THE MICROBIOLOGY OF SAFE FOOD

S.J. Forsythe
Department of Life Sciences
Nottingham Trent University

Blackwell
Science

© 2000 by Blackwell Science Ltd, a Blackwell Publishing company

Editorial offices:
Blackwell Science Ltd, 9600 Garsington Road, Oxford OX4 2DQ, UK
    Tel: +44 (0) 1865 776868
Blackwell Publishing Inc., 350 Main Street, Malden, MA 02148-5020, USA
    Tel: +1 781 388 8250
Blackwell Science Asia Pty Ltd, 550 Swanston Street, Carlton, Victoria 3053, Australia
    Tel: +61 (0)3 8359 1011

First published 2000
Reprinted 2002, 2006

ISBN-10: 0-632-05487-5
ISBN-13: 978-0-632-05487-9

Library of Congress Cataloging-in-Publication Data

Forsythe, S.J. (Steve J.)
    The microbiology of safe food/S.J. Forsythe.
        p. cm.
    Includes bibliographical references and index.
    ISBN 0-632-05487-5 (pb)
    1. Food—Microbiology. I. Title.
    QR115+
                        00-029784

A catalogue record for this title is available from the British Library

Set in 10.5/12.5 pt Garamond Book
by DP Photosetting, Aylesbury, Bucks
Printed and bound by Replika Press Pvt. Ltd, India

The publisher's policy is to use permanent paper from mills that operate a sustainable forestry policy,
and which has been manufactured from pulp processed using acid-free and elementary chlorine-free
practices. Furthermore, the publisher ensures that the text paper and cover board used have met
acceptable environmental accreditation standards.

For further information on Blackwell Publishing, visit our website:
www.blackwellpublishing.com

THE MICROBIOLOGY OF SAFE FOOD

CONTENTS

Preface   *ix*

1   Introduction to Safe Food   **1**
   1.1   What is safe food?   1
   1.2   The manufacture of hygienic food   3
   1.3   Functional foods   5

2   Basic Aspects   **10**
   2.1   The microbial world   10
   2.2   Bacterial cell structure   12
   2.3   Microbial growth cycle   16
   2.4   Death kinetics   18
   2.5   Factors affecting microbial growth   23
   2.6   Preservatives   31
   2.7   Microbial response to stress   36
   2.8   Predictive modelling   42

3   Foodborne Illness   **53**
   3.1   The size of the food poisoning problem   53
   3.2   Consumer pressure effect on food processing   61
   3.3   Testing foods for the presence of pathogens   63
   3.4   Control of food poisoning   64
   3.5   Surveillance programmes   66
   3.6   Causes of food poisoning   72
   3.7   The microbial flora of the human gastrointestinal
       tract   74
   3.8   The mode of action of bacterial toxins   79
   3.9   Virulence factors of foodborne pathogens   87

4   The Microbial Flora of Food   **96**
   4.1   Spoilage microorganisms   96
   4.2   Shelf life indicators   104

4.3	Methods of preservation and shelf life extension	107
4.4	Fermented foods	118
4.5	Prebiotics, probiotics and synbiotics	134
4.6	Microbial biofilms	138
5	**Food Poisoning Microorganisms**	**142**
5.1	Indicator organisms	143
5.2	Foodborne pathogens: bacteria	146
5.3	Foodborne pathogens: viruses	172
5.4	Seafood and shellfish poisoning	177
5.5	Foodborne pathogens: eucaryotes	181
5.6	Mycotoxins	184
5.7	Emerging and uncommon foodborne pathogens	189
6	**Methods of Detection**	**193**
6.1	Prologue	193
6.2	Conventional methods	195
6.3	Rapid methods	198
6.4	Rapid end-detection methods	203
6.5	Specific detection procedures	218
6.6	Accreditation schemes	254
7	**Food Safety Management Tools**	**256**
7.1	Microbiological safety of food in world trade	257
7.2	The management of hazards in food which is in international trade	258
7.3	Hazard Analysis Critical Control Point (HACCP)	259
7.4	Outline of HACCP	261
7.5	Microbiological criteria and HACCP	268
7.6	Microbiological hazards and their control	270
7.7	HACCP plans	276
7.8	Sanitation Standard Operating Procedures (SSOPs)	279
7.9	Good Manufacturing Practice (GMP) and Good Hygiene Practice (GHP)	293
7.10	Quality Systems	293
7.11	Total Quality Management (TQM)	294
7.12	ISO 9000 Series of standards	294
8	**Microbiological Criteria**	**296**
8.1	International Commission on Microbiological Specifications for Foods	296
8.2	Codex Alimentarius principles for the establishment and application of microbiological criteria for foods	297

8.3   Sampling plans   298
8.4   Variables plans   303
8.5   Attributes sampling plan   306
8.6   Principles   307
8.7   Microbiological limits   315
8.8   Examples of sampling plans   317
8.9   Implemented microbiological criteria   320
8.10   Public Health (UK) Guidelines for Ready-To-Eat Foods 334

9   Microbiological Risk Assessment   335
9.1   Risk assessment (RA)   337
9.2   Risk management   344
9.3   Risk communication   345
9.4   Food Safety Objectives   346
9.5   Application of MRA   347

10   Regulations and Authorities   361
10.1   Regulations in international trade of food   361
10.2   Codex Alimentarius Commission   363
10.3   Sanitary and Phytosanitary measures (SPS),
       Technical Barriers to Trade (TBT) and the World
       Health Organisation (WHO)   364
10.4   Food authorities in the United States   365
10.5   European Union legislation   367
10.6   Food safety agencies   369

Glossary of Terms   *371*

Appendix: Food Safety Resources on the World Wide Web *376*

References   *380*

Index   *400*

PREFACE

Throughout the world, food production has become more complex. Frequently raw materials are sourced globally and the food is processed through an increasing variety of techniques. No longer does the local farm serve the local community through a local shop, nowadays there are international corporations adhering to national and international regimes. Therefore approaches to safe food production are being assessed on an expanding platform from national, European, transatlantic and beyond. Against this backdrop there have been numerous highly publicised food safety issues such as BSE and *E. coli* O157:H7 which has caused the general public to become more vociferous concerning food issues. The controversy in Europe over genetically modified foods is perceived by the general public within the context of 'food poisoning'.

This book aims to review the production of food and the level of microorganisms which humans ingest. Certain circumstances require zero tolerances for pathogens, whereas more frequently there are acceptable limits set, albeit with statistical accuracy or inaccuracy depending upon whether you subsequently suffer from food poisoning. Microbes are traditionally ingested in fermented foods and this has developed into the subject of pre- and probiotics with disputed health benefits. Whether engineered 'functional foods' will be able to attain consumer acceptance remains to be seen.

Food microbiology covers both food pathogens and food spoilage organisms. This book aims to cover the wide range of microorganisms occurring in food, both as contaminants and deliberate inoculation. Due to the heightened public awareness over food poisoning it is important that all companies in the food chain maintain high hygienic standards and assure the public of the safety of the produce. Obviously over time there are technological changes in production methods and in methods of microbiological analysis. Therefore the food microbiologist needs to know the effect of processing changes (pH, temperature, etc.) on the microbial load. To this end this book reviews the dominant foodborne microorganisms, the means of their detection, microbiological criteria as the numerical means of interpreting end-product testing, predictive microbiology as a tool to understanding the consequences of processing changes, the role of Hazard Analysis Critical Control Point (HACCP), the objectives of Microbial Risk Assessment (MRA) and the setting of Food Safety Objectives which have recently become a focus of attention. In recent years the Web has become an invaluable source of information and to reflect this a range of useful food safety resource sites is given in the back to encourage the reader to boldly go and surf. Although primarily aimed for undergraduate and postgraduate courses, I hope the book will also be of use to those working in industry.

The majority of this book was written during the last months of 1999, a time when France was being taken to the European Court over its refusal to sell British beef due to BSE/nvCJD and there had been riots in Seattle concerning the World Trade Organisation. While large organisations were wondering about the impact of the 'millennium bug', in the UK the public were waiting to see the impact of the BSE 'bug' (a few hundred or a few thousand cases?).

As usual no book can be achieved without assistance and special thanks are due to Phil Vosey concerning MRA, Ming Lo for considerable help with the computer packages, Alison Howie at Oxoid Ltd for the invaluable information on microbiological testing procedures around the world, Pete Silley and Andrew Pridmore at Don Whitley Scientific Ltd for the RABIT diagrams and Garth Lang at Biotrace Ltd for the ATP bioluminescence data. Not forgetting of course Debbie and Cathy for reading through the draft copy; nevertheless all mistakes are the author's fault.

This book is especially dedicated to Debbie, James and Rachel, Mum and Dad for their patience while I've been burning the midnight oil.

Steve Forsythe
June 2000

1

INTRODUCTION TO SAFE FOOD

This chapter is largely an overview of the book in order for the reader to understand the approach being taken. The key topics will be covered in greater depth in each of the specific chapters. Definition of terms will be found in the glossary at the end of the book, where there is also a listing of useful hypertext links.

1.1 What is safe food?

The increasing number and severity of food poisoning outbreaks world-wide has considerably increased public awareness about food safety. Public concern on food safety has been raised due to well publicised incidences such as food irradiation, BSE, *E. coli* O157:H7 and genetically modified foods.

'What is "safe" food?' invokes different answers depending upon whom is asked. Essentially, the different definitions would be given depending upon what consitutes a significant risk. The general public might consider that 'safe food' means zero risk, whereas the food manufacturer would consider 'what is an acceptable risk?'. The opinion expressed in this book is that *zero risk is not feasible* given the range of food products available, the complexity of the distribution chain and human nature. Nevertheless, the risks of food poisoning should be reduced during food manufacture to an 'acceptable risk'. Unfortunately there is no public consensus on what constitutes an acceptable risk. After all how can one compare the risk of hang gliding with eating rare beef? Also hang gliding has known risks which can be evaluated and a decision 'to glide or not to glide' taken. In contrast, the general public (rightly or wrongly) often feels it is not informed of relevant risks. Table 1.1 shows the possible causes of death in the next 12 months. Many of these risks are acceptable to the general public; people continue to drive cars and cross the road. This table can be compared with Tables 3.2 and 3.3 in Chapter 3 which give the recent data

Table 1.1   Risk of death during the next twelve months.

Event	Chance of one in
Smoking 10 cigarettes a day	200
Natural causes, middle-aged	850
Death through influenza	5 000
Dying in a road accident	8 000
Dying in a domestic accident	26 000
Being murdered	100 000
Death in a railway accident	500 000
Struck by lightning	10 000 000

for the USA and UK on the chances of food poisoning. The USA data indicate that each year 0.1% of the population will be hospitalised due to food poisoning. Food scares cause public outcries and can give the industry an undeserved bad reputation. Whereas, in fact, the majority of the food industry has a good safety record and is in the business to stay in business, not to go bankrupt due to adverse publicity. See Table 1.2 for some business philosopy on consumer confidence.

Table 1.2   Loss of consumer confidence in your product (adapted with permission from Corlett, 1998 *HACCP User's Manual*, Aspen Publishers Inc.).

- In the average business, for every customer who bothers to complain, 26 remain silent.
- The 'wronged' customer will tell 8 to 16 people of the problem, and more than 10% will tell 20 more.
- Of the unhappy customers, 91% will never purchase the offending goods or service again.
- If 26 people tell 8 people = 208 lost customers.
- If 26 people tell 16 people = 416 lost customers.
- It costs five times as much to attract a new customer as it costs to retain an old one.

A difficulty that arises in manufacturing 'safe' food is that the consumer is a mixed population with varying degrees of susceptibility and general life style. Additionally, food with 'high' levels of preservatives to reduce microbial growth are undesirable by the consumer and perceived as 'over processed' with 'chemical additives'! The consumer pressure is for greater varieties of fresh and minimally processed foods, natural preservatives with a *guarantee of absolute safety*.

The manufacture of safe food is the responsibility of everyone in the food chain, and food factory, from the operative on the conveyor belt to the higher management. It is not the sole responsibilty of the food microbiologist. Nevertheless, the food microbiologist in industry will need not only to know which food pathogens are likely to occur in the ingredients, but also the affect of the food matrix and processing steps on cell survival in order to give the best advice on the most appropriate manufacturing regimes. The best methods for microbiological analysis are still being developed. It is obvious from the plethora of differing methods adopted by different countries that food poisoning statistics cannot be directly compared between countries due to the differing methods of analysis applied.

1.2   The manufacture of hygienic food

Is there a way forward for the manufacture of food which is nutritious and appetising, yet meets the expectations of the consumer regarding risk? This book aims to cover the microbiological aspects of safe hygienic food manufacture. It is only one aspect of the whole jigsaw, but should direct the reader to sources of supplemental information where necessary. It also points towards areas of future expansion such as functional foods.

The production of safe food requires:

- Control at source;
- Product design and process control;
- Good hygienic practice during production, processing, handling and distribution, storage, sale, preparation and use;
- A preventative approach because effectiveness of microbial end-product testing is limited.

Taken from CAC Alinorm 97/13 (Codex Alimentarius Commission 1995).

Control of foodborne pathogens at source is not always easy. Many pathogens survive in the environment for long periods of time (Table 1.3). They can be transmitted to humans by a variety of routes (Fig. 1.1).

Production processes can be very complicated. A general flow diagram for poultry processing is given in Fig. 1.2. This is a relatively simple flow diagram with only one branching. It does not indicate the temperature and time at each step which affect microbial growth. Therefore safe food production requires an all-encompassing approach involving the food operatives at the shop floor through to the management. Hence a number of mangement safety tools such as Good Hygienic Practices (GHP), Good Manufacturing Practice (GMP), Total Quality Management (TQM) and Hazard Analysis Critical Control Point (HACCP) need to be implemented. Only an outline of some of these tools are given in this book as detailed

Table 1.3   Survival of pathogenic microorganisms in sewage sludge, soil and on vegetables.

Organism	Conditions	Survival
Coliforms	Soil	30 days
Mycobacterium tuberculosis	Soil	up to 2 years
	Soil	5–15 months
	Radish	3 months
Salmonella spp.	Soil	72 weeks
	Potatoes at soil surface	40 days
	Vegetables	7–40 days
	Beet leaves	21 days
	Carrots	10 days
	Cabbage/gooseberries	5 days
	Apple juice, pH 3.68	>30 days and multiplies
	Apple juice pH <3.4	2 days
S. typhi	Soil	30 days
	Vegetables/fruit	1–69 days
	Water	7–30 days
Shigella spp.	Tomatoes	2–7 days
Vibrio cholerae	Spinach, lettuce, non-acid vegetables	2 days
Enterovirus group (polio, echo, coxsackie)	Soil	150–170 days
	Cucumber, tomato, lettuce, at 6–10°C	>15 days
	Radish	>2 months

descriptions are outside its scope. Regarding legislation, the reader should always seek the appropriate regional authority.

For information on hygienic food production with an emphasis on the factory layout, equipment design and staff training the reader is recommended Forsythe & Hayes (1998) *Food Hygiene, Microbiology and HACCP* and Shapton & Shapton (1991) *Principles and Practices for the Safe Processing of Foods*. Detailed information on individual microorganisms can be found in the International Commission on Microbiological Specifications for Foods (ICMSF) book series, especially *Microorganisms in Foods 5: Characteristics of Microbial Pathogens* (1996) and *Microorganisms in Foods 6: Microbial Ecology of Food Commodities* (1998), which should be regarded as essential requirements on the bookshelf (or library) of any food microbiologist.

The whole issue of safe food manufacturing comes within the umbrella of quality control and quality assurance. Hence it requires the hygienic

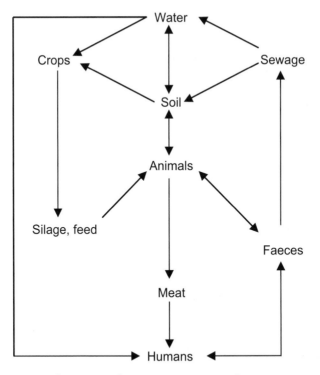

Fig. 1.1 Routes of enteric pathogen transmission to humans.

design of equipment and factory, and managerial commitment to safety and quality. Diagrammatically this can be perceived in Fig. 1.3. The current issue concerning food safety is microbiological risk assessment and the development of *food safety objectives*. These are governmental activities that eventually may decide the permissible level of foodborne pathogens, etc. (Fig. 1.4). This issue is covered in Chapter 9.

There are some foods which are currently difficult to produce without a significant risk of foodborne infection. Outbreaks of *Salmonella* spp. and *E. coli* O157:H7 associated with raw seed sprouts have occurred in several countries and currently the elderly, children and those with compromised immune systems are advised not to eat raw sprouts (such as alfalfa) until effective measures to prevent sprout-associated illness are identified (Taormina & Beuchat 1999).

1.3 Functional foods

In direct contrast to the general concern on the presence of bacteria in food, there are in fact a number of foods which deliberately contain

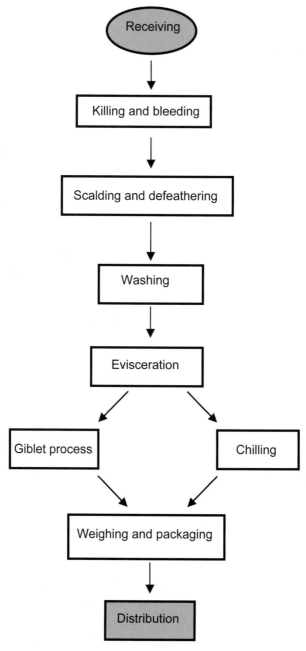

Fig. 1.2   Flowsheet for poultry processing.

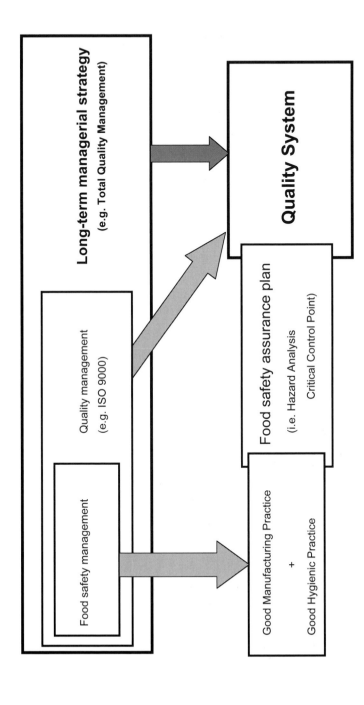

Fig. 1.3 Food safety management tools (adapted from Jouve *et al.* 1998).

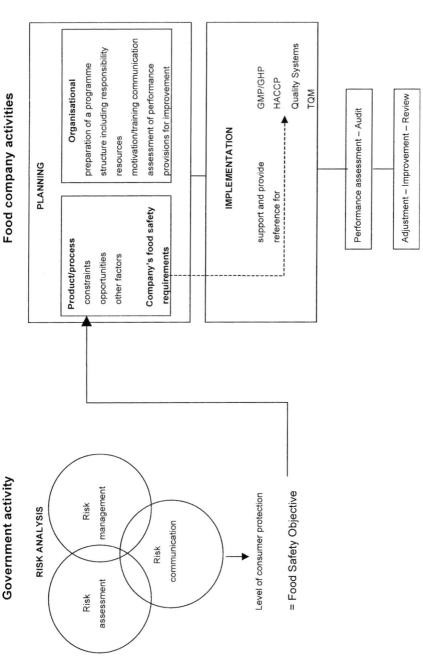

Fig. 1.4   Interaction between governments' and companies' food safety activities (adapted from Jouve *et al.* 1998).

bacteria and fungi. These are the 'fermented' foods (Section 4.4) and these foods have been produced since the early era of civilisation. The taste, texture and flavour of the food are due to microbial metabolism. Hence they are regarded as microbiologically safe. These foods now form the basis over the past 10–12 years for the development of 'functional foods'.

In Japan, and to a lesser extent in the USA, research into functional food has expanded greatly in recent years. Functional foods are expected to have a health-related or physiological effect, such as reduce the risk of disease. Most functional foods currently approved contain either oligo-saccharides or lactic acid bacteria for promoting intestinal health. In 1998 the Food and Drug Adminsitration (FDA) had recognised 11 foods or food components as showing correlation between intake and health benefits (Diplock *et al.* 1999). A significant portion of function foods is concerned with lactic acid bacteria ingestion ('probiotics'). Therefore this book will review probiotics (Section 4.5) as an extension of the age-old practice of fermented foods. The legislation regarding genetically modified foods will not be covered as there are more suitable texts available (IFBC 1990; WHO 1991; OECD 1993; Jonas *et al.* 1996; FAO 1996; SCF 1997; Tomlinson 1998; Mosely 1999).

2

BASIC ASPECTS

2.1 The microbial world

The world of microbiology covers a wide range of life. Using the definition that microbiology studies life forms that are not clearly visible to the naked eye means that it includes protozoa, fungi, bacteria, viruses and prions. The major organisms studied have been bacteria, because of their medical importance and because they are easier to cultivate than other organisms such as viruses, and the more recently recognised 'prions' which are uncultivatable infectious organisms. Despite the predominance of bacteria in our understanding of microscopic life there are numerous important organisms in the other microbial categories.

The cell structure reveals whether an organism is 'eucaryotic' (also spelt eukaryotic) or 'procaryotic' (also spelt prokaryotic). Eucaryotes contain cellular organelles such as mitochondria, endoplasmic reticulum and a defined nucleus, whereas procaryotes have no obvious organelle differentiation and are in fact similar in size to the organelles of eucaryotes. Analysis of the genetic information in the ribosome (16S rRNA analysis) has revealed a plausible relationship and evolution of life from procaryotes to eucaryotes through intracellular symbiotic relationships.

A brief survey (see Table 2.1) of food poisoning microorganisms can start with eucaryotic organisms such as helminths. These include the cestode worms that are responsible for taeniasis, *Taenia solium*, the pork tapeworm, and *T. saginata*, the beef tapeworm. Both have worldwide distribution. Infection results from the ingestion of undercooked or raw meats containing the cysts. The mature worms can infect the eye, heart, liver, lungs and brain. A third tapeworm is *Diphyllobothrium latum* which is found in a variety of freshwater fish including trout, perch and pike. Contaminated water supplies can carry infectious organisms including pathogenic protozoa.

Fungi (a term which includes yeast and moulds) are eucaryotic. They are members of the plant kingdom that are not differentiated into the

Table 2.1 Microbiological contaminants.

	Where found	Sources
Viruses		
A wide range which cause diseases, including hepatitis A	Most common in shellfish, raw fruit and vegetables	Associated with poor hygiene and cultivation in areas contaminated with untreated sewage and animal and plant refuse
Bacteria		
Includes *Bacillus* spp. *Campylobacter*, *Clostridium*, *Escherichia coli*, *Salmonella*, *Shigella*, *Staphylococcus* and *Vibrio*	Raw and processed foods: cereal, fish and seafood, vegetables, dried food and raw food of animal origin (including dairy products)	Associated with poor hygiene and unclean conditions generally: carried by animals such as rodents and birds, and human secretions
Moulds		
Aspergillus flavis and related fungi	Nuts and cereals	Products stored in high humidity and temperature
Protozoa		
Amoebae and Sporidia	Vegetables, fruits and raw milk	Contaminated production areas and water supplies
Helminths		
A group of internal parasites including *Ascaris*, *Fasciola*, *Opisthorchis*, *Taenia*, *Trichinella* and *Trichuris*	Vegetables and uncooked or undercooked meat and raw fish	Contaminated soil and water in production areas

usual roots, stems and leaves; they also do not produce the green photosynthetic pigment chlorophyll. They may form a branching mycelium with differentiating cells producing hyphae to dispere spores or exist as single cell forms commonly referred to as yeast. Yeasts are very important in food microbiology. In fact the brewing and bakery industries are dependent upon yeast metabolism of sugars to generate ethanol (and other alcohols) for beer and wine production, and carbon dioxide for bread manufacture.

Mycotoxicoses are caused by the ingestion of poisonous metabolites (mycotoxins) which are produced by fungi growing in food. Aflatoxins are produced by the fungi *Aspergillus flavus* and *A. parasiticus*. There are four main aflatoxins designated B1, B2, G1 and G2 according to the blue (B) or green (G) fluorescence given when viewed under a UV lamp. Ochratoxins are produced by *A. ochraceus* and *Penicillium viridicatum*. Ochratoxin A is the most potent of these toxins.

The bacteria are procaryotes and are divided into the eubacteria (true bacteria) and archaea organisms (old term 'Archaebacteria') according to 16S rRNA analysis and detailed cell composition and metabolism studies. There are few *Archae* organisms of importance in the food industry, the vast majority are eubacteria. As a generalisation, the size of a rod-shaped bacterium is about $2\,\mu m \times 1\,\mu m \times 1\,\mu m$. Although they are very small, even 500 cells of *Listeria monocytogenes* can be an infectious dose to a pregnant women resulting in a stillbirth. Familiar foodborne pathogens such as *Salmonella* spp., *Escherichia coli* and *Campylobacter jejuni* are eubacteria which are able to grow at body temperatures (37°C) and damage human cells resulting in 'food poisoning' symptoms such as diarrhoea and vomiting.

Viruses are very much smaller than bacteria. The large ones, such as the cowpox virus, are about $0.3\,\mu m$ in diameter; the smaller ones, such as the foot and mouth disease virus, are about $0.1\,\mu m$ in diameter. Because of their small size, viruses pass through bacteriological filters and are invisible under the light microscope. Bacterial viruses are termed bacteriophages (or 'phages'). They can be used to 'fingerprint' bacterial isolates which is necessary in epidemiological studies; this is termed phage typing.

Prions (short for proteinaceous infectious particles) have a very long incubation period (months or even years) and resistance to high temperature, formaldehyde and UV irradiation. With regard to sheep and cattle the isomer of the normal cellular protein PrP^C, termed PrP^{SC}, accumulates in the brain causing holes or plaques. This leads to the symptoms of scrapie in sheep and BSE in cattle. The equivalent disease in humans (new variant Creutzveld Jacob Disease, nvCJD) is probably due to ingestion of infectious agents from cattle.

2.2   Bacterial cell structure

2.2.1   Morphology

Bacteria are characteristically unicellular organisms. Their morphology can be straight or curved rods, cocci or filaments depending upon the organism concerned. The morphology of the organism is consistent with the type of organism and to a lesser extent to the growth conditions. Bacteria in the genera *Bacillus*, *Clostridium*, *Desulphotomaculum*,

Sporolactobacillus and *Sporosarcina* form spores in the cytoplasm under certain environmental conditions (stress related). The spore is more resistant to heat, drying, pH, etc. than the vegetative cell and hence enables the organism to persist until more favourable conditions when the spore can subsequently germinate and grow into a vegetative cell.

2.2.2 Gram stain

Christian Gram was a Danish microbiologist who wanted to visualise bacteria in muscle tissue. He tried a range of histological staining protocols and eventually came up with the, nowadays ubiquitous, 'Gram stain' procedure. It was noted by Christian Gram that the bacteria stained either dark blue or red. This was due to the precipitation of crystal violet with Lugol's iodine in the cell cytoplasm which could not be extracted using solvents such as ethanol or acetone from certain organisms (Gram positives) but was extracted from others (Gram negatives). Since the latter cells were no longer stained by the crystal violet, counterstaining was required with safranine or basic fuschin. More recently, the reason for the differential extraction of the crystal violet–iodine complex was found to be due to differences in the cell wall structure (Fig. 2.1).

Gram-positive organisms have a thick cell wall surrounding the cytoplasmic membrane. This is composed of peptidoglycan (also known as murein) and teichoic acids. The Gram-negative organisms have a thinner cell wall which is surround by an outer membrane. Hence Gram-negative organisms have two membranes. The outer membrane differs from the inner membrane and contains the molecule known as lipopolysaccharide (LPS). Since peptidoglycan is the site of action for the original penicillin antibiotic (penicillin G) this explains why the antibiotic was initially so effective against streptococci and staphylococci (both Gram positives) rather than Gram negatives such as *Escherichia coli*. Consequently the semi-synthetic penicillins were developed to widen the range of sensitive organisms to penicillin.

2.2.3 Lipopolysaccharide (LPS, O antigen)

The outer membrane of Gram-negative organisms contains the molecule lipopolysaccharide (LPS). It is composed of three regions: lipid A, core and O antigen. Lipid A anchors the molecule in the outer membrane and is toxic (Fig. 2.2). It is a virulence factor for organisms such as *Salmonella* and *Chlamydia*. The core region is composed of sugar molecules, the sequence of which reflects the organism's identity. The O region is more variable. In some organisms the O region may only contain a few sugar residues, whereas in others there are repeating units of sugars. Isolates of a

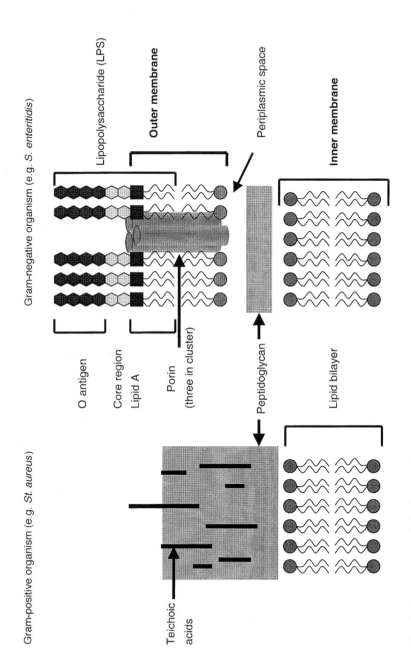

Fig. 2.1   Structure of the bacterial cell wall.

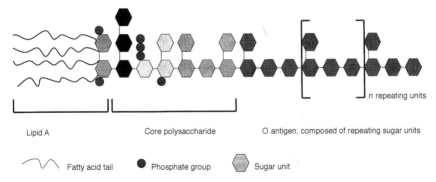

Fig. 2.2 Structure of the lipopolysaccharide molecule of Gram-negative organisms.

single species vary in the amount of O region present and hence this gives rise to the terms 'smooth' and 'rough' variants. The antigens are on the body of the organism and are known as the 'somatic' or 'O antigens'. 'O' is for 'ohne', the German for 'without', which originally referred to non-swarming or non-flagellated forms. The LPS structure is resistant to boiling for 30 minutes and is therefore also referred to as a 'heat-stable' antigen.

One of the factors causing the fatality of Gram-negative septicaemia is that the presence of LPS causes the overproduction of tumour necrosis factor (TNF), which leds to overstimulation of nitric oxide synthase.

2.2.4 Flagella (H antigen)

Most rod-shaped (and a few coccoid) bacteria are motile in liquid media. Motility occurs due to the rhythmic movement of thin fila-mentous structures called flagella. The cell may have a single flagellum (monotrichus) or a tuft of flagella (lophotrichus) at one or both poles, or many flagella (peritrichus) over the entire surface. The proteinaceous nature of the flagellum gives rise to its antigenicity, called the H anti-gen. 'H' is from the German word 'hauch' meaning 'breath' which origi-nated from a description of the appearance of *Proteus* swarming on moist agar plates being similar to the light mist caused by breathing on cold glass. Flagella are denatured by heat (100°C, 20 minutes) and there-fore the H antigen is referred to as heat-labile. Flagella are also dena-tured by acid and alcohol.

Salmonella species may express two flagellar antigens: phase 1, which is possessed by only a few other serotypes of *Salmonella*, and phase 2, which is less specific. Phase 1 antigen is represented by letters, phase 2 antigen by numbers. A culture may entirely express one phase (mono-phasic culture) which will give rise to mutants in the other phase (diphasic culture), especially if the culture is incubated for more than 24 hours. The serotyping of some *Salmonella* serovars is given in Table 2.2.

E. coli has a total of 173 different O antigens and 56 different H antigens. *E. coli* O157:H7 is a very pathogenic strain which is recognised by serotyping the O and H antigens.

Table 2.2 *Salmonella* serotyping.

Serotype	Group	O antigen	H antigen Phase 1	Phase 2
S. paratyphi A	A	(1), 2, 12	a	—
S. typhimurium	B	(1), 4, (5), 12	i	1, 2
S. paratyphi C	C₁	6, 7, Vi	c	1, 5
S. newport	C₂	6, 8	e, h	1, 2
S. typhi	D	9, 12, Vi	d	—
S. enteritidis	D	(1), 9, 12	g, m	
S. anatum	E₁	3, 10	e, h	1, 6
S. newington	E₂	3, 15		
S. minneapolis	E₃	(3), (15), 34		

Parentheses indicate antigen determinant which may be difficult to detect. Dominant antigenic determinants are underlined.

2.2.5   Capsule (Vi antigen)

Some bacteria secrete a slimy polymeric material composed of polysaccharides, polypeptides or polynucleotides. If the layer is very dense then it may be visualised as a capsule. The possession of the capsule endows a resistance to white blood cell engulfment. Subsequently the capsule's antigenicity is called the Vi antigen and was originally thought to be responsible for the virulence of *S. typhi* (see Table 2.2).

2.3   Microbial growth cycle

The microbial growth cycle is composed of six phases (Fig. 2.3):

(1)   Lag phase
      Cells are not multiplying, but are synthesising enzymes appropriate for the environment.

(2)   Acceleration phase
      An increasing proportion of the cells are multiplying.

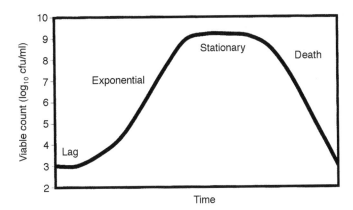

Fig. 2.3 The microbial growth curve.

(3) Exponential (or log) phase

The cell population are multipying by doubling (1-2-4-8-16-32-64, etc.). The cell numbers are increasing at such a rate that to represent them graphically it is best to use exponential values (logarithms). This results in a straight line, the slope of which represents the μ_{max} (rate of maximum growth), and the doubling time t_d (time required for the cell mass to increase two-fold).

(4) Deceleration phase

An increasing proportion of cells are no longer multiplying.

(5) Stationary phase

The rate of growth equals the rate of death, resulting in equal numbers of cells at any given time. Death is due to the exhaustion of nutrients, the accumulation of toxic end products and/or other changes in the environment, such as pH changes. The length of the stationary phase is dependent upon a number of factors such as the organism and environmental conditions (temperature, etc.). Spore forming organisms will develop spores due to the stress conditions.

(6) Death phase

The number of cells dying is greater than the number of cells growing. Cells which form spores will survive longer than non-sporeformers.

The length of each phase is dependent upon the organism and the growth environment, temperature, pH, water activity, etc. The growth cycle can be modelled using sophisticated computer programs and leads to the area of microbial modelling and predictive microbiology (see Section 2.8).

2.4   Death kinetics

2.4.1   *Expressions*

There are a number of expressions used to describe microbial death:

- D value: decimal reduction time. Defined as the time at any given temperature for a 90% reduction (=1 log value) in viability to be effected.
- Z value: temperature increase required to increase the death rate 10-fold, or in other words reduce the D value 10-fold.
- P value: time at 70°C. A cook of 2 minutes at 70°C will kill almost all vegetative bacteria. For a shelf life of 3 months the P value should be 30 to 60 minutes, according to other risk factors.
- F value: this value is the equivalent time, in minutes at 250°F (121°C), of all heat considered, with respect to its capacity to destroy spores or vegetative cells of a particular organism.

Since these values are mathematically derived they can be used in predictive microbiology and Section 2.8 and Chapter 9, Microbial Risk Assessment (MRA).

2.4.2   *Decimal reduction times (D values)*

In order to design an effective heating treatment time and temperature regimes it is imperative to have an understanding of the effects of heat on microorganisms. The thermal destruction of microorganisms (death kinetics of vegetative cells and spores) can be expressed logarithmically. In other words for any specific organism, in a specific substrate and at a specific temperature, there is a certain time required to destroy 90% (=1 log reduction) of the organism. This is the decimal reduction time (D value). The rate of death depends upon the organism, including the ability to form spores, and the environment (Table 2.3). Free (or planktonic) vegetative cells are more sensitive to detergents than fixed cells (biofilms or slime). The heat sensitivity of an organism at any given temperature varies according to the suspending medium. For example, the presence of acids and nitrite will increase the death rate whereas the presence of fat may decrease it.

Plotting the D values for an organism in a substrate against the heating temperature should give a straight-line relationship (Fig. 2.4). The straight-line relationship does not always occur due to cell clumping. The slope of the line (in degrees Fahrenheit or Celsius) can be used to determine the change in temperature resulting in a 10-fold increase (or decrease) in the D value. This coefficient is called the Z value (Fig. 2.5). The integrated

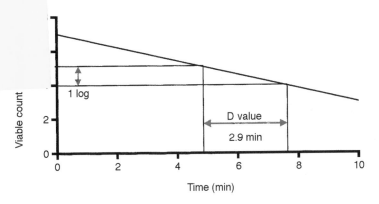

Fig. 2.4a Death rate of *E. coli* O157:H7 in beef at 60°C.

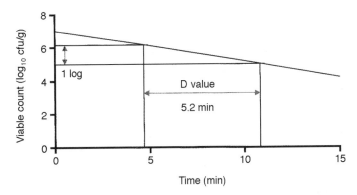

Fig. 2.4b Death rate of *St. aureus* in liquid egg at 60°C.

lethal value of heat received by all points in a container during processing is designated F_s or F_o. This represents a measure of the capacity of a heat process to reduce the number of spores or vegetative cells of a given organism per container. When we assume instant heating and cooling throughout the container of spores, vegetative cells, or food, F_o may be derived as follows:

$$F_o = D(\log a - \log b)$$

where a is the number of cells in the initial population and b is the number of cells in the final population.

See Table 7.9 for an example of the effect of cooking temperature on the survival of *E. coli* O157:H7 in beef using D and Z values. The decimal reduction in viability can also be determined following irradiation of microorganisms (Fig. 2.6).

Table 2.3 Variation in microbial heat resistance of microorganisms according to test conditions. (Various sources, including Mortimore and Wallace, 1994; Borche *et al.*, 1996 and ICMSF, 1996.)

Organism	Medium	pH	Temperature (°C)[a]	D value (min)	Z value
A. hydrophila	Saline		51.0	8.08–122.8	5.22–7.69
Brucella spp.	—	—	65.5	0.1–0.2	—
B. cereus	—	—	100	5.0	6.9
(spores)			100	2.7–3.1	6.1
(toxin destruction; diarrhoeal/emetic)			56.1/121	5/stable	—
B. coagulans	Buffer	4.5	110	0.064–1.46	—
	Red pepper	4.5	110	5.5	—
B. licheniformis	Buffer	7.0	110	0.27	—
	Buffer	4.0	110	0.12	—
B. stearothermophilus			120	4.0–5.0	10
B. subtilis	—	—	100	11.0	
C. jejuni	Buffer	7.0	50	0.88–1.63	6.0–6.4
	Beef		50	5.9–6.3	
Cl. butyricum	—	—	100	0.1–0.5	
Cl. perfringens	—	—	100	0.3–20.0	3.8
(spores)			98.9	26–31 minutes	7.2
(toxin)	—	—	90	4	5.5
Cl. botulinum					
(type A and B proteolytic strain's spores	—	—	121.1	0.21	10
type E and non-proteolytic types B and F)			82	0.49–0.74	5.6–10.7
(toxin destruction)			85	2.0	4.0–6.2
Cl. thermosaccharolyticum	—	—	120	3–4	7.2–10
D. nigrificans	—	—	120	2–3	2–3

(Contd)

Table 2.3 *(Contd)*

Organism	Medium	pH	Temperature (°C)[a]	D value (min)	Z value
E. coli O157:H7	Beef		62.8	0.47	4.65
	Apple juice	3.6	58	1.0	4.8
		4.5	58	2.5	4.8
	Growth, 23°C		58	1.6	4.8
	Growth, 37°C		58	5.0	4.8
L. monocytogenes	Beef		62.0	2.9–4.2	5.98
S. enteritidis	Liquid whole egg		62.8	0.06	3.30
S. senftenberg	Beef		62.0	2.65	5.91
St. aureus	—	—	65.5	0.2–2.0	4.8–5.4
(toxin destruction)			98.9	>2 hours	(approx. 27.8)
Streptococcus Group D	Cured meat		70	2.95	10
Y. enterocolitica	Saline	4.5	60.0	0.4–0.51	4.0–5.2
Sac. cerevisiae	Saline	4.5	60	22.5[b]	5.5
Z. bailii	Saline	4.5	60	0.4[c]	3.9
	Saline		60	14.2[b]	—

[a] To convert to °F use the equation °F = (9/5)°C + 32. As a guidance: 0°C = 32°F, 4.4°C = 40°F, 60°C = 140°F.
[b] Ascospores.
[c] Vegetative cells.

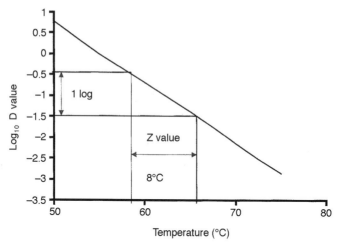

Fig. 2.5   *C. jejuni* Z value in lamb cubes.

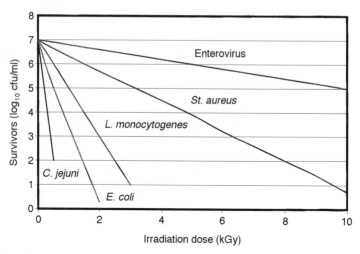

Fig. 2.6   Survival of microorganisms following irradiation.

12-D concept

The 12-D concept refers to the process lethality requirement which has for a long time been used in the canning industry. It implies that the minimum heat process should reduce the probability of survival of the most resistant *Cl. botulinum* spores to 10^{-12}. Since *Cl. botulinum* spores do not germinate and produce toxin below pH 4.6, this concept is observed only for foods above this pH value.

$$F_o = D(\log a - \log b)$$

$$F_o = 0.21(\log 1 - \log 10^{-12})$$

$$F_o = 0.21 \times 12 = 2.52$$

Processing for 2.52 minutes at 250°F (121°C) should reduce the *Cl. botulinum* spores to one spore in one million million containers (10^{12}). When it is considered that some flat-sour spores have D values of about 4.0 and some canned foods receive F_o treatments of 6.0–8.0, the potential number of *Cl. botulinum* spores is reduced even more (see Table 2.3).

2.4.3 Pasteurisation

Pasteurisation treatment aims to reduce the number of pathogenic-spoilage organisms by a set amount (frequently a 6 log reduction) and to ensure that the product formulation and storage conditions inhibit the growth of any surviving cells during the intended shelf life of the product. For example, if the D value of a target organism is 0.5 minutes at 70°C and the Z value is 5°C, a process of 3 minutes at 70°C would give a 6 log reduction. At 75°C the organism would die 10 times faster, hence 0.3 minutes (20 seconds) at 75°C is an equivalent treatment giving a 6 log reduction.

Since the determination of actual Z values for each organism and product would require considerable work, assumed appropriate values are often used. Although this may be satisfactory for general Quality Assurance purposes, it is important to establish accurate values for the most appropriate microorganism present in the product.

2.5 Factors affecting microbial growth

Traditional ways to control microbial spoilage and safety hazards in foods include:

- Freezing
- Blanching
- Pasteurisation
- Sterilisation
- Canning
- Curing
- Syruping
- Inclusion of preservatives

Table 2.4   Methods of food preservation (ICMSF, 1988).

Operation	Intended effect
Cleaning, washing	Reduces microbial load
Cold storage (below 8°C)	Prevents the growth of most pathogenic bacteria; slows the growth of spoilage microbes
Freezing (below −10°C)	Prevents growth of all microbes
Pasteurising (60–80°C)	Kills most non-sporing bacteria, yeast and moulds
Blanching (95–110°C)	Kills surface vegetative bacteria, yeast and moulds
Canning (above 100°C)	'Commercially sterilises' food; kills all pathogenic bacteria
Drying	Stops growth of all microbes when $a_w < 0.60$
Salting	Stops growth of most microbes at *ca*. 10% salt
Syruping (sugars)	Halts growth when $a_w < 0.70$
Acidifying	Halts growth of most bacteria (effects depend on acid type)

a_w denotes water activity.

Table 2.5   Most common faults in handling food which allow growth or formation of toxins by microorganisms (Sinell *et al.*, 1995).

Temperature abuse	Other process parameters
No or not sufficient chilling	Not properly controlled
Inadequate heat processing	a_w
temperature too low	pH
process time too short	NO_2^-/NO_3^- and concentration of
holding time too long at $\leq 65°C$	other preservatives not adequate

These methods (Table 2.4) prevent microbial growth due either to unfavourable temperatures or to the presence of compounds toxic to microorganisms. In order to design adequate treatment processes an understanding of the factors affecting microbial growth is necessary. Table 2.5 lists the common faults during processing which result in the growth of foodborne pathogens or toxin production.

2.5.1   *Intrinsic and extrinsic factors affecting microbial growth*

Food is a chemically complex matrix, and predicting whether, or how fast, microorganisms will grow in any given food is difficult. Most foods contain sufficient nutrients to support microbial growth. Several factors encourage, prevent or limit the growth of microorganisms in foods; the most important are a_w, pH, and temperature.

Factors affecting microbial growth are divided into two groups: intrinsic and extrinsic parameters (Table 2.6). These factors affect the growth of microorganisms on foods.

Table 2.6 Intrinsic and extrinsic parameters affecting microbial growth.

Intrinsic parameters	Extrinsic parameters
Water activity, humectant identity	Temperature
Oxygen availability	Relative humidity
pH, acidity, acidulant identity	Atmosphere composition
Buffering capacity	Packaging
Available nutrients	
Natural antimicrobial substances	
Presence and identity of natural microbial flora	
Colloidal form	

2.5.2 a_w *(water activity or water availability)*

When other substances (solutes) are added to water, water molecules orient themselves on the surface of the solute and the properties of the solution change dramatically. The microbial cell must compete with solute molecules for free water molecules. Except for *St. aureus*, bacteria are rather poor competitors for free water, whereas moulds are excellent competitors (Chirife & del Pilarbuera 1996).

Water activity (a_w) is a measure of the available water in a sample. The a_w is the ratio of the water vapour pressure of the sample to that of pure water at the same temperature:

$$a_w = \frac{\text{water vapour pressure of sample}}{\text{pure water vapour pressure}}$$

A solution of pure water has an a_w of 1.00. The addition of solute decreases the a_w to less than 1.00. The a_w varies very little with temperature over the range of temperatures that support microbial growth.

The a_w of a solution may dramatically affect the ability of heat to kill a bacterium at a given temperature. For example, a population of *S. typhimurium* is reduced 10-fold in 0.18 minutes at 60°C if the a_w of the suspending medium is 0.995. If the a_w is lowered to 0.94, 4.3 minutes are required at 60°C to cause the same 10-fold reduction.

An a_w value stated for a bacterium is generally the minimum a_w which supports growth. At the minimum a_w growth is usually minimal, increasing as the a_w increases. At a_w values below the minimum for growth, bacteria do not necessarily die, although some proportion of the population does die. The bacteria may remain dormant, but infectious. Most importantly, a_w is only one factor, and the other factors (e.g. pH and temperature) of the food must be considered. It is the interplay between factors that ultimately determines whether a bacterium will grow or not. The a_w of a food may not be a fixed value; it may change over time, or may vary considerably between similar foods from different sources. The minimum water activity supporting growth for various microorganisms is given in Table 2.7.

The water activity has been widely used as the preservation factor by the addition of salt and sugar. Sugar has been traditionally used in the preservation of fruit products (jams and preserves). In contrast, salt has been used for the preservation of meat and fish. The water activity of a range of foods is given in Table 2.8.

2.5.3　pH (hydrogen ion concentration, relative acidity or alkalinity)

The pH range of a microorganism is defined by a minimum value (at the acidic end of the scale) and a maximum value (at the basic end of the scale). There is a pH optimum for each microorganism at which growth is maximal (Table 2.7). Moving away from the pH optimum in either direction slows microbial growth. A range of food pH values is given in Table 2.9. Shifts in pH of a food with time may reflect microbial activity, and foods that are poorly buffered (i.e. do not resist changes in pH), such as vegetables, may shift pH values considerably. For meats, the pH of muscle from a rested animal may differ from that of a fatigued animal.

A food may start with a pH which precludes bacterial growth, but as a result of the metabolism of other microbes (yeasts or moulds), pH shifts may occur and permit bacterial growth.

2.5.4　Temperature

Temperature values for microbial growth, like pH values, have a minimum and maximum range with an optimum temperature for maximal growth. The optimum growth temperature determines its classification as a thermophile, mesophile or psychrophile (Table 2.10). A thermophile cannot grow at room temperature and therefore canned foods can be stored at room temperature even though they may contain thermophilic microorganisms which survived the high processing temperature.

Time and temperature guidance has been given for the major foodborne pathogens in seafoods with regard to the total time at a given

Table 2.7 Limits of microbial growth; water activity (a_w), pH and temperature.

Organism	Minimal water activity (a_w)	pH range	Temperature range (°C)[a]	Growth rate[b] (t_d)
A. hydrophila	0.970	(7.2 optimum)	−0.1–42	12 h, 4°C
B. cereus	0.930	4.3–9.3	4–52	4 h/generation, 8°C
B. stearothermophilus	–	5.2–9.2	28–72	6 h/generation, 32°C
C. jejuni	0.990	4.9–9.5	30–45	(8 d, 10°C)[c]
Cl. botulinum type A and proteolytic B and F	0.935	4.6–9.0	10–48	(8 d, 10°C)[c]
Cl. botulinum type E and non-proteolytic B and F	0.965	5.0–9.0	3.3–45	12 h, 12°C
Cl. perfringens	0.945	5.0–9.0	10–52	25 h/generation, 8°C
E. coli	0.935	4.0–9.0	7–49.4	1 d, 4.4°C
Lactobacillus spp.	0.930	3.8–7.2	5–45	
L. monocytogenes	0.920	4.4–9.4	−0.4–45	(60 h),[d] 10 h, 10°C
Salmonella spp.	0.940	3.7–9.5	5–46	(3.6 d, 8°C)[e]
Shigella spp.	0.960	4.8–9.3	6.1–47.1	(2.8 d),[d] 1 d, 10°C
St. aureus	0.830	4.0–10	7–50	(4 h),[d] 98 min, 20°C
(toxin production)	0.850	4.0–9.8	10–48	60 min, 18°C
V. cholerae	0.970	5.0–10.0	10–43	
V. parahaemolyticus	0.936	4.8–11	5–44	
V. vulnificus	0.960	5.0–10	8–43	
Y. enterocolitica	0.945	4.2–10	−1.3–45	17 h, 5°C
Saccharomyces spp.	0.85	2.1–9.0	—	
Asp. oryzae	0.77	1.6–13.0	10–43	
F. miniliforme	0.87	<2.5–>10.6	2.5–37	
Pen. verrucosum	0.79	<2.1–>10.0	0–31	

[a] To convert to °F use the equation °F = (9/5)°C + 32. As a guidance: 0°C = 32°F, 4.4°C = 40°F, 60°C = 140°F.

[b] These are only examples of doubling time (t_d). The values will vary according to food composition.

[c] Time to toxin production.

[d] Lag time.

[e] Average time to turbidity (inoculation from 1:1 dilution of 5 h, 37°C culture).

Various sources were used, principally ICMSF (1996), Corlett (1998), Mortimore & Wallace (1994); where data differed between sources the wider growth range was quoted.

Table 2.8 Water activity of various foods.

Water activity (a_w) range	Foods
1.00–0.95	Highly perishable foods: meat, vegetables, fish, milk Fresh and canned fruit Cooked sausages and breads Foods containing up to 40% sucrose or 7% NaCl
0.95–0.91	Some cheeses (e.g. Cheddar) Cured meat (e.g. ham) Some fruit juice concentrates Foods containing 55% sucrose or 12% NaCl
0.91–0.87	Fermented sausages (salami), sponge cakes, dry cheeses, margarine Foods containing 65% sucrose or 15% NaCl
0.87–0.80	Most fruit juice concentrates, sweetened condensed milk, chocolate syrup, maple and fruit syrups, flour, rice, pulses containing 15–17% moisture, fruit cake
0.80–0.75	Jam, marmalade, marzipan, glace fruits
0.75–0.65	Rolled oats, fudge, jelly, raw cane sugar, nuts
0.65–0.60	Dried fruits, some toffees and caramels, honey
0.60–0.50	Noodles, spaghetti
0.50–0.40	Whole egg powder
0.40–0.30	Cookies, crackers
0.30–0.20	Whole milk powder, dried vegetables, cornflakes

temperature (Table 2.11). For example, to control the growth of *Salmonella* species the food must not be exposed to temperatures between 5.2 and 10°C for more than 14 days, between 11 and 21°C for more than 6 hours or above 21°C for more than 3 hours. Synder (1995) has calculated the total time for food quality (defined as the growth of pathogens through five multiplications) and food safety (defined as the growth of food pathogens through 10 multiplications) at a range of temperatures (Table 2.11). These times and temperatures are based on the known growth rates of common food pathogens and give a danger zone of microbial pathogen growth as a lower limit of −1.5°C (29.3°F) to control *L. monocytogenes*, *Y. enterocolitica* and *A. hydrophila* and an upper limit of 53°C (127.5°F), which is the limit of *C. perfringens* growth (Table 2.7). However, these do not necessarily concur with the national standards of 4 to 60°C (40 to 140°F) which were set before the emergence of certain

Table 2.9   pH values of various foods (ICMSF, 1988).

pH range	Food	pH
Low acid (pH 7.0–5.5)	(Whole eggs	7.1–7.9)
	(Frozen eggs	8.5–9.5)
	Milk	6.3–8.5
	(Camembert cheese	7.44)
	Cheddar cheese	5.9
	Roquefort cheese	5.5–5.9
	Bacon	6.6–5.6
	Carcass meat	7.0–5.4
	Red meat	6.2–5.4
	Ham	5.9–6.1
	Canned vegetables	6.4–5.4
	Poultry	5.6–6.4
	Fish	6.6–6.8
	Crustaceans	6.8–7.0
	Butter	6.1–6.4
	Potatoes	5.6–6.2
	Rice	6.0–6.7
	Bread	5.3–5.8
Medium acid (pH 5.3–4.5)	Fermented vegetables	5.1–3.9
	Cottage cheese	4.5
	Bananas	4.5–5.2
	Green beans	4.6–5.5
Acid (pH 4.5–3.7)	Mayonnaise	4.1–3.0
	Tomatoes	4.0
High acid (<pH 3.7)	Canned pickles and fruit juice	3.9–3.5
	Sauerkraut	3.3–3.1
	Citrus fruits	3.5–3.0
	Apples	2.9–3.3

Foods in parentheses are outside the pH range, but are included for comparison.

Table 2.10   Grouping of microorganisms according to temperature growth range.

Group	Minimum ($^\circ$C)	Optimum ($^\circ$C)	Maximum ($^\circ$C)
Psychrophiles	−5	12–15	20
Mesophiles	5	30–45	47
Thermophiles	40	55–75	60–90

Note: To convert to $^\circ$F use the equation $^\circ F = (9/5)^\circ C + 32$. As a guidance: $0^\circ C = 32^\circ F$, $4.4^\circ C = 40^\circ F$, $60^\circ C = 140^\circ F$.

Table 2.11 Safe food holding times at specified temperatures (adapted from Snyder 1995).

Temp. (°C)	Safety[a]	Quality[b]
54.4		
↑	4 h	2 h
37.8		
35.0	5 h	2.5 h
32.2	6 h	3 h
29.4	7 h	3.5 h
26.7	8.5 h	4 h
23.9	11 h	5.5 h
21.1	14 h	7 h
18.3	18 h	9 h
15.6	1 day	12 h
12.8	1.5 days	18 h
10.0	2.5 days	30 h
7.2	5 days	2.5 days
5.0	10 days	5 days
4.4	13 days	6.5 days
1.7	30 days	15 days
0.0	60 days	30 days
−1.1	Safe chilled food holding	
−2.2	Meat, poultry, fish thaw	
−5.0	Spoilage bacteria begin to multiply	
−10.0	Yeasts and moulds begin to multiply	

[a] 10 multiplications of pathogens. [b] 5 multiplications of pathogens.

foodborne pathogens such as *L. monocytogenes*. Regrettably legislation is not always updated promptly.

2.5.5 Interplay of factors affecting microbial growth in foods

Although each of the major factors listed above plays an important role, the interplay between the factors ultimately determines whether a microorganism will grow in a given food. Often, the results of such interplay are unpredictable, as poorly understood synergism or antagonism may occur. Advantage is taken of this interplay with regard to preventing the outgrowth of *Cl. botulinum*. Food with a pH of 5.0 (within the range for *Cl. botulinum*) and an a_w of 0.935 (above the minimum for *Cl. botulinum*) may not support the growth of this bacterium. Certain processed cheese spreads take advantage of this fact and are therefore shelf stable at room temperature, even though each individual factor would permit the outgrowth of *Cl. botulinum*.

Therefore, predictions about whether or not a particular microorganism will grow in a food can, in general, only be made through experimentation. Also, many microorganisms do not need to multiply in food to cause disease.

2.6 Preservatives

Preservative agents are required to ensure that manufactured foods remain safe and unspoiled during their whole shelf-life. A range of preservatives is used in food manufacture including traditional foods (Table 2.12). Many preservatives are effective under low pH conditions: benzoic

Table 2.12 Antimicrobial food preservatives (adapted from Gould and reprinted with permission of Elsevier Science from the *International Journal of Food Microbiology*, 1996, **33**, 51–64).

Preservative (typical concentration range, mg/kg)	Examples of use
Weak organic and ester preservatives	
Propionate (1–5000)	Bread, bakery and cheese products
Sorbate (1–2000)	Fresh and processed cheese, dairy products, bakery products, syrups, jams, jellies, soft drinks, margarines, cakes, dressings
Benzoate (1–3000)	Pickles, soft drinks, dressings, semi-preserved fish, jams, margarines
Benzoate esters (parabens, 10)	Marinaded fish products
Organic acid acidulants	
Lactic, citric, malic, acetic acids (no limit)	Low pH sauces, mayonnaises, dressings, salad creams, drinks, fruit juices and concentrates, meat and vegetable products
Inorganic acid preservatives	
Sulphite (1–450)	Fruit pieces, dried fruit, wine, meat sausages
Nitrate and nitrite (50)	Cured meat products
Mineral acid acidulants	
Phosphoric acid, hydrochloric acid	Drinks
Antibiotics	
Nisin	Cheese, canned foods
Natamycin (pimaricin)	Soft fruit
Smoke	Meat and fish

acid (pH < 4.0), propionic acid (pH < 5.0), sorbic acid (pH < 6.5), sulfites (pH < 4.5). The parabens (benzoic acid esters) are more effective at neutral pH conditions.

2.6.1 Organic acids

Acetic, lactic, benzoic and sorbic acids are weak acids which are commonly used as preservative agents. These molecules inhibit the outgrowth of both bacterial and fungal cells. Sorbic acid also inhibits the germination and outgrowth of bacterial spores (Sofos & Busta 1981; Blocher & Busta 1985). The addition of 0.2% calcium propionate to bread dough delays *B. cereus* germination sufficiently to reduce the risk of food poisoning to negligible (Kaur 1986). Benzoic acid concentration is usually *ca* 500 ppm in order to preserve fruit juice-based beverages. Sulphur dioxide concentrations are controlled in European regulations to a limit of 10 ppm.

In solution, weak acid preservatives exist in a pH-dependent equilibrium (measured as the pK value) between the undissociated and dissociated state. Optimal inhibitory activity is at low pH because this favours the uncharged, undissociated state of the molecule which is freely permeable across the plasma membrane (lipophilic) and is thus able to enter the cell (Fig. 2.7). The molecule will dissociate after entering the cell and this results in the release of charged anions and protons which cannot cross the plasma membrane. Hence the preservative molecule diffuses into the cell until an equilibrium is achieved. This results in the accumulation of anions and protons inside the cell (Booth & Kroll 1989). Therefore, inhibition of growth by weak acid preservatives has been proposed to be due to a number of actions, including membrane disruption (Bracey *et al.* 1998), inhibition of essential metabolic reactions, stress on intra-

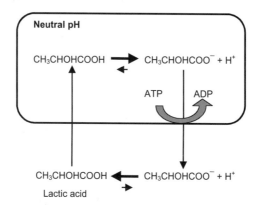

Fig. 2.7 Mode of action of organic acids.

cellular pH homeostasis and the accumulation of toxic anions (Eklund 1985). The concentrations of undissociated acid that inhibit micro-organisms are given in Table 2.13. An additional inhibitory action on yeasts is the induction of an energetically expensive stress response that attempts to restore homeostasis and results in the reduction of available energy pools for growth and other essential metabolic functions (Holyoak *et al.* 1996; Bracey *et al.* 1998).

Table 2.13   Inhibitory concentration (%) of dissociated organic acids (Various sources including Mortimore and Wallace, 1994; Borch *et al.*, 1996 and ICMSF, 1996).

Acid	Enterobacteriaceae	Bacillaceae	Yeasts	Moulds
Acetic	0.05	0.1	0.5	0.1
Benzoic	0.01	0.02	0.05	0.1
Sorbic	0.01	0.02	0.02	0.04
Propionic	0.05	0.1	0.2	0.05

2.6.2   Hydrogen peroxide

The lactoperoxidase system, found in milk, has profound antimicrobial effects against both bacteria and fungi (Reiter & Harnulv 1984; Russel 1991; de Wit & van Hooydonk 1996). The system requires hydrogen peroxide and thiocyanate for optimal activity and is primarily active against microorganisms producing H_2O_2. Alternatively, hydrogen per-oxide can be added to the foods that are to be preserved. Under suitable experimental conditions, the reaction generates a short-lived singlet oxygen which is extremely biocidal (Fig. 2.8; Tatsozawa *et al.* 1998). Furthermore, during incomplete reduction of molecular oxygen, the superoxide radical is generated. H_2O_2 may lead together with the pur-peroxide radical and trace amounts of transition metal ions (e.g. Fe (II)) in

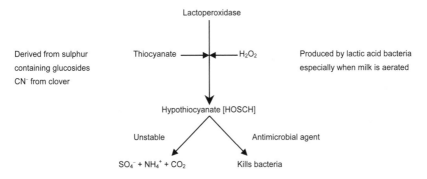

Fig. 2.8   The lactoperoxidase system.

the so-called Fenton reaction to the formation of the extremely biocidal hydroxyl radical (Luo *et al.* 1994). A wide range of both Gram-negative and Gram-positive bacteria are inhibited by the lactoperoxidase system. However, studies have shown that Gram-negative bacteria are generally more sensitive to lactoperoxidase-mediated (food) preservation than Gram-positive species. Hydrogen peroxide on its own is also known to be bactericidal depending on the concentrations applied and on environmental factors such as pH and temperature (Juven & Pierson 1996).

Temperature is also an extremely important parameter in determining the sporicidal efficacy of hydrogen peroxide. H_2O_2 was found to be weakly sporicidal at room temperatures but very potent at higher temperatures. While the mechanism by which hydrogen peroxide kills spores is not known, killing of vegetative bacteria and fungi is known to involve DNA damage. In the USA, the regulatory authorities allow the direct addition of hydrogen peroxide to food products such as raw milk for the preparation of certain cheese variants; whey intended for use in modified whey preparation; corn starch and dried eggs; and for the decontamination of packaging material (Juven & Pierson 1996). Several other procedures where H_2O_2 is used as a preservative have been reported, such as fruit and vegetable disinfection and raisin decontamination (Falik *et al.* 1994).

2.6.3 Chelators

Chelators that can be used as food additives include the naturally occurring acid, citric acid, and the disodium and calcium salts of ethylenediaminetetraacetic acid (EDTA; Russel 1991). EDTA is known to potentiate the effect of weak acid preservatives against Gram-negative bacteria, while citric acid inhibits growth of proteolytic *Cl. botulinum* due to its Ca^{2+} chelating activity (Graham & Lund 1986).

2.6.4 Small organic biomolecules

In addition to the weak organic acids, H_2O_2 and certain chelators, a few other antimicrobial compounds are permitted by the regulatory authorities for inclusion in foods. In fact, some of these compounds are naturally present in spices (Table 2.14): eugenol in cloves, allicin in garlic, thymol in rosemary, cinnamic aldehyde and eugenol in cinnamon and allyl isothiocyanate in mustard. These compounds are generally hydrophobic and can have membrane perturbing or even membrane rupturing characteristics. However, the concentrations at which the latter occurs, and thus cell death is achieved, are above tolerable taste thresholds. Also it is debatable whether the levels in fresh garlic are significant enough to exert

Table 2.14 Concentration of essential oils in some spices and antimicrobial activity of active components (adapted from ICMSF 1998).

Spice	Essential oil in whole spice (%)	Antimicrobial compounds in distillate or extract	Antimicrobial concentration (ppm)	Organisms
Allspice (*Piementa dioica*)	3.0–5.0	Eugenol Methyl eugenol	1000 150	Yeast *Acetobacter* *Cl. botulinum* 67B
Cassis (*Cinnamomum cassis*)	1.2	Cinnamic aldehyde Cinnamyl acetate	10–100	Yeast *Acetobacter*
Clove (*Syzgium aromaticum*)	16.0–19.0	Eugenol Eugenol acetate	1000 150	Yeast *Cl. botulinum* *V. parabaemolyticus*
Cinnamon Bark (*Cinnamomum zeylanicum*)	0.5–1.0	Cinnamic aldehyde Eugenol	10–1000 100	Yeast, *Acetobacter* *Cl. botulinum* 67B *L. monocytogenes*
Garlic (*Allium sativum*)	0.3–0.5	Allyl sulphonyl Allyl sulphide	10–100	*Cl. botulinum* 67B *L. monocytogenes* Yeast, bacteria
Mustard (*Sinapis nigra*)	0.5–1.0	Allyl isothionate	22–100	Yeast, *Acetobacter* *L. monocytogenes*
Oregano (*Origanum vulgare*)	0.2–0.8	Thymol Carvacrol	100 100–200	*V. parabaemolyticus* *Cl. botulinum* A, B, E
Paprika (*Capsicum annuum*)		Capsicidin	100	*Bacillus*
Thyme (*Thymus vulgaris*)	2.5	Thymol Carvacrol	100 100	*V. parabaemolyticus* *Cl. botulinum* 67B Gram-positive bacteria *Asp. parasiticus* *Asp. flavus* aflatoxin B_1 and G_1

any antibacterial affect; there is no allicin in cooked garlic. Hence, in practice most of the above mentioned compounds are used at concentrations that may have only growth inhibitory effects on microorganisms. An extensive overview of the antimicrobial (and antioxidant) properties of spices can be found in Hirasa & Takemasa (1998). Essential oils from plants such as basil, cumin, caraway and coriander have inhibitory efffects on organisms such as *A. hydrophila, Ps. fluorescens* and *St. aureus* (Wan *et al.* 1998).

2.6.5 Hurdle concept

Several factors can be used as a series of 'hurdles' to prevent microbial growth. Each hurdle represents a barrier that must be overcome by the bacteria to initiate food spoilage and food poisoning. Modified atmosphere packaging, high hydrostatic pressure, ultraviolet light, ethanol and bacteriocins are examples of secondary preservation methods. Additional precautions can be the rotation of antimicrobials and sanitisers to prevent the build-up of resistant strains. The interaction of factors can best be seen in the effect on microbial growth according to growth conditions (see Fig. 2.12 in Section 2.8.4).

2.7 Microbial response to stress

The common means of food preservation (freezing, etc., see Table 2.4) are being replaced by new techniques such as:

- Mild heating
- Modified atmosphere and vacuum packaging
- Inclusion of natural antimicrobial agents
- High hydrostatic pressure
- Pulse electric field and high intensity laser

The development of these techniques is partially due to consumer demand for less processed, less heavily preserved yet higher quality foods. They are 'milder' than the traditional methods and commonly rely on storage and distribution at refrigeration temperatures for their preservation. These are commonly called 'minimal processed foods'.

The microorganisms of concern in minimal processed foods are psychrotrophic and mesotrophic microorganisms. The psychrotrophic organisms, by definition, can grow at refrigeration temperatures, whereas mesophilic pathogens can survive under refrigeration and grow during periods of temperature abuse (see Table 2.15).

Table 2.15 Food poisoning microorganisms of concern in minimally processed foods (adapted from Abee & Wouters 1999).

Minimum growth temperature (°C)	Heat resistance	
	Low[a]	High[b]
0-5	*L. monocytogenes* *Y. enterocolitica* *A. hydrophila*	*Cl. botulinum* type E and non-proteolytic type B *B. cereus* *B. subtilis* *B. licheniformis*
5-10	*Salmonella* spp. *V. parahaemolyticus* Pathogenic strains of *E. coli* *St. aureus*	
10-15		*Cl. botulinum* type A and proteolytic type B *Cl. perfringens*

[a] Organisms undergo a 6 log kill following heat treatment at 70°C for 2 minutes.
[b] Organisms require heat treatment at 90°C or above to destroy spores.
Note: To convert to °F use the equation °F = (9/5)°C + 32. As a guidance: 0°C = 32°F, 4.4°C = 40°F, 60°C = 140°F.

Microorganisms are able, within limits, to adapt to stress conditions such as acidity and cold (Table 2.16). The mechanism of adaptation is by signal transduction systems which control the coordinated expression of genes involved in cellular defence mechanisms (Huisman & Kolter 1994; Rees *et al.* 1995; Kleerebezem *et al.* 1997). Microbial cells are able to adapt to many processes used to retard microbial growth starvation, that is cold shock, heat shock, (weak) acids, high osmolarity and high hydrostatic pressure. The easiest mechanism of survival to recognise is in *Bacillus* and *Clostridium* species. These organisms form spores under stress conditions which can germinate under favourable conditions later. Other organisms, such as *E. coli*, undergo significant physiological changes to enable the cell to survive environmental stresses such as starvation, near-UV radiation, hydrogen peroxide, heat and high salt.

The regulatory mechanism involves the modification of sigma (δ) factors whose primary role is to bind to core RNA polymerase conferring promoter specificity (Haldenwang 1995). The sigma factors of *B. subtilis* and *E. coli* have been extensively studied. The main sigma factor is responsible for the transcription from the majority of the promoters. Alternative sigma factors have different promoter specificities, directing expression of specific regulons involved in heat-shock response, the

Table 2.16 Response mechanisms in microorganisms (adapted from Gould and reprinted with permission from Elsevier Science from the *International Journal of Food Microbiology*, 1996, **33**, 51-64).

Environmental stress	Stress response reaction
Low nutrient levels	Nutrient scavenging, oligotrophy, generation of viable non-culturable forms
Low pH, presence of weak organic acids	Extrusion of hydrogen ions, maintenance of cytoplasmic pH and membrane pH gradient
Reduced water activity	Osmoregulation, avoidance of water loss, maintenance of membrane turgor
Low temperature – growth	Membrane lipid changes, cold shock response
High temperature – growth	Membrane lipid changes, heat shock response
High oxygen levels	Enzymic protection from oxygen-derived free radicals
Biocides and preservatives	Phenotypic adaptation and development of resistance
Ultraviolet radiation	Excision of thymine dimers and repair of DNA
Ionizing radiation	Repair of DNA single strand breaks
High temperature – survival	Low water content in the spore protoplast
High hydrostatic pressure – survival	Low spore protoplast water content?
High voltage electric discharge	Low conductivity of spore protoplast
Ultrasonication	Structural rigidity of cell wall
High levels of biocides	Impermeable outer layers of cells
Competition	Formation of biofilms, aggregates with some degree of symbiosis

chemotactic response, sporulation and general stress response (Haldenwang 1995; Abee & Wouters 1999).

2.7.1 Response to pH stress

Acidification of food is a commonly used method of preserving food, such as dairy products. This method is very effective since the main foodborne pathogens grow best at neutral pH values (Table 2.7). These organisms, however, are able to tolerate and adapt to weak acid stresses and this possibly enables them to survive passage through the acidic human stomach. Additionally acid resistance can cross-protect against heat treatment (see *E. coli*, Table 2.3; Buchanan & Edelson 1999).

Acid stress is the combined effect of low pH and weak (organic) acids such as acetate, propionate and lactate (see Section 2.6.1). Weak acids in their unprotonated form can diffuse into the cell and dissociate. Consequently they lower the intracellular pH (pH_{in}) resulting in the inhibition of various essential cytoplasmic enzymes. In response microorganisms have inducible acid survival strategies. Regulatory features include an alternative sigma factor sigma^s (an acid shock protein), two-component signal transduction systems (e.g. PhoP and PhoQ) and the major iron regulatory protein Fur (ferric uptake regulator, Bearson *et al.* 1997).

The important aspect of the acid tolerance response (ATR) is the induction of cross-protection to a variety of stresses (heat, osmolarity, membrane active compounds) in exponentially grown (log phase) cells. Acid-adapted cells are those that have been exposed to a gradual decrease in environmental pH, whereas acid-shocked cells are those which have been exposed to an abrupt shift from high pH to low pH. It is important to differentiate between these two conditions since acid-adapted, but not acid-shocked *E. coli* O157:H7 cells in acidified tryptone soya broth and low pH fruit juices have enhanced heat tolerance in TSB at 52°C and 54°C, and in apple cider and in orange juice at 52°C. Acid-induced general stress resistance may reduce the efficiency of hurdle technologies which are dependent upon multiple stress factors (Section 2.6.5).

Acid resistance can be induced by factors other than acid exposure (Baik *et al.* 1996; Kwon & Ricke 1998). *S. typhimurium* acid tolerance can be induced by exposure to short-chain fatty acids, which are used as food preservatives and also occur in the intestinal tract (see Sections 2.6.1 and 3.7). Subsequently the virulence of *S. typhimurium* may be enhanced by increasing acid resistance upon exposure to short-chain fatty acids, such as propionate, and further enhanced by anaerobiosis and low pH.

2.7.2 Response to heat shock

Food preservation by heat is a common method, for example blanching, pasteurisation and sterilisation. Other means of reducing or inactivating the microbial population use water, steam, hot air, electrical, light, ultrasound or microwave energy (Table 2.16; Heldman & Lund 1992). The targets for heat inactivation of microbes are the intrinsic stability of macromolecules, i.e. ribosomes, nucleic acids, enzymes and intracellular proteins and the membrane. The exact primary cause for cell death due to heat exposure is not fully understood (Earnshaw *et al.* 1995). Bacterial

thermotolerance increases upon exposure to sublethal heating temperatures, phage infection and chemical compounds such as ethanol and
streptomycin.

Microbes can adapt to mild heat treatment in a variety of ways:

- Cell membrane composition changes by increasing the saturation and
  the length of the fatty acids in order to maintain the optimal membrane
  fluidity and the activity of intrinsic proteins.
- Accumulation of osmolytes may enhance protein stability and protect
  enzymes against heat activation.
- *Bacillus* and *Clostridium* species produce spores.
- Production of heat shock proteins (HSPs).

When bacteria cells are exposed to higher temperatures, a set of HSPs is
rapidly induced. The primary structure of most HSPs appears to be highly
conserved in a wide variety of microorganisms. HSPs involve both
chaperones and proteases which act together to maintain quality control of
cellular proteins. Both types of enzymes have as their substrates a variety of
misfolded and partially-folded proteins that arise from slow rates of folding
or assembly, chemical or thermal stress, intrinsic structural instability and
biosynthetic errors. The primary function of classical chaperones, such as
the *E. coli* DnaK (Hsp70) and its co-chaperones, DnaJ and GrpE, and GroEL
(Hsp60) and its co-chaperone, GroES, is to modulate protein folding
pathways, thereby preventing misfolding and aggregation and promoting
refolding and proper assembly. HSPs are induced by several stress situations, for example heat, acid, oxidative stress and macrophage survival,
which suggests that HSPs contribute to bacterial survival during infection
(see Section 3.9). In addition, HSPs may enhance the survival of foodborne
pathogens in foods during exposure to high temperatures. It is important
to note that microorganisms develop a complicated, tightly regulated
response upon an upshift in temperature. Different stressors can activate
(parts of) this stress regulon by which they can induce an increased heat
tolerance. The process of adaptation and initiation of defence against
elevated temperature is an important target when considering food preservation and the use of hurdle technology (Section 2.6.5).

2.7.3 Response to cold shock

Cold adaptation by microorganisms is of particular importance due to the
increased use of frozen and chilled foods and the increased popularity of
fresh or minimally processed food, with little or no preservatives. The
relevant growth ranges of *L. monocytogenes*, *Y. enterocolitica*, *B. cereus*
and *Cl. botulinum* are of particular importance (see Table 2.7).

Mechanisms of cold adaptation are:

- Membrane composition modifications, to maintain membrane fluidity for nutrient uptake.
- Structural integrity of proteins and ribosomes.
- Production of cold shock proteins (CSPs).
- Uptake of compatible solutes (i.e. betaine, proline and carnitine).

To maintain membrane fluidity and function at low temperatures, microorganisms increase the proportion of shorter and/or unsaturated fatty acids in the lipids. In *E. coli* the proportion of cis-vaccenic acid (C18:1) increases at low temperature at the expense of palmitic acid (C16:0). The increase in average chain length has the opposite effect on membrane fluidity, but is outweighed by the greater fluidity of increased unsaturation. In *L. monocytogenes*, the fatty acid composition is altered in response to low temperature by an increase of a C15:0 fatty acid and a decrease of a C17:0 fatty acid. Compatible solutes (such as betaine, proline and carnitine) may play a role in osmoprotection and in cold adaptation. At 7°C, *L. monocytogenes* increases its uptake of betaine 15-fold compared to growth at 30°C. Cold shock proteins are small (7 kDa) proteins which are synthesised when bacteria are subjected to a sudden decrease in temperature. They are involved in protein sysnthesis and mRNA folding.

Different cold shock treatments prior to freezing have different effects on survival of bacteria after freezing. This might result in a high survival rate of bacteria in frozen food products. Furthermore, low temperature adapted bacteria may be relevant to food quality and safety.

2.7.4 *Response to osmotic shock*

Lowering water activity (a_w) is one of the common ways of preserving food (Section 2.5.2). This works by increasing the osmotic pressure on the microbial cell. This is achieved by the addition of high amounts of salts and sugar (osmotically active compounds) or desiccation. The internal osmotic pressure in bacterial cells is higher than that of the surrounding medium. This results in an outward pressure called 'turgor pressure', which is necessary for cell elongation.

The microbial response to loss of turgor pressure is the cytoplasmic accumulation of 'compatible solutes' that do not interfere too seriously with cellular functions (Booth *et al.* 1994). These compounds are small organic molecules, which share a number of common properties:

- Soluble to high concentrations.
- Can be accumulated to very high levels in the cytoplasm.

- Neutral or zwitterionic molecules.
- Specific transport mechanism is present in the cytoplasmic membrane.
- Do not alter enzyme activity.
- May protect enzymes from denaturation by salts or protect them against freezing and drying.

The adaptation of *E. coli* O157:H7, *S. typhimurium*, *B. subtilis*, *L. monocytogenes* and *St. aureus* to osmotic stress is mainly through the accumulation of betaine (*N,N,N*-trimethylglycine) via specific transporters. Other compatible solutes include: carnitine, trehalose, glycerol, sucrose, proline, mannitol, glucitol, ectoine and small peptides.

2.7.5   Response to high hydrostatic pressure

The use of pressure technology is a novel means of food preservation (Table 2.16; Knorr 1993; Hendrickx *et al.* 1998). The key objective is to reduce foodborne pathogens by about 8 log cycles (8D). Elevated pressure can exert detrimental effects on microbial physiology and viability (Kalchayanand *et al.* 1998). Growth of microorganisms is generally inhibited at pressures in the range of 20 to 130 MPa, while higher pressures of between 130 and 800 MPa may result in cell death; the maximum pressure allowing for growth or survival depends on the species and medium composition. The site of cell damage is possibly the cytoplasmic membrane and ribosomes. Exposure of *E. coli* to high pressure results in the synthesis of HSPs, CSPs and proteins only associated with high pressure exposure.

Pressure cycling has been proposed to inactivate bacterial spores. Spore germination is induced at low pressure (100 to 250 MPa) followed by inactivation at high pressure (500 to 600 MPa). A combination of pressure cycling and other preservation methods may be required for the control of *Clostridium* species as the spores of these organisms have greater pressure tolerance than the *Bacillus* species spores. Moderate hydrostatic pressure (270 MPa) at 25°C only reduces the viable count by approximately 1.3 log cycles. However 5 log cycle kills can be achieved by combined treatment conditions of 35°C and the inclusion of bactericidal compounds such as pediocin and nisin (Fig. 2.9; Kalchayanand *et al.* 1998).

2.8   Predictive modelling

The major factors affecting microbial growth are:

- pH
- Water activity

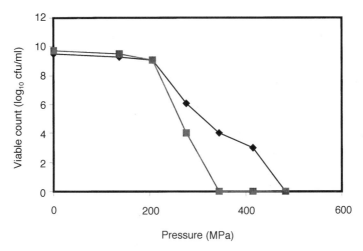

Fig. 2.9 *St. aureus* survival following high pressure treatment. Plated onto selective (■) and non-selective (◆) media (Kalchayanard *et al.*, 1998).

- Atmosphere
- Temperature
- Presence of certain organic acids, such as lactate

However our ability to predict accurately the subsequent growth of microbes in food with regard to the safety and shelf life of food is very limited. Obviously if food is subjected to temperature fluctuations during distribution and storage then the rate of microbial growth will be affected.

Predictive food microbiology is a field of study that combines elements of microbiology, mathematics and statistics to develop models that describe and predict the growth and decline of microbes under prescribed (including varying) environmental conditions (Whiting 1995). There are 'kinetic' models which model the extent and rate of microbial growth and 'probability' models which predict the likelihood of a given event occurring, such as sporulation (Ross & McMeekin 1994).

The main objective is to describe mathematically the growth of microorganisms in food under prescribed growth conditions. Initially the data for models are collected using a range of bacterial strains to represent the variation of the target organism present in the commercial situation. Ideally this will include strains associated with outbreaks, the fastest growing strains and the most frequently isolated strain. Although there is a considerable wealth of knowledge from the growth of microbes in bio-reactors (fermenters) this is not directly applicable to the food industry. In food the environmental factors are more varied and fluctuate and one is often dealing with a mixed population. A detailed study of predictive microbiology is given by McMeekin *et al.* (1993) and McDonald & Sun (1999).

2.8.1 Predicting modelling development

Predictive models have various applications (Whiting 1995):

- *Prediction of risk:* the model can estimate the likelihood of pathogen survival during storage.
- *Quality control:* predicting the effect of environmental factors can help in deciding on Critical Control Points in a Hazard Analysis Critical Control Point (HACCP) plan (see Section 7.4).
- *Product development:* microbial survival can be predicted for changes in processing and new product formulation without the need for extensive laboratory analysis.
- *Education:* the implications of changes in temperature, pH, etc., can be visualised during a staff training programme.

The origin of predictive microbiology is in the canning industry and the D values used to describe the rate of microbial death (Section 2.4). However, the ability to solve complex mathematical equations has required the revolution in computing power and this has greatly facilitated the development of predictive modelling. Models developed to predict microbial survival and growth may become an integral tool to evaluate, control, document and even defend the safety designed into a food product (Baker 1995). The application of microbial models in HACCP and Microbiological Risk Assessment are covered in Section 7.4 and Chapter 9. A form of predictive microbiology is the development of expert systems which can quantify the safety risks of food products without the necessity of extensive laboratory work (Schellekens *et al.* 1994; Wijtzes *et al.* 1998).

Models can be thought of as having three levels (Whiting 1995; McDonald & Sun 1999):

(1) Primary level models describe changes in microbial numbers (or equivalent) with time.
(2) Secondary level models show how the parameters of the primary model vary with environmental conditions.
(3) Tertiary level combines the first two types of models with user-friendly application software or expert systems that calculate microbial behaviour under changing environmental conditions.

2.8.2 *Primary models and the Gompertz and Baranyi equations*

Primary models may quantify the increase in microbial biomass as colony forming units per millilitre or absorbance; alternatively it could be by changes in media composition, such as metabolic end products, conductivity and toxin production.

According to Whiting (1995) the first primary models were simple equations such as growth verses no-growth conditions, for example acetic acid and sugar concentrations preventing the growth of spoilage yeast. Subsequent models described the time between inoculation and growth (or equivalent). This approach was used to model the time required for toxin production by *Cl. botulinum*. Primary models describing growth parameters started by plotting the growth curve and determining the rate of growth from the exponential phase:

$$N_t = N_0 e^{kt/\ln 2}$$

where

N_t = population size (logarithm) at a specified time
N_0 = initial population size (logarithm) at time 0
k = slope
t = time

This simple modelling was applied to the growth of *L. monocytogenes* in milk in the presence of pseudomonads (Marshall & Schmidt 1988) and in various meats (Yeh *et al.* 1991; Grau & Vanderlinde 1992). Further models were developed which included the lag phase.

The Gompertz function has become the most widely used primary model (Fig. 2.10; Pruitt & Kamau 1993). The Gompertz function can be represented as:

$$\log(N_t) = A + Ce^{-e(\exp(-B(t-M)))}$$

where :

N_t = population density (cfu/ml) at time t (hours)
A = initial population density (log(cfu/ml))
C = difference in initial and maximum population densities (log(cfu/ml))
M = time of maximum growth rate (hours)
B = relative maximum growth rate at M (log(cfu/ml)/hour)

This Gompertz function produces a sigmodial curve that consists of four phases comparable to the phases of microbiological growth: lag, acceleration, deceleration and stationary.

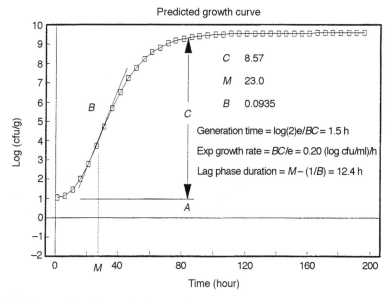

Fig. 2.10　Predicted microbial growth curve using the Gompertz function (from Pathogen Modelling Program Website).

Baranyi and co-workers developed an alternative equation based on the basic growth model which incorporated the lag, exponential and stationary phases and the specific growth rate:

$$N_t = N_{max} - \ln[1 + (\exp(N_{max} - N_0) - 1\exp(-\mu_{max}A(t))]$$

where:

N_t　　= population size (logarithm)
N_0　　= initial population size (logarithm)
N_{max} = maximum population (logarithm)
μ_{max} = maximum specific growth rate
$A(t)$　= integral of the adjustment function

The model fitted experimental data better than the Gompertz function with regard to predicting the lag time and exponential growth phase.

The Baranyi equation has been used as the basis (with modification) for the MicroFit software. MicroFit is a freeware program (see Appendix on Web resources) developed by MAFF (UK) and four other partners. Microbiological data are analysed according to the Baranyi growth model (Baranyi & Roberts 1994). The program enables microbiological growth data to be easily analysed:

- To determine μ_{max}, doubling time, lag time, initial cell count and final cell count.
- To estimate confidence intervals on the above parameters.
- To analyse simultaneously two data sets and compare them graphically.
- To perform statistical testing on difference between two data sets.

An example for *C. jejuni* is given in Fig. 2.11. The model cannot analyse the decline phase as this is not described by the Baranyi growth model.

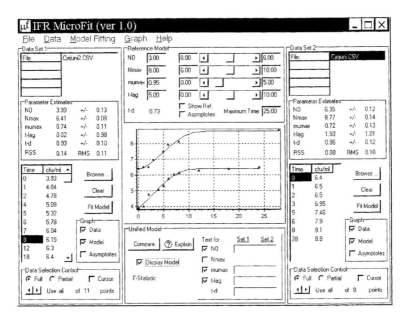

Fig. 2.11 An example of the MicroFit screen shot for the *C. jejuni* growth curve.

2.8.3 Secondary models

Commonly used secondary models describe the responses to changes in an environmental factor, for example temperature, pH and a_w. There are three types of models: the second-order response surface equation, the square root model (Belehardek) and Arrhenius relationships. The Arrhenius equation is applicable if the growth rate is determined by a single rate-limiting enzymatic reaction. Broughall *et al.* (1983) used the Arrhenius equation to describe the lag and generation time of *St. aureus* and *S. typhimurium*. This model was later modified to take into account the pH value (Broughall & Brown 1984). Initially the response surface equation was used when a number of factors affected the primary model. The square root model (Ross 1993) is based on the linear relationship between the square root of growth rate and temperature. An important feature is that the equation includes the concept of a 'biological zero' which is the temperature when the growth rate is zero. The square root model has been applied to the growth of *E. coli*, *Bacillus* species, *Y. enterocolitica* and *L. monocytogenes* (Gill & Phillips 1985; Wimptheimar *et al.* 1990; Heitzer *et al.* 1991; Adams *et al.* 1991). In order to have sufficient data for model fitting a large number of data points must be collected. Growth can be measured by a number of methods such as turbidity and plate counts. Automatic data collection is of obvious benefit since it is less laborious and the data are digitised. The Bioscreen (Labsystems) automatically records turbidity of a large number of samples at any given time. Alternative measurements include changes in the media conductivity (see Section 6.4.1; Borch & Wallentin 1993). The growth rate of *Y. enterocolitica* in pork was modelled using conductance microbiology and the data closely fitted to the Gompertz function.

A listing of selected modelling papers is given in Table 2.17.

Table 2.17   Examples of predictive modelling of microbial growth and toxin production.

Organism	Comments	Reference
B. cereus		Zwietering *et al.* 1996
Brochothrix thermosphacta		McClure *et al.* 1993
Cl. botulinum types A and B	Growth and toxin production	Lune *et al.* 1990; Roberts & Gibson 1986; Robinson *et al.* 1982
Cl. botulinum	Toxin production	Lindroth & Genigeorgis 1986; Baker & Genigeorgis 1990; Hauschild *et al.* 1982; Meng & Genigeorgis 1993, 1994
E. coli O157:H7		Sutherland *et al.* 1995
L. monocytogenes		McClure *et al.* 1997
St. aureus		Broughall *et al.* 1983; Sutherland *et al.* 1994
S. typhimurium		Broughall *et al.* 1983
Yersinia enterocolitica		Sutherland & Bayliss 1994

2.8.4 Tertiary models

Tertiary models use the primary and secondary model to generate models used to calculate the microbial response to changing conditions and to compare the effects of different conditions. The Pathogen Modelling Program and the FoodMicro model program are two easily available tertiary models. The Pathogen Modelling Program was developed by the USDA food safety group as a spreadsheet software-based system (see Appendix on Web resources for the downloading address). It includes models for the effect of temperature, pH, water activity, nitrite concentration and atmospheric composition on the growth and lag responses of major food pathogens (Fig. 2.12). The Food MicroModel was developed in the UK (Campden Food and Drink Research Association, UK) and includes environmental parameters similar to the Pathogen Modelling Program (Jones 1993). It has predictive equations for growth, survival and death of pathogens. Panisello & Quantick (1998) demonstrated the application of the Food MicroModel for risk assessment within the development of an HACCP plan and this is decribed in Section 7.4.

2.8.5 Application of predictive microbial modelling

Predictive microbial modelling has many applications:

- Product research and development
- Shelf life studies (Section 4.3)
- HACCP (Section 7.4)
- Risk assessment studies (Chapter 9)
- Education and training

Zwietering *et al.* (1992) describe an expert system which models bacterial growth in food production and distribution chains. The system combines two databases of physical parameters of the food (temperature, a_w, pH and oxygen level) with the physiological data (growth kinetics) of spoilage organisms. This enables predictions of possible spoilage types and kinetics of deterioration to be made.

Currently both predictive microbiology and HACCP computer programs are still in the developmental stages as food safety tools. However, predictive models are available that are potentially useful in the development and maintenance of HACCP systems. When conducting an HACCP study, models can be used to assess the risk or probability and determine the consequence of a microbiological hazard in food. By using predictive models, ranges and combinations of process parameters can be established as critical limits for CCPs. Models can also be used in combi-

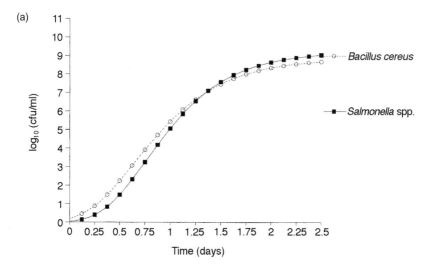

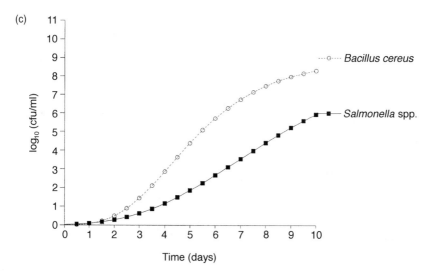

Fig. 2.12   The growth of *Salmonella* species and *Bacillus cereus* under different growth conditions as predicted using the pathogen modelling program. (a) Temperature 20°C, pH 6.8, a_w 0.997. (b) As per (a) except pH 5.6. (c) As per (a) except temperature is 10°C. (d) As per (a) except a_w 0.974. Note the change in scale for (c).

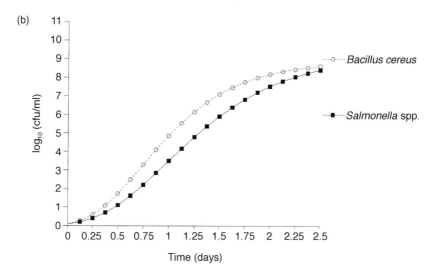

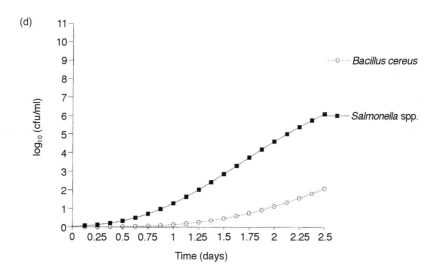

nation with sensors and microprocessors for real-time process control. This has the advantage of providing more processing options while maintaining a degree of safety equivalent to that of a single set of critical limits. Validation testing of individual CCPs can be reduced if the CCP models were developed with a similar food type. Microbiological as well as mechanical and human reliability models may be used to establish sets of rules for rule-based expert computer systems in an effort to automate the development of HACCP plans and evaluate the status of process deviations. Models could also be used in combination with sensors and microprocessors for real-time process control. Since HACCP is a risk-reduction tool, then predictive microbiological models are tools used to aid in the decision-making processes of risk assessment and in describing process parameters necessary to achieve an acceptable level of risk (Elliott 1996). Using a combination of risk assessment and predictive modelling it is possible to determine the effect of processing modifications on food safety (Buchanan & Whiting 1996).

Growth models have been generated for most food pathogens. Linton *et al.* (1996) have shown that the Gompertz equation and non-linear regression can be used to predict the survival curve shape and response to heat of *L. monocytogenes* under many environmental conditions.

3

FOODBORNE ILLNESS

3.1 The size of the food poisoning problem

Foodborne illness occurs when a person gets sick by eating food that has been contaminated with an unwanted microorganism or toxin. This condition is often called 'food poisoning'. Many cases of foodborne illness go unreported because their symptoms often resemble influenza. The most common symptoms of foodborne illness include stomach cramps, nausea, vomiting, diarrhoea and fever. It is accepted that only a small proportion of cases of foodborne illness are brought to the attention of food inspection, control and health agencies. This is partially because many foodborne pathogens cause mild symptoms and the victim may not seek medical help. Hence the notified number of cases is just the 'tip of the iceberg' with regard to true numbers of food poisoning cases (Fig. 3.1). Recently in the USA and England there have been studies to estimate the proportion of cases which are not recorded and hence obtain a more

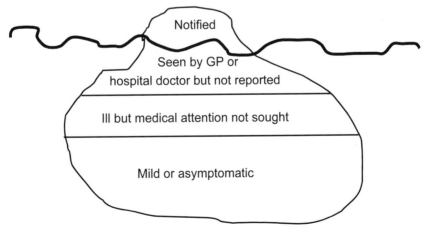

Fig. 3.1 The reporting pyramid – 'tip of the iceberg'.

accurate figure of food poisoning numbers. Table 3.1 gives the proportion of under-reporting for the USA.

The range of food products available offers the consumer a wide choice. However despite the progress in medicine, food science and the technology of food production, illness caused by foodborne pathogens has continued to present a major problem of both health and economic

Table 3.1 Under-reporting of foodborne pathogens in the USA (Mead *et al.* 1999).

Organism	Under-reporting factor
Bacterial pathogens	
B. cereus	38
Cl. botulinum	2
Brucella spp.	14
Campylobacter spp.	38
Cl. perfringens	38
E. coli O157:H7	20
STEC (VTEC) other than O157	Half as common as *E. coli* O157:H7 cases
E. coli enterotoxigenic (ETEC)	10
E. coli, other diarrhoegenic	Assumed to be as common as ETEC
L. monocytogenes	2
S. typhi	2
Salmonella, non-typhoid spp.	38
Shigella spp.	20
St. aureus	38
Streptococcus Group A	38
V. cholerae O1 or O139	2
V. vulnificus	2
Vibrio spp. other than those above	20
Y. enterocolitica	38
Parasitic pathogens	
Cry. parvum	45
Cyc. cayetanensis	38
G. lamblia	20
Tox. gondii	7
Tri. spiralis	2
Viral pathogens	
Rotavirus	Not given (number of cases taken as equal to birth cohort)
Astrovirus	Not given (number of cases taken as equal to birth cohort)
Norwalk-like virus (SRSV)	11% of all acute primary gastroenteritis
Hepatitis A	3

significance. In 1990, an average of 120 cases of foodborne illness per 100 000 population were reported from 11 European countries and more recent estimates, based on a 'sentinel' study, indicate that in some European countries there are at least 30 000 cases of acute gastroenteritis per 100 000 population yearly (Notermans & van der Giessen 1993), much of which is thought to be foodborne.

Mead *et al.* (1999) reported that the total burden of foodborne illness per year caused approximately 76 million illnesses, 323 000 hospitalisations and 5000 deaths in the USA (Table 3.2). Three pathogens (*Salmonella*, *Listeria* and *Toxoplasma*) were responsible for 1500 deaths per year which is more than 75% of those caused by known pathogens. Unknown agents caused 62 million illnesses, 265 000 hospitalisations and 3200 deaths. Using a population size of 270 299 000 (US Census Bureau 1998) this equates to 28% of the population suffering from food poisoning each year and 0.1% being hospitalised due to food poisoning.

The USA study can be compared with the recent sentinel study in England (Table 3.3; Wheeler *et al.* 1999; also Sethi *et al.* 1999 and Tompkins *et al.* 1999). This was undertaken to determine a more accurate estimate of the incidence of food poisoning in England. The study estimated the overall extent of under-reporting and that for every case detected by laboratory surveillance, there were 136 in the community (Fig. 3.2). Hence the scale of infectious intestinal disease in England was estimated at 9.4 million cases occurring annually, and 1.5 million cases presented to general practitioners. Under-reporting for individual organisms varied, most likely due to the severity of the illness: salmonella (3.2:1), campylobacter (7.6:1) compared with rotavirus (35.1:1) and Norwalk-like viruses (approximately 1562:1). These under-reporting

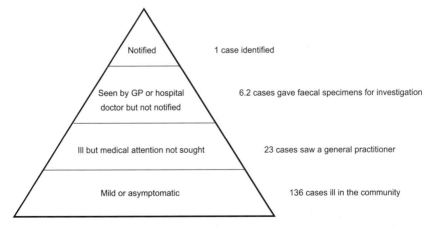

Fig. 3.2 The under-reporting pyramid (data from Wheeler *et al.* 1999).

Table 3.2 Estimated illnesses, hospitalisations, and deaths caused by known foodborne pathogens in the USA (adapted from Mead *et al.* 1999).

Disease or agent	Illnesses				Hospitalisations		Deaths	
	Total	Foodborne	% Foodborne transmission	% of total foodborne	Foodborne	% of total foodborne	Foodborne	% of total foodborne
Bacterial								
Bacillus cereus	27 360	27 360	100	0.2	8	0.0	0	0.0
Botulism, foodborne	58	58	100	0.0	46	0.1	4	0.2
Brucella spp.	1 554	777	50	0.0	61	0.1	6	0.3
Campylobacter spp.	2 453 926	1 963 141	80	14.2	10 539	17.3	99	5.5
Clostridium perfringens	248 520	248 520	100	1.8	41	0.1	7	0.4
Escherichia coli O157:H7	73 480	62 458	85	0.5	1843	3.0	52	2.9
E. coli, non-O157 STEC	36 740	31 229	85	0.2	921	1.5	26	1.4
E. coli, enterotoxigenic	79 420	55 594	70	0.4	15	0.0	0	0.0
E. coli, other diarrhoegenic	79 420	23 826	30	0.2	6	0.0	0	0.0
Listeria monocytogenes	2518	2493	99	0.0	2 298	3.8	499	27.6
Salmonella typhi	824	659	80	0.0	494	0.8	3	0.1
Salmonella, non-typhoidal	1 412 498	1 341 873	95	9.7	15 608	25.6	553	30.6
Shigella spp.	448 240	89 648	20	0.6	1 246	2.0	14	0.8
Staphylococcus food poisoning	185 060	185 060	100	1.3	1 753	2.9	2	0.1
Streptococcus, foodborne	50 920	50 920	100	0.4	358	0.6	0	0.0
Vibrio cholerae, toxigenic	54	49	90	0.0	17	0.0	0	0.0
V. vulnificus	94	47	50	0.0	43	0.1	18	1.0
Vibrio, other	7 880	5 122	65	0.0	65	0.1	13	0.7
Yersinia enterocolitica	96 368	86 731	90	0.6	1 105	1.8	2	0.1
Subtotal	5 204 934	4 175 565		30.2	36 466	59.9	1297	71.7

(Contd)

Table 3.2 (*Contd*)

Disease or agent	Illnesses				Hospitalisations		Deaths	
	Total	Foodborne	% Foodborne transmission	% of total foodborne	Foodborne	% of total foodborne	Foodborne	% of total foodborne
Parasitic								
Cryptosporidium parvum	300 000	30 000	10	0.2	199	0.3	7	0.4
Cyclospora cayetanensis	16 264	14 638	90	0.1	15	0.0	0	0.0
Giardia lamblia	2 000 000	200 000	10	1.4	500	0.8	1	0.1
Toxoplasma gondii	225 000	112 500	50	0.8	2 500	4.1	375	20.7
Trichinella spiralis	52	52	100	0.0	4	0.0	0	0.0
Subtotal	2 541 316	357 190		2.6	3 219	5.3	383	21.2
Viral								
Norwalk-like viruses	23 000 000	9 200 000	40	66.6	20 000	32.9	124	6.9
Rotavirus	3 900 000	39 000	1	0.3	500	0.8	0	0.0
Astrovirus	3 900 000	39 000	1	0.3	125	0.2	0	0.0
Hepatitis A	83 391	4 170	5	0.0	90	0.9	4	0.2
Subtotal	30 833 391	9 282 170		67.2	21 167	34.8	129	7.1
Grand total	38 629 641	13 814 924		100.0	60 854	100.0	1809	100.0

Table 3.3 Incidence of infectious intestinal disease in community and reported to general practice by organism (Wheeler *et al.*, 1999).

	Community		General practice		No. of community cases/GP cases
	No. of cases[a]	Rate/1000 person years	No. of cases	Rate/1000 person years	
Bacteria					
Aeromonas spp.	46	12.4	165	1.88	6.7
Bacillus spp ($>10^4$/g)	0	0	4	0.05	—
Campylobacter spp.	32	8.7	354	4.14	2.1
Clostridium difficile cytotoxin	6	1.6	17	0.20	8.0
Clostridium perfringens enterotoxin	9	2.4	114	1.30	1.9
E. coli O157	0	0	3	0.03	—
E. coli DNA probes:					
Attaching and effacing	20	5.4	119	1.32	4.1
Diffusely adherent	23	6.2	103	1.18	5.3
Enteroaggregative	18	4.9	141	1.62	3.0
Enteroinvasive	0	0	0	0	—
Enteropathogenic	1	0.27	4	0.05	5.4
Enterotoxigenic	10	2.7	52	0.59	4.6
Verocytotoxigenic (non-O157)	3	0.82	6	0.06	13.4
Salmonella spp.	8	2.2	146	1.57	1.4
Shigella spp.	1	0.27	23	0.27	1.0
Staphylococcus aureus ($>10^6$/g)	1	0.27	10	0.11	2.5
Vibrio spp.	0	0	1	0.01	—
Yersinia spp.	25	6.8	51	0.58	11.7

(Contd)

Table 3.3 *(Contd)*

	Community		General practice		No. of community cases/GP cases
	No. of cases[a]	Rate/1000 person years	No. of cases	Rate/1000 person years	
Protozoa:					
Cryptosporidium parvum	3	0.81	39	0.43	1.9
Giardia intestinalis	2	0.54	28	0.28	1.9
Viruses:					
Adenovirus group F	11	3.0	81	0.88	3.4
Astrovirus	14	3.8	77	0.86	4.4
Calicivirus	8	2.2	40	0.43	5.1
Rotavirus group A	26	7.1	208	2.30	3.1
Rotavirus group C	2	0.54	6	0.06	8.9
Small round structured viruses	46	12.5	169	1.99	6.3
No organism identified	432	117.3	1305	14.82	7.9
Total	781	194	8770[b]	33.1	5.8

[a] Excluding cases where individual follow-up was not known.
[b] Total cases are greater than the sum of individual organisms due to cases for which a stool sample was not sent for testing. The general practice total includes cases from the enumeration arm, for which full stool testing was not carried out.

values differ from those of the USA (Table 3.1). Nevertheless, the total burden of gastroenteritis is that 1 in 5 of people in the general population of England develops the diseases each year. This proportion is similar to the estimate of 28% in the USA. It should be noted that, across Europe, the number of food poisoning cases has been decreasing for the past 2 years for all foodborne pathogens except *C. jejuni* (Anon. 1999d).

In England the organisms most commonly detected in patients with infectious intestinal disease were *Campylobacter* spp. (12.2% of stools tested), rotavirus group A (7.7%) and small round structured virus (6.5%). No pathogen or toxin was detected in 45.1 to 63.1% of cases. Surprisingly, *Aeromonas* spp., *Yersinia* spp. and some enterovirulent groups of *E. coli* were detected as frequently from controls as from cases.

It is evident that the causes of gastroenteritis vary with age (Fig. 3.3). SRSV, caliciviruses and rotaviruses probably cause the majority of gastroenteritis in children under four, whereas bacteria (*Campylobacter* and *Salmonella* spp.) are the major cause of gastroenteritis in other age groups. Figure 3.4 indicates that males suffer from gastroenteritis more than females, except for one age group (>74, probably because of the lower ratio of men to women in this age group). A possible reason for part of this difference is that fewer men than women wash their hands after using the lavatory (33% compared with 66%, taking an average of 47 seconds, against 79 seconds).

Estimates have been made of the economic consequences of foodborne illness, where costs are incurred by individuals who become ill and by their employees and by families, health care agencies and the food companies and businesses involved. For instance, in England and Wales in

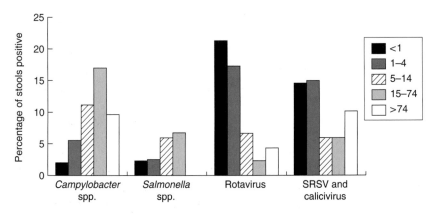

Fig. 3.3 Variation in causative organism of gastroenteritis with age.

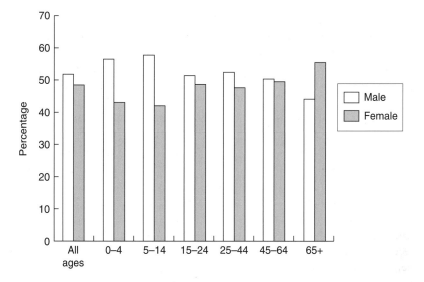

Fig. 3.4 Comparison of reported cases of gastroenteritis between males and females.

1991, some 23 000 cases of salmonellosis were estimated to have resulted in an overall cost of between £40 million and £50 million (Sockett 1991). This is comparable with the £28 million Sweden spends on the treatment of salmonella infections. The cost of campylobacter infections in the USA has been estimated at $1.5 to $8.0 billion. The total economical aspect of foodborne diseases is a loss of $5 to $17 billion by the US Food and Drug Administration (Table 3.4).

3.2 Consumer pressure effect on food processing

In the production of food it is crucial that proper measures are taken to ensure the safety and stability of the product during its whole shelf life. In particular, modern consumer trends and food legislation have made the successful attainment of this objective much more of a challenge to the food industry. First, consumers require more high quality, preservative-free, safe but mildly processed foods with extended shelf life. For example, this may mean that foods have to be treated at mild-pasteurisation rather than sterilisation temperatures. As acidity and sterilisation are two crucial factors in the control of outgrowth of pathogenic spore-forming bacteria, such as *Cl. botulinum*, addressing this consumer need calls for innovative approaches to ensure preservation of products (Gould

Table 3.4 Medical costs and productivity losses estimated for selected human pathogens, 1993.

Pathogen	Foodborne illness		Foodborne costs (bil $)	Per cent from meat/poultry (%)	Meat/poultry related		Total costs meat/poultry (bil $)
	Cases	Deaths			Cases (#)	Deaths (#)	
Bacteria:							
Campylobacter jejuni or coli	1375000–1750000	110–511	0.6–1.0	75	1031250–1312500	83–383	0.5–0.8
Clostridium perfringens	10000	100	0.1	50	5000	50	0.1
Escherichia coli O157:H7	8000–16000	160–400	0.2–0.6	75	6000–12000	120–300	0.2–0.5
Listeria monocytogenes	1526–1767	378–485	0.2–0.3	50	763–884	189–243	0.1–0.2
Salmonella	696000–3840000	696–3840	0.6–3.5	50–75	348000–2880000	348–2610	0.3–2.6
Staphylococcus aureus	1513000	1210	1.2	50	756500	605	0.6
Subtotal	3603526–7130767	2654–6546	2.9–6.7	N/A	2147513–4966884	1395–4191	1.8–4.8
Parasite:							
Toxoplasma gondii	3056	41	2.7	100	2056	41	2.7
Total	3606582–7133823	2695–6587	5.6–9.4	N/A	2149569–4968940	1436–4232	4.5–7.5

Source: USDA, FSIS, Pathogen Reduction; Hazard Analysis and Critical Control Point (HACCP) Systems; Proposed Rule (13).

1995; Schellekens 1996; Peck 1997). Second, legislation has restricted the use and permitted levels of some currently accepted preservatives in different foods. This has created problems for the industry because the susceptibility of some microorganisms to most currently used preservatives is falling. For example, recent work has identified that resistance to weak acid preservatives in spoilage yeast is mediated by a multidrug resistance protein (Piper *et al.* 1998).

The consumer has demanded convenience foods which require minimal further processing before consumption. They also require fewer 'additives' and consequently fewer preservatives which affect product shelf life (Sections 4.2 and 4.3). Therefore new processing techniques are being introduced to increase product quality such as milder thermal processing, microwave heating, ohmic heating and high-pressure processing techniques. All these processes need to be evaluated fully regarding safe food production, especially since 'mild' treatment might confer resistance to the stomach acid and hence lower the infectious dose (Section 2.7). Additionally there has been an increase in the consumption of food outside the home and an increase in the population of the vulnerable elderly.

There has been the dramatic appearance of emerging pathogens such as shiga toxin producing *E. coli* (STEC, also known as verotoxigenic *E. coli* or VTEC), multiantibiotic resistant *S. typhimurium* DT104 and a greater awareness of viral gastroenteritis. The link between new variant CJD and BSE has been established and drastically affected abattoir procedures in certain countries and production methods (Section 5.7.1).

3.3 Testing foods for the presence of pathogens

Note: End-product testing in itself does not guarantee a safe food product.

Testing food at the end of production for microorganisms ('end-product testing') has been standard practice in the food industry for decades. However, a statistical appreciation of its usefulness has been largely overlooked (Section 8.6.3). In 1974 the International Commission (ICMSF) wrote an excellent text regarding the setting of microbiological criteria (Chapter 8). This book was written at a time of increasing global food transportation and its primary application is to food entering a country (port of entry) with no known history. Nevertheless, these criteria have been applied within industry for their own products despite in-house knowledge of the product. This book was revised in 1986 and is currently being revised again and will be republished in 2000.

It is very important nowadays to recognise that microbiological testing of foods should be carried out under the umbrella of Hazard Analysis Critical Control Point (HACCP) as part of the verification principle (Section 7.4). In other words, *end-product testing in itself does not guarantee a safe food product*, but supports the HACCP plan implementation. Nevertheless, microbiological criteria (levels of microbes acceptable in a particular food) are required by governments (for example the European Union Vertical Directives) and between companies in a supply chain. The details of these criteria are often historical, not necessarily the most appropriate and in the case of criteria between companies they are often confidential. These criteria may eventually be replaced by 'food safety objectives' which are currently being evaluated (Section 9.4).

3.4 Control of food poisoning

Pathogenic microbes are present in soil and hence on crops and on farming livestock and fish. It is inevitable therefore that raw ingredients entering a process can contain pathogenic organisms. Therefore to control food poisoning the pathogens associated with an ingredient and food product need to be recognised and controlled. Control programmes then need to be in place and monitored to assess their efficacy, reviewed and modified if necessary. As an example of a control programme the case of salmonella in Sweden is given.

3.4.1 The control of salmonella in Sweden

In the 1950s, Sweden had a severe salmonella food poisoning outbreak. It caused about 9000 cases and 90 people died. This resulted in the implementation of an effective salmonella control programme. Salmonella is now found in less that 1% of animal products for human consumption produced in the country. The programme has compulsory reporting of all cases of infection and subsequent elimination measures (Weirup 1992).

The importation of animal feeds is handled under licence. At least 10 samples, or a 100 g sample for every 2 tonnes of feed, are sent to the Swedish National Food Administration for bacterial evaluation. If salmonella is recovered, importation is not permitted. In domestic bonemeal and animal protein plants, at least five samples of 100 g are collected each day. These samples are pooled and examined for salmonella. When salmonella is recovered, an inspection is conducted. If needed, design modifications are recommended. The Swedish government has authority to investigate a feed operation if a farm it services is found to have salmonella contamination. Samples of raw materials, dust and finished rations are cultured. Feed contaminated with salmonella is subject to two sequences of heat treatment. If salmonella is found in finished feed, the processing line is cleaned and fumigated. When necessary the line is modified so that feed reaches 70 to 75°C. Because of *S. enteritidis* (Section 5.2.2), regulators now require heat treatment of feed and approved storage methods for grain to be used for laying hen operations.

Only day-old grandparent breeding chicks may be imported into Sweden. The birds are quarantined for 5 weeks. The liver, yolk sac and intestines of all chicks that die in transit are examined for salmonella. Cloacal swabs are taken from 100 birds and pooled to 20 swabs per sample at arrival. A total of 60 dead and culled birds are tested at 1-2, 4-6, 12-14 and 16-18 weeks. A total of 0.5% of the birds is tested, using dead and stunted birds, and composites of five faecal swabs if there is low mortality. The flock is destroyed if salmonella is detected. Egg-type multiplier flocks are tested on a voluntary basis. Participating flocks are tested for salmonella at 2, 6-10 and 14-18 weeks of age. During rearing, two composite faecal samples containing 30 droppings each and two composite necropsy samples containing liver and caecum from five birds each are cultured. After the multiplier flock is in production, two composite faecal samples and one composite necropsy sample are collected monthly. Salmonella-positive flocks are destroyed. Sanitation is the principal control for salmonella in hatcheries. Eggs are disinfected before incubation and fumigated on day 3 of incubation. Eggs and chicks are separated by source. Hatcheries are bacteriologically monitored every 3 months.

Pullets are voluntarily monitored at 2-3 weeks of age and 2 weeks before placement. Three samples containing 30 faecal droppings each are collected for bacterial isolation; this procedure is repeated at 25 and 55 weeks of age. Producers that participate in voluntary monitoring are eligible for insurance to defray losses caused by the regulation of salmonella. Egg-laying flocks with *S. enteritidis* or *S. typhimurium* are depopulated without processing or compensation. After flock depopulation, the poultry house must be cleaned and disinfected. Layer flocks with serotypes other than *S. enteritidis* and *S. typhimurium* are permitted to sanitary slaughter, followed by heat treatment of carcasses. Recovery of *S. enteritidis* or *S. typhimurium* from articles associated with a flock permits inspection of the farm. Articles may include diagnostic specimens, liquid or powdered egg. This authority may be used if a human isolate of *S. enteritidis* implicates eggs.

These measures remained in place when Sweden joined the European Union in 1995. However despite the strict control, reported salmonella food poisoning has risen in Sweden. This is partially due to the increase in packaged holidays resulting in Swedish tourists bringing salmonella back to the country. There are about 5500 salmonella cases per year, 85% of which are due to travel outside the country. The Swedish programme is financed by food producers and the state. It costs about $8 million per year. This amount is small compared to the cost of medical treatment for the 5500 annual cases, which is estimated at $28 million. The control programme costs less than 11 cents per kilo of chicken.

3.5   Surveillance programmes

A foodborne-disease surveillance programme is an essential part of a food safety programme.

Todd 1996

Surveillance refers to the systematic collection and use of epidemiological information for the planning, implementation and assessment of disease control. With the exception of cholera (which is subject to the International Health Regulations), there is no obligation to report foodborne diseases internationally. Attempts to obtain a global picture of foodborne disease are usually hindered by differences in national surveillance systems.

Global surveillance of foodborne problems can play an important role in the early detection, early warning, rapid investigation and control of such problems. It could also help limit the extent and distribution of the problem, thereby preventing many cases of illness and minimising the negative impact on trade and the economics of individual countries. Due to the increase in world trade the events within one country might also affect many other nations. International outbreaks can be recognised in two ways: one country recognising an outbreak, passing this information on to the network for information, and other countries recognising a similar occurrence in their country, or by analysing pooled, international databases for unusually high levels of infection.

The study by Todd (1996) surveyed the incidence of food poisoning in 17 countries (Table 3.5, see also Tables 3.2 and 3.3). The survey demonstrated considerable regional differences, for example *St. aureus* is the major food poisoning organism in Cuba and Japan, whereas salmonella dominates in most other countries. The most frequent vector of foodborne disease was meat in 13 countries and foodborne illness was most frequently acquired in the home.

3.5.1   FoodNet in the United States

The Foodborne Disease Active Surveillance Network (FoodNet) in the United States was established in 1995. It is the principal foodborne disease

Table 3.5 Annual number of outbreaks by country (Todd 1996, reprinted with permission from *Journal of Food Protection*, International Association for Food Protection, Des Moines, Iowa).

Country	B. cereus	C. perfringens	Salmonella	S. aureus	Total	Population in millions
Israel	0.0	3.0	4.4	4.0	11.4	4.4
Finland	3.8	8.8	7.8	5.6	26.0	5.0
Denmark	2.0	5.8	5.2	0.0	13.0	5.1
Scotland	3.0	6.6	152.0	2.4	164.0	5.2
Sweden	0.6	4.6	7.0	2.8	15.0	8.4
Hungary	5.2	5.0	131.2	16.0	157.4	10.6
Portugal	0.0	0.3	6.8	7.3	14.4	10.4
Cuba	4.8	13.4	6.8	60.8	85.8	10.4
Netherlands	4.8	2.4	8.0	0.0	15.2	14.8
Yugoslavia	0.4	2.8	46.0	13.0	62.2	24.0
Canada	14.0	18.0	49.4	18.8	100.2	26.0
Spain	0.6	5.0	467.6	0.0	473.2	39.3
England/Wales	28.3	53.3	450.0	9.5	541.1	50.4
France	0.0	17.6	177.0	12.8	207.4	56.0
Fed. Rep. of Germany	0.4	1.4	3.0	3.2	8.0	61.4
Japan	10.5	14.0	84.0	128.0	236.5	123.0
USA	3.2	4.8	68.4	9.4	85.8	247.4

[a] No. of outbreaks per year (mean 2–5 years).

component of the Emerging Infections Program (EIP) of the Centers for Disease Control and Prevention (CDC) of the United States. It is a collaborative project of the CDC, nine EIP state health department sites, the Food Safety and Inspection Service (FSIS), US Department of Agriculture (USDA) and the Food and Drug Administration (FDA). It covers areas of Minnesota, Oregon, Colorado, Tennessee, Georgia, California, Connecticut, Maryland and New York. The total population of this area is 25.4 million, which is 10% of the US population.

It is a sentinel network designed to:

- Produce national estimates of the burden and sources of specific foodborne diseases in the United States through active surveillance and epidemiological studies.
- Determine how much foodborne illness results from eating specific foods, such as meat, poultry and eggs.
- Document the effectiveness of new food safety control measures, such as the USDA Pathogen Reduction and HACCP Rule, in decreasing the number of cases of major foodborne diseases in the US each year.
- Describe the epidemiology of new and emerging bacterial, parasitic and viral foodborne pathogens.
- Respond rapidly to new and emerging foodborne pathogens.

FoodNet has five components:

- Active laboratory-based surveillance at over 300 clinical laboratories that test stool samples. Information is collected on every laboratory-diagnosed case of bacterial pathogens including *Salmonella*, *Shigella*, *Campylobacter* (including Guillain-Barre syndrome), *E. coli* O157 (including HUS), *L. monocytogenes*, *Y. enterocolitica* and *Vibrio*. The parasitic organisms *Cryptosporidium* and *Cyclospora* are also included.
- Survey of clinical laboratories to give baseline information on which pathogens to include in routine bacterial stool cultures and to standardise procedures of sample collection and examination.
- Survey of physicians to obtain information on physician stool culturing practices to determine how often samples are requested and for what reason.
- Survey (by telephone questionnaire) of the population to determine the number of cases of diarrhoea in the general population and how often medical advice is sought.
- Epidemiological studies of *E. coli* O157, *Campylobacter* and *Salmonella* serogroups B and D, which cause 60% of salmonella infections in the US. The aim is to determine which foods or other exposures might be risk factors for these bacterial infections. Isolates are subjected to

antibiotic resistance testing, phage typing and molecular subtyping (PFGE, see PulseNet). In the future sporadic outbreaks of *E. coli* O157, *Cryptosporidium* and *L. monocytogenes* will be included.

FoodNet has recorded the decrease in *Campylobacter*, *Salmonella* and *Cryptosporidium* infections (see also Section 3.1) and the decrease in meat and poultry products contaminated with *Salmonella*. It has reported high isolation rates of *Y. enterocolitica* in Georgia and *Campylobacter* in California. An outbreak of *Salmonella* infections was detected in Oregon which was due to alfalfa sprouts. It has also detected two *E. coli* O157:H7 outbreaks in Connecticut which were due to lettuce and apple cider. FoodNet has also assisted in the surveillance of vCJD (linked to BSE in the UK). It has contributed to the investigations into multistate outbreaks of *Listeria* and *Cyclospora*. The latter outbreak was associated with raspberries from Guatemala and resulted in restrictions on the importation of raspberries into the United States.

3.5.2 *PulseNet: US* E. coli O157:H7 *and* S. typhimurium *detection network*

PulseNet is based on pulse-field gel electrophoresis (PFGE) which has been standardised in the US Department of Health and Human Services to identify distinctive DNA fingerprint patterns of *E. coli* O157:H7 and *S. typhimurium*. Pulse-field gel electrophoresis (PFGE) involves cutting bacterial DNA into shorter segments which are separated (in hours) in a polyacrylamide gel due to a pulsing electric field. The DNA fragments separate according to size and generate a barcode-like appearance which is a 'finger-print' of the organism. Using a computer network across 16 states, PFGE patterns from humans and suspected foods can be shared across the country. Hence multistate foodborne disease outbreaks can be recognised rapidly and investigated. In the future, PulseNet will include other foodborne pathogens.

3.5.3 *Enter-Net: European surveillance network for salmonellosis and shiga toxin-producing* E. coli *(STEC)*

Enter-Net is an international surveillance network for human gastro-intestinal infections. The network involves the 15 countries of the European Union, plus Switzerland and Norway. It conducts international surveillance of salmonellosis and shiga toxin-producing *E. coli* O157 (STEC, also known as verotoxigenic *E. coli* or VTEC), including anti-microbial (antibiotic) resistance. It is a continuation of the Salm-Net surveillance network which operated from 1994 to 1997.

The objectives of Enter-Net are:

(1) To collect standardised data on the antimicrobial resistance patterns of salmonellae isolated.
(2) To facilitate the study of resistance mechanisms and their genetic control by arranging the collection of representative strains of multiple drug resistant salmonellae and co-ordinating the required research work between specialised centres, and where available compare the resistances of animal isolates.
(3) To extend the typing of STEC for surveillance purposes by:
   • Extending the availability of phage-typing for *E. coli* O157
   • Using poly- and monovalent antisera to identify common non-O157 serogroups.
(4) To pilot an international quality assessment scheme for laboratory methods used in the identification/typing of STEC.
(5) To establish a core set of data items to accompany, where possible, each laboratory typed STEC isolate.
(6) To create an international database of STEC isolates which is updated regularly and is readily available to each participating team.
(7) To detect clusters of STEC isolate types in time, place and person and to bring such clusters to the attention of collaborators rapidly.
(8) To support the above objectives by continuing the existing Salm-Net surveillance system consisting of regular, frequent data exchange on salmonellae.

From Fisher (1997a, 1997b, 1999); Anon (1997a, 1999a).

A number of international outbreaks have been recognised by the Salm-Net and subsequently by Enter-Net (Table 3.6).

3.5.4   Antibiotic resistance

Resistance to medically important antibiotics has increased in *Salmonella* and *Campylobacter* to a disturbing degree. A strain of *S. typhimurium* DT104 resistant to ampicillin, chloramphenicol, streptomycin, sulphonamides and tetracyline (R-type ACSSuT) was first isolated in 1984. This microorganism has been isolated from cattle, poultry, sheep, pigs and horses. Antimicrobial therapy is used extensively to combat *S. typhimurium* infection in animals, and the evolution of a strain resistant to the commonly used antibiotics has made infections with *S. typhimurium* in food animals difficult to control. The primary route by which humans acquire infection is by the consumption of a large range of contaminated foods of animal origin.

Table 3.7 shows the significant increase in isolates in recent years for England and Wales. In addition to the antibiotic resistant type ACSSuT,

Table 3.6 International outbreaks recognised and investigated by Enter/Salm-Net.

Outbreak	No. of cases	Countries with cases
S. newport	100+	England & Wales and Finland
S. livingstone	100+	Austria, Czech Republic, Denmark, England & Wales, Finland, France, Germany, Netherlands, Norway and Sweden
E. coli O157(HUS)	15	Denmark, England & Wales, Finland and Sweden
S. anatum	19	England & Wales, Israel and USA
S. agona	4000+	Canada, England & Wales, Israel and USA
S. dublin	30+	France and Switzerland
S. stanley	100+	Finland and USA
S. tosamanga	28	Eire, England & Wales, France, Germany, Sweden and Switzerland
Sh. sonnei	100+	England & Wales, Germany, Norway, Scotland and Sweden

Source: Enter-Net home page (see Appendix for address).

Table 3.7 Antibiotic resistance of *Salmonella typhimurium* DT104 from humans in England and Wales, 1990–1996.

Antibiotic	1990	1991	1992	1993	1994	1995	1996
Ampicillin	37[a]	50	72	85	88	90	95
Chloramphenicol	32	49	60	83	87	89	94
Streptomycin	38	52	75	85	92	97	97
Sulphonamides	37	53	76	86	93	90	97
Tetracyclines	36	50	74	83	88	90	97
Trimethoprim	0.4	3	3	2	13	30	24
Ciprofloxacin	0	0	0.2	0	1	7	14

[a] Per cent resistant.

many isolates are also resistant to trimethoprim and ciprofloxacin. Ciprofloxacin (a medically important antibiotic) resistance has been most notable in *S. hadar* (39.6%) compared to other *Salmonella* serovars. *Campylobacter* species have also shown a rise in ciprofloxacin resistance. The source of antibiotic resistance is possibly the veterinary use of enrofloxacin (a fluoroquinolone antibiotic). The systemic use of this antibiotic may be imposing a selective pressure on the microbial flora which has resulted in the selection of DNA gyrase mutants (gyrA).

Fluoroquinolone resistance has also been reported in *C. jejuni*. Molecular typing has revealed an association between resistant *C. jejuni* strains from chicken products and *C. jejuni* strains from human cases of campylobacteriosis (Jacobs-Reitsma *et al.* 1994; Smith *et al.* 1999). A Microbiological Risk Assessment (Chapter 9) of fluoroquinolone-resistant *C. jejuni* has been published on the Web (see the Appendix on the Web).

Several countries are monitoring the occurrence of antimicrobial resistance and establishing guidelines (either through voluntary or regulatory means) to control their increase (Bower & Daeschel 1999):

• WHO: Division of Emerging and other Communicable Diseases Surveillance and Control (EMC)
• USA: Alliance for the Prudent Use of Antibiotics (APUA)
• Denmark: Danish Integrated Antimicrobial Resistance Monitoring and Research Programme (DANMAP)
• European Antimicrobial Resistance Surveillance System (EARSS); Web address in Appendix.

Control of salmonellae is achieved through a number of mechanisms: the requirement for the absence (<1 salmonella cell in 25 g food) in ready-to-eat foods, temperature control during storage and a processing step (i.e. cooking) to eliminate salmonellae from raw meats.

3.6   Causes of food poisoning

There are a number of factors which contribute to food being unsafe and causing food poisoning (Table 3.8). The principal causes can be summarised as:

• Inadequate control of temperature during cooking, cooling and storage
• Poor personnel hygiene
• Cross-contamination of raw and processed products
• Inadequate monitoring of processes

These contributing factors can be considerably reduced by adequate training of staff, implementation of HACCP combined with risk assessment (Sections 7.4 and 9.1). As Chapter 8 will explain, it is now widely accepted that it is inadequate to rely on testing the end-product for the presence of microorganisms and as a means of controlling the hygienic status of the process.

Table 3.8 Factors contributing to outbreaks of foodborne disease (various sources).

Contributing factors	Percentage[a]
Factors relating to microbial growth	
Storage at ambient (room) temperature	43
Improper cooling	32
Preparation too far in advance of serving	41
Improper warm holding	12
Use of leftovers	5
Improper thawing and subsequent storage	4
Extra large quantities prepared	22
Factors relating to microbial survival	
Improper reheating	17
Inadequate cooking	13
Factors relating to contamination	
Food workers	12
Contaminated processed non-canned foods	19
Contaminated raw foods	7
Cross-contamination	11
Inadequate cleaning of equipment	7
Unsafe source	5
Contaminated canned foods	2

[a] Percentages exceed a total of 100 since multiple factors often contribute to foodborne illness.

The key to the production of safe food is producing food which is microbiologically stable. In other words, to ensure that any intrinsic microbes are unable to multiply to an infectious dose (see Section 9.1.3) – ideally they die off – and that toxins are absent.

Essentially the cooking and cooling temperature profile should be aimed at:

- Reducing the number of infectious organisms by 6 log orders (i.e. from 10^6 cells/g to 1 cells/g).
- Not providing suitable conditions for the outgrowth of microbial spores which survive cooking.
- Avoiding conditions which enable heat-stable toxins to be produced; by definition these toxins are resistant to 100°C for 30 minutes and hence are not destroyed during cooking.

Cross-contamination causes post-processing (i.e. after the cooking step) contamination of the food. This can be avoided through:

- Careful design of the factory layout.
- Control of personnel movement.
- Good personnel hygiene habits.

Foods which do not undergo a cooking process are normally acidified (for example fermented foods, Section 4.4) and stored under chilled conditions. These practices rely upon the pH and temperature of the food stopping microbial growth. The growth range of the major food poisoning organisms has been documented (see ICMSF 1996 for details) and the general growth parameters are given in Table 2.7. Hence it is possible to predict the pH and storage temperature of the food which will restrict the growth of foodborne pathogens.

3.7   The microbial flora of the human gastrointestinal tract

The human intestinal tract is divided into a number of regions: oesophagus, stomach, small intestine, large intestine and anus (Fig. 3.5). The stomach produces gastric enzymes (endopeptidases, gelatinase and lipase) and has a low acidic pH (pH 2). Most digestion and absorption occur in the small intestine. Pancreatic enzymes include trypsin, chymotrypsin, carboxypeptidase, amylase, lipases, ribonuclease, deoxyribonuclease, collagenase and elastase. Bile salts secreted from the liver aid the absorption of fats.

The structure of the small intestine maximises the area available for absorption. The surface of the mucosa is convoluted and folded. It is covered with finger-like projections called villi, which are, in turn, covered with absorptive cells. The effective surface area of the mucosal cells is further increased by the microvilli that occur on the luminal membrane of the enterocyte (Fig. 3.6). The brush border of the enterocytes contains various enzymes including many disaccharidases such as maltase, isomaltase, sucrase and lactase. These are involved in both the digestion and absorption of carbohydrates. Mucosal enzyme levels are affected by bacterial activity and lactose deficiency is a sensitive indicator of the colonisation of the small intestine by pathogenic bacteria. Interactions between the cell membranes and the luminal contents are facilitated by the glycocalyx, a complex mucous layer overlaying the enterocytes.

The large intestine has a neutral pH and has a slow transit time (up to 60 hours). It contains a complex microbial flora which is approximately 55% of the solids and is dominated by strict anaerobes (Table 3.9). The initial 'inoculum' is from the mother at the time of birth and subsequently from

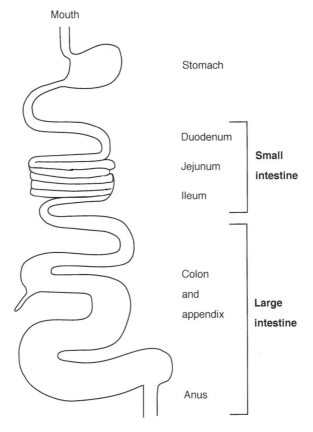

Fig. 3.5   The human intestinal tract.

ingested food and the environment. The indigenous flora is composed of the normal flora, microorganisms which persist in the large intestines and the transient autochthonous flora.

The intestinal flora is complex and is primarily composed of a number of strictly anaerobic bacteria. Commonly isolated genera are:

(1) *Bacteroides* spp. such as *B. fragilis*. These are Gram-negative, non-sporeforming rods. They produce volatile and non-volatile fatty acids (VFA and n-VFA); acetic, succinic, lactic, formic, propionic, *N*-butyric, isobutyric and isovaleric acids.
(2) *Bifidobacterium* spp., such as *Bif. bifidium*. These are Gram-positive, non-sporeforming rods with characteristic club-shaped ends. They produce acetic and lactic acids (3:2 ratio).
(3) *Clostridium* spp., such as *Cl. innocuum*. These are Gram-positive sporeforming rods.

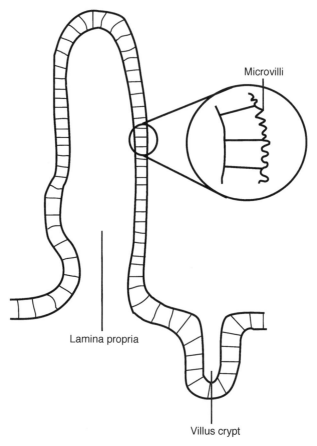

Fig. 3.6   Gut mucosal structure.

(4) *Enterococcus* spp., such as *Ent. faecalis*. These are Gram-positive cocci which are aerotolerant. They are in Lancefield group D and can grow in 6.5% NaCl and under alkaline conditions, <pH 9.6.
(5) *Eubacterium* spp. are Gram-positive non-sporeforming rods. They produce butyric, acetic and formic acids.
(6) *Fusobacterium* spp. are Gram-negative non-sporeforming rods which produce N-butyric acid.
(7) *Peptostreptococcus* spp. are Gram-positive cocci that can degrade peptone and amino acids.
(8) *Ruminococcus* are Gram-positive cocci that produce acetic, succinic and lactic acids, ethanol, carbon dioxide and hydrogen from carbohydrates.

The dominant genus of intestinal bacteria in animals and humans is *Bacteroides* (10^{11}cfu/g). The intestinal microflora also contains Gram-

Table 3.9 Microbial flora of the human intestinal tract.

Organism	Cell density (cfu/g)[a]
Protozoan parasites	10^6-10^7
Ascaris	10^4-10^5
Enteric viruses	
Enteroviruses	10^3-10^7
Rotaviruses	10^{10}
Adenovirus	10^{12}
Enteric bacteria	
Acidaminococcus fermentans	10^7-10^8
Bacteroides ovatus	
B. uniformis	
B. coagulans	10^9-10^{10}
B. eggerthii	
B. merdae	
B. stercoris	
Bifidobacterium bifidum	10^8-10^9
Bif. breve	
Clostridium cadaveris	
Cl. clostridioforme	
Cl. innocuum	10^8-10^9
Cl. paraputrificum	
Cl. perfringens	
Cl. ramosum	
Cl. pertium	
Coprococcus cutactus	10^7-10^8
Enterobacter aerogenes	10^5-10^6
Enterococcus faecalis	10^5-10^6
Escherichia coli	10^6-10^7
Eubacterium limosum	10^8-10^9
E. tenue	
Fusobacterium mortiferum	
F. naviforme	
F. necrogenes	10^6-10^7
F. nucleatum	
F. prousnitzil	
F. varium	
Klebsiella pneumoniae	10^5-10^6
K. oxytoca	
Lactobacillus acidophilus	
Lb. brevis	
Lb. casei	
Lb. fermentum	
Lb. leichmannii	10^7-10^8
Lb. minutus	
Lb. plantarum	

(Contd)

Table 3.9   *(Contd)*

Organism	Cell density (cfu/g)[a]
Lb. rogosa	
Lb. ruminis	
Lb. salovarius	
Megamonas hypermegas	10^7–10^8
M. elsdenii	10^7–10^8
Methanobrevibacter smithii	Undetectable–10^9
Methanosphaerae stadtmaniae	Undetectable–10^9
Morganella morgannii	
Peptostreptococcus assccharolyticus	
P. magnus	10^8–10^9
P. productus	
Proteus mirabilis	10^5–10^6
Salmonella spp.	10^4–10^{11}
Shigella spp.	10^5–10^9
Veillonella parvula	10^5–10^6
Indicator bacteria	
Coliforms	10^7–10^9
Faecal coliform	10^6–10^9

Sources: Tannock (1995) and Haas *et al.* (1999).
[a] Values are given as concentration per gram of faeces as representative of the large intestinal tract.

positive non-sporing strictly anaerobic rods: *Eubacterium*, *Propionibacterium* and *Bifidobacterium* species. There are many other genera of organisms present (a minimum of 400 species), including *Lactobacillus*, and *Clostridium* spp. Gram-positive cocci are numerically important in the intestinal tract, including *Peptostreptococcus* and *Enterococcus* species. The facultative Gram-negative rods such as *Proteus*, *Klebsiella* and *E. coli* are not numerically important, being outnumbered by *Bacteroides* species by about 1000:1.

E. coli are Gram-negative, catalase-positive, oxidase-negative, facultatively anaerobic short rods. They are members of the family Enterobacteriaceae. Most isolates ferment lactose. The majority of *E. coli* serotypes are not pathogenic and are part of the normal intestinal flora (about 10^6 organisms/g). Subsequently they are used as indicator organisms as one of the 'coliforms' to indicate faecal pollution of water, raw ingredients and foods. Non-pathogenic strains of *E. coli* typically colonise the infant gastrointestinal tract within a few hours after birth. The presence of this bacterial population in the intestine suppresses the growth of harmful bacteria and is important for synthesizing appreciable amounts of B vitamins. *E. coli* usually remains harmless when confined to the intestinal lumen. However, in debilitated or immunosuppressed humans, or when gastrointestinal barriers are violated, even normal, 'non-pathogenic' strains of *E. coli* can cause infection.

3.8   The mode of action of bacterial toxins

There are a number of different bacterial toxin definitions:

(1)   *Exotoxins* are essentially the same as 'bacterial protein toxins', and are also called 'secreted toxins'. The term exotoxin emphasises the nature of the substance, being extracellular, excretory, heat-labile and antigenic. They exert their biological activity (often lethal) in minute doses and are released during the decline phase of batch culture.

(2)   *Enterotoxins* are exotoxins which result in extremely watery diarrhoea. They are further categorised by their mode of action:
   • Exotoxins which induce by direct action on gut tissue, bio-chemical and/or structural lesions which lead to diarrhoea.
   • Exotoxins which cause specific action (e.g. net fluid loss) in the gut.
   • Exotoxins responsible for elevating cAMP levels which subse-quently causes ion flux changes and excess fluid secretion.

(3)   *Cytotonic enterotoxins* is the term applied to those toxins, such as cholera toxin, that induce net fluid secretion by interfering with biochemical regulatory mechanisms without causing overt histo-logical damage.

(4)   *Cytotoxic enterotoxins* is used to describe toxins which induce actual damage to intestinal cells as a necessary prelude to onset of net fluid secretion. Cytotoxins can be a protein or LPS (endotoxin) and may also be called 'cell-associated' toxins and 'cytolysins'. They have different mode of action to enterotoxins; essentially they are invasive and they kill target cells. Their mode of action is either intracellular or by formation of pores within cells. Intracellular cytotoxins inhibit cellular protein synthesis and/or actin filament formation. Cyto-toxins which cause pore formation in target cells can also be detected by their lytic activity upon erythrocytes, hence they are also known as 'haemolysins'. The result of cytotoxin invasion is inflam-matory diarrhoea which often contains blood and leukocytes.

(5)   *Endotoxins* are heat-stable, cell-associated, complex lipopoly-saccharide (LPS) structures (Section 2.2.3). The LPS is part of the outer membrane of Gram-negative bacteria. Lipid A is the part of the LPS which is responsible for toxicity. The endotoxins cause toxic

shock, inflammation and fever, as occurs with *Salmonella* spp. infections. They elicit cytotoxic activity upon cells.

(6)   *Cytolethal distending toxins (CLDT)* affect host cell-cycle regulation. These have been found in a variety of unrelated organisms: *C. jejuni, E. coli, Sh. dysenteriae, Haemophilus ducreyi* and *Actinobacillus actinomycetemcomitans* (Picket & Whitehouse 1999). The toxin blocks cells in G_2. CLDT inhibits the dephosphorylation of the protein kinase cdc2 and this prevents the cells entering into mitosis.

3.8.1   Exotoxins

Exotoxins are toxic bacterial proteins. The term is derived from the early observation that many bacterial exotoxins were excreted into the medium during growth. This differentiated them from endotoxins. This differentiation is not exact since some exotoxins are localised in the cytoplasm or periplasm and are released upon cell lysis. The naming of exotoxins is not systematic. Some toxins are named to indicate the type of host cell. Exotoxins that attack a variety of different cell types are called cytotoxins, whereas exotoxins that attack specific cell types can be designated by the cell type or organ affected, such as neurotoxin, leukotoxin, hepatotoxin and cardiotoxin. Exotoxins can also be named after their mode of action, for example lecithinase produced by *Cl. perfringens*. However exotoxins can also be named after the producer organism, or the disease which they cause. Two examples are cholera toxin, which causes cholera and is produced by *Vibrio cholerae*, and shiga toxin, which causes bacterial dysentery and is produced by *Shigella* species. Some toxins have more than one name, such as the shiga toxin of *E. coli* O157:H7 which is also called a verotoxin since it affects the mammalian Vero cell culture line. This confusion is reflected by the name changes for VTEC and STEC pathogenic *E. coli* (Section 5.2.3). The term 'enterotoxin' refers to protein toxins that cause diarrhoea or vomiting.

Exotoxins can be divided into four groups (Henderson *et al.* 1999):

(1)   Those that act at the cell membrane: type I.
(2)   Those that attack the membrane: type II.
(3)   Those that penetrate the membrane to act inside the cell: type III.
(4)   Those that are directly transported from the bacterium into the target eucaryotic cell by type III secretion.

Toxins that act at the cell membrane (type I toxins)

Although proteins are normally sensitive to high temperatures (such as 100°C) there are a few small exotoxins that are heat stable (ST toxins). The stable toxin of *E. coli* is a peptide of 18 to 19 amino acids with three disulphide bonds. This structure is responsible for the toxin's resistance to heat denaturation. The toxin causes diarrhoea since it binds to the endogenous ligand guanylin, a peptide hormone that regulates salt and water homeostasis in the gut and kidney. The *B. cereus* emetic toxin (also called cereulide) is a ring of three repeats of four amino acids (Table 3.10). It is plausible that this toxin binds to the 5-HT$_3$ receptor to stimulate the vagus afferent nerve, leading to vomiting.

Table 3.10 Distinguishing characteristics of *Cl. botulinum*.

	Group			
	I	II	III	IV
Toxin produced	A, B, F	B, E, F	C1, C2, D	G
Proteolysis	+	−	+ or −	+
Lipolysis	+	+	+	−
Glucose fermentation	+	+	+	−
Mannose fermentation	−	+	+	−
Minimum growth temp.	10–12°C	3.3°C	15°C	12°C
Salt inhibition (%)	10	5	3	>3
Volatile fatty acids produced[a]	Ac, iB, B, iV, Ph	A, B	A, P, B	A, iB, iV, Pa

[a] Ac, acetic acid; iB, isobutyric acid; B, butyric acid; iV, isovaleric acid; Ph, phenylpropionic acid; Pa, phenylacetic acid.

Superantigens are toxins which lack an AB structure and act by stimulating T cells to release cytokines and so induce an inappropriate immune response. Superantigens are produced by *St. aureus* and *Strep. pyogenes*. One of the most studied is TSST-1 (approximately 22 kDa) from *St. aureus* which causes toxic shock syndrome (associated with the use of tampons). *St. aureus* also produces the enteroxins A–E (approximately 27 kDa) which are responsible for food poisoning (Section 5.2.7).

Membrane-damaging toxins (type II toxins)
These toxins have A and B subunits which do not separate and act by disrupting the host cell membranes, for example listeriolysin O. This group is subdivided into channel-forming (pore-forming) toxins which allow cytoplasmic contents to leak out of the host cell. The thiol-activated

cholesterol binding cytolysins (52 to 60 kDa) bind to the cholesterol of the host cell to form pores of about 30 to 40 nm. Examples include listeriolysin O (*L. monocytogenes*), perfringolysin O (*Cl. perfringens*) and cereolysin O (*B. cereus*). These may sometimes be called 'haemolysins' due to the common use of blood cells to detect the presence of a toxin. In the past they have also been termed 'oxygen labile' and 'sulphydryl activated' toxins, although these terms are now recognised as misleading. The second type is phospholipase (PLC) which removes the charged head group from the lipid portion of phospholipids. A number of both Gram-positive and Gram-negative bacteria produce phospholipases. *L. monocytogenes* produces a phophoinositol-specific PLC which enables the organism to escape the vacuole after engulfment (Section 3.9.6) and a broad-range PLC involved in cell-to-cell spread. *Cl. perfringens* produces phospholipase C also known as α-toxin which has necrotic and cytolytic activity (Titbull *et al.* 1999). *St. aureus* produces a β-haemolysin.

Membrane-penetrating toxins (type III toxins)
The membrane-penetrating toxins have a so-called AB structure, where A is the catalytic subunit and the B subunit binds to the host cell receptor. There are two common types of AB toxins.

In the first type, A and B are joined in one molecule, although the structure can be disrupted to yield two subunits joined by a disulphide bridge. Examples include botulinum toxins types A–F (Section 5.2.9) and tetanus toxin. These are large (>150 kDa) single-chain toxins that are proteolytically cleaved during activation and cellular entry. The botulinum toxin blocks the action of peripheral nerves (Figs 3.7 and 3.8, Table 3.10).

In the second type the structure is AB_5, where five B subunits (pentamer) form a doughnut-like ring structure and possibly the A subunit enters the cell through the central hole (Fig. 3.9). Examples include cholera toxin, *E. coli* heat-labile toxin and the shiga toxins. Cholera toxins and shiga toxins bind to glycolipid ganglioside receptors on the host cell. The shiga toxins attack 28S rRNA to depurinate adenine 4324 which is involved in elongation factor 1 mediated binding of tRNA to the ribosomal complex and subsequently inhibits protein synthesis. The A subunit of shiga toxin has sequence and structural homology with the ricin family of plant toxins which have an identical mode of action. The A subunit is activated by proteolytic cleavage on cell entry to yield two fragments joined by a disulphide bridge which is subsequently reduced. Cholera toxin and *E. coli* heat-labile (LT) toxin ADP-ribosylate G_3, a heterotrimeric G protein involved in stimulation of adenylate cyclase (Fig. 3.10). The A subunit of G_s is modified at Arg201 which inhibits the GTPase activity. This keeps the G_s protein in the 'on' position, leading

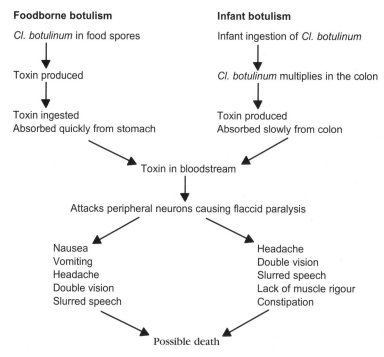

Fig. 3.7 *Cl. botulinum* intoxication.

to permanent activation of adenylate cyclase. The resultant high concentration of cAMP in the gut epithelial cells causes massive fluid accumulation in the gut lumen and watery diarrhoea, which can be fatal. Cholera toxin is chromosomally encoded on a phage, whereas *E. coli* LT and ST are plasmid encoded and shiga toxin is chromosomally encoded.

Transported toxins
Toxins in this group do not display toxic activity when purified from bacteria. They are transported directly from the bacterial cytoplasm to the eucaryotic cytoplasm by a complex array of proteins that bridge the membranes of the two cell types. This secretion system is often encoded on pathogenicity islands. Bacteria known to produce such toxins include *Salmonella*, *Shigella* and *Yersinia* species. Pathogenicity islands (PAIs) are large (>30 kb) distinct chromosomal elements encoding virulence-associated genes (Table 3.11). Because the G + C ratios of PAIs are often distinct from the rest of the bacterial DNA G + C ratios, it has been proposed that they may have been acquired in the past by horizontal gene transfer from other bacterial species.

Toxin mode of action

150 kDa toxin activated by protease activity, either gastric or clostridial

↓

Toxin is a zinc-requiring endoprotease

↓

Nicked AB toxin
A, 50 kDa (light chain)
B, 100 kDa (heavy chain)

↓

B subunit binds to sialic acid containing glycoprotein on peripheral neurones

↓

Toxin internalised into neurone

↓

Toxin prevents release of the neurotransmitter acetylcholine

↓

Consequently, nerve pulse transmission stops, causing flaccid paralysis

Cl. botulinum

Cl. botulinum is classified into four groups using Table 3.10

Eight toxin serotypes: A, B, C1, D, E, F & G
A & B most common cause of human
   botulism
C1 – farm animal botulism
D & G – infrequent cases of human
   botulism
E & G produced by non-*Cl. botulinum*
   species

BoNTB = Botulinum toxin serotype B, etc.

Fig. 3.8a   *Cl. botulinum* and toxin mode of action.

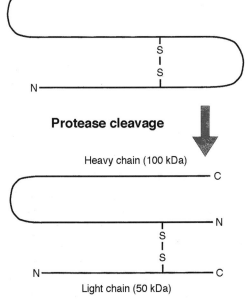

Fig. 3.8b   Structure and activation of *Cl. botulinum* toxin.

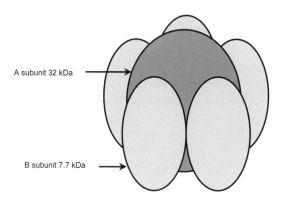

A subunit 32 kDa

B subunit 7.7 kDa

Fig. 3.9 AB$_5$ structure of shiga toxin.

Table 3.11 Pathogenicity islands of three important foodborne pathogens.

Pathogen	PAI designation	Size (kb)	G + C ratio (PAI/host)	Phenotype
E. coli	Pai I	70	40/51	Haemolysin production
	Pai II	190	40/51	Haemolysin, P-fimbriae production
	LEE (Pai III)	35	39/51	Induction of attaching and effacing lesions on enterocytes
S. typhimurium	SPI-1	40	42/52	Invasion of non-phagocytic cells
	SPI-2	40	45/52	Survival in macrophages
	SPI-3	17	Unknown	Survival in macrophages
V. cholerae	VPI	39.5	35/46	Colonisation, expression of phage CTXφ receptor

PAI = pathogenicity island.
Adapted from Henderson *et al.* 1999.

The locus of enterocyte effacement (LEE) is a PAI (35 kb) of entero-pathogenic *E. coli* strains. It induces the attaching and effacing lesions on enterocytes (Fig. 3.6) and encodes the secretion system for transfer of toxin from *E. coli* cell to host cell. This results in cytoskeletal rearrangements and the formation of a pedestal on which the *E. coli* cell is located (see later Fig. 3.16). *Salmonella* species have about 100 genes required for virulence and contain at least four PAIs. Salmonella pathogenicity island (SPI)-1 is associated with the toxin secretion system and encodes for around 25 genes (Table 3.11). *V. cholerae* pro-

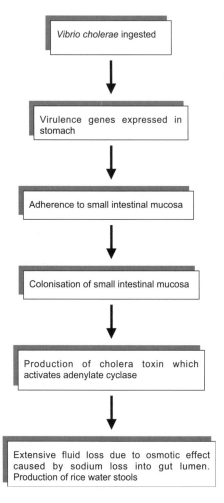

Fig. 3.10   Mode of action of cholera toxin.

duces cholera toxin encoded by the *ctx*A and *ctx*B genes that are encoded by the filamentous phage CTX. The bacterial receptor for phage
infection is the toxin co-regulated pilus (TCP) that is also an important
adherence determinant. TCP is found on the 39.5 kb PAI called VPI in
V. cholerae. It is believed that the acquisition of VPI enables aquatic *V.
cholerae* strains to colonise the human intestine and the subsequent
generation of epidemic and pandemic *V. cholerae* strains. A possible
Gram-positive PAI is found in pathogenic strains of *L. monocytogenes*.
The 10 kb element encodes for genes required for listeriolysin O (*hly*),
*act*A and *plc*B (responsible for intra- and intercellular movement).

3.9 Virulence factors of foodborne pathogens

3.9.1 Campylobacter jejuni

Campylobacter jejuni colonises the distal ileum and colon of the human intestinal tract. After colonising the mucus and adhering to intestinal cell surfaces, the organisms perturb the normal absorptive capacity of the intestine by damaging epithelial cell function. This is due either to direct action, by cell invasion or the production of toxin(s) (Section 5.2.1), or indirectly, following the initiation of an inflamatory response (Fig. 3.11). Intracellular survival within the host cell may be aided by the production of catalase to protect against oxidative stress from lysosomes. The organism may also enter a viable but non-culturable state (VNC, Section 6.2.2) which may be of importance in the organism's virulence.

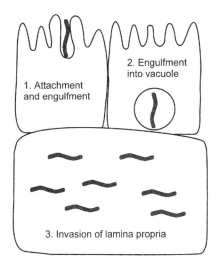

Fig. 3.11 *C. jejuni* infection of gut mucosa.

3.9.2 Vibrio cholerae

In the small intestine, *V. cholerae* attaches to the mucosal surface and produces an exotoxin, cholera toxin, that acts on intestinal mucosal cells. The toxin is 84 kDa in size and is composed of an A1 subunit (21 kDa) and an A2 subunit (7 kDa) covalently linked to five B subunits (10 kDa each). The B subunit binds to a receptor ganglioside on the mucosal cells and the A subunit enters the cell, activating adenylate cyclase and subsequently increasing the concentration of cAMP.

Mucosal cells have a set of ion transport pumps (for Na^+, Cl^-, HCO_3^- and K^+) that normally maintain a tight control over ion fluxes across the intestinal mucosa. Because water can pass freely through membranes, the only way to control the flow of water into and out of tissue is to control the concentration of ions in different body compartments. Under normal conditions, the net flow of ions is from lumen to tissue (Fig. 3.10) resulting in a net uptake of water from the lumen. The effect of cholera toxin on mucosal cells is to alter this balance. Cholera toxin causes no apparent damage to the mucosa, but the increased level of cAMP decreases the net flow of sodium into tissue and produces a net flow of chloride (and water) out of tissue and into the lumen, resulting in massive diarrhoea and electrolyte imbalance (Fig. 3.10).

3.9.3   Shigella dysenteriae

Shigella dysenteriae has a low infectious dose (Tables 7.7 and 9.1) probably because the organism is not readily killed by the acidity of the stomach. The organism then colonises the colon and enters the mucosal cells, where it multiplies rapidly. The bacteria inside the mucosal cells move laterally to infect adjacent mucosal cells (Fig. 3.12). The organism provokes an intense inflammatory response in the lamina propria and the mucosal layer. The inflammatory response causes the blood and mucus to be found in the victim's faeces.

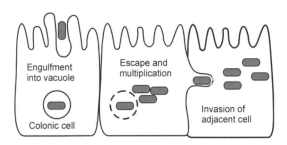

Fig. 3.12   *Shigella* spp. infection of gut mucosa.

The organism produces shiga toxin, an exotoxin (Section 3.8.1) with AB_5 structure (Fig. 3.7; A subunit is 32 kDa, B subunit 7.7 kDa) and a poorly characterised cytolethal distending toxin (CLDT, Section 3.8). The AB_5 toxin is structurally similar to cholera toxin, although there is little similarity in amino acid sequences. The B subunits of shiga toxin, like those of cholera toxin, recognise a host cell surface glycolipid (Gb3), but the carbohydrate moiety on this lipid is Gal α-1,4-Gal, not sialic acid as for cholera toxin receptor G_{M1}. The shiga toxin first binds to the surface of mammalian cells and is then internalised by endocytosis. Nicking to activate and translocate the A subunit takes place inside the host cell. In contrast to other AB toxins (Section 3.8.4), the A subunit of shiga toxin does not ADP-ribosylate a host cell protein, but stops host cell protein synthesis. It inactivates the 60S subunit of host cell ribosomes by cleaving the *N*-glycosidic bond of a specific adenosine residue in 28S rRNA, a component of the 60S ribosomal subunit. Cleavage of 23S rRNA at this site prevents binding of aminoacyl-tRNAs to the ribosome and stops elongation of proteins.

3.9.4 Pathogenic strains of E. coli

The differences in the pathogenic strains of *E. coli* are summarised in Table 3.12 and Figs 3.13–18 (Section 5.2.3).

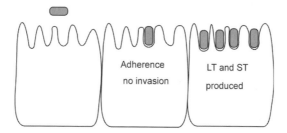

Fig. 3.13 ETEC infection of gut mucosa.

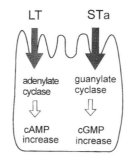

Fig. 3.14 Effect of LT and STa toxins on gut cells.

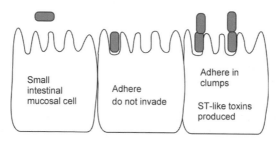

Fig. 3.15 EAggEC infection of gut mucosa.

Table 3.12 Pathogenic strains of *E. coli* (adapted from Post).

Group	Serotypes	Adhesion and invasion characters	Toxins	Disease symptoms
Enterotoxigenic (ETEC)	O6, O8, O15, O20, O27, O63, O78, O80, O85, O115, O128, O139, O148, O153, O159, O167	Adhere uniformly but do not invade	Heat-labile (LT). Heat-stable (ST). LT is similar to cholera toxin and acts on mucosal cells	Cholera-like diarrhoea but generally less severe
Enteropathogenic (EPEC)	O18, O44, O55, O86, O111, O112, O114, O119, O125, O126, O127, O128, O142	Adhere in clumps. Invade host cells, attach and efface	Not apparent	Infantile diarrhoea, vomiting
Enteroinvasive (EIEC)	O124, O143, O152	Invade cells of colon. Spread laterally to adjacent cells	No Shiga-like toxin has been detected	Cell-to-cell spread and disease is similar to dysentery
Enterohaemorrhagic (EHEC)	O6, O26, O46, O48, O91, O98, O111, O112, O146, O157, O165	Adhere tightly. Attach and efface host cells. Invade	Verocytotoxic Shiga-like toxin	Bloody diarrhoea. Haemorrhagic colitis. May progress to haemolytic uraemic syndrome (HUS) and thrombotic thrombocytopenic purpura (TTP)
Enteroaggregative (EAggEC)	Wide range of serotypes, recent outbreaks O62, O73, O134	Adhere in clumps but do not invade	ST-like toxin. Haemolysin. Verocytotoxin reported in some strains	Diarrhoea. Some strains have been reported to cause haemolytic uraemic syndrome (HUS)

Enterotoxigenic *E. coli* (ETEC) strains resemble *V. cholerae* in that they adhere to the small intenstinal mucosa and produce symptoms not by invading the mucosa but by producing toxins that act on mucosal cells to cause diarrhoea (Fig. 3.13). ETEC strains produce two types of entero-toxin: a cholera-like toxin called heat-labile toxin (LT) and a second diarrhoeal toxin called heat-stable toxin (ST, Fig. 3.14). Heat-stable is defined as retaining activity after heat treatment at 100°C for 30 minutes. There are two main types of LT: LT-I and LT-II. LT-I shares about 75% amino acid sequence identity with cholera toxin and has an AB_5 structure. The A subunit catalyses the ADP-ribosylation of G_s, which raises host cell cAMP levels, identical to the cholera toxin mode of action. The B subunits of LT-I interact with the same receptor as cholera toxin (G_{M1}). However, LT-I is not excreted (as per cholera toxin), but is localised in the periplasm. The toxin leaks out of the organism on exposure to bile acids and low iron concentrations (as would occur in the small intestine).

ST is a family of small (approximately 2 kDa) toxins which can be divided into two groups: methanol soluble (STa) and methanol insoluble (STb). STa causes an increase in cGMP (not cAMP) levels in the host cell cytoplasm, which leads to fluid loss. cGMP, like cAMP, is an important signalling molecule in eucaryotic cells, and changes in cGMP affect a number of cellular processes, including activities of ion pumps. LT-I and STa are encoded on plasmids.

The enteroaggregative *E. coli* (EAggEC) resemble ETEC strains in that they bind to small intestinal cells, are not invasive, and cause no obvious histological changes in the intestinal cells to which they adhere (Fig. 3.15). They differ from ETEC strains primarily in that they do not adhere uni-formly over the surface of the intestinal mucosa but tend to clump in small aggregates. EAggEC strains produce an ST-like toxin (EAST) and a hae-molysin-like toxin (120 kDa in size). Some strains are reported to produce a Shiga-like (verocytotoxin) toxin.

Enteropathogenic *E. coli* (EPEC) do not produce any enterotoxins or cytotoxins. The cells do invade, attach and efface. These bacteria induce a characteristic attaching and effacing (A/E) lesion (Fig. 3.6, Fig. 3.16) in epithelial cells in which microvilli are lost and the underlying cell mem-brane is raised to form a pedestal which can extend outward for up to 10 μm. This phenomenon is the result of extensive rearrangement of host cell actin in the vicinity of the adherent bacteria that results in formation of a cuplike pedestal structure under the bacteria. The genes encoding for this virulence are encoded on the locus of enterocyte effacement (LEE), a pathogenicity island (Table 3.11). Contact with epithelial cells results in the secretion of a number of proteins, including EspA (25 kDa) and EspB (37 kDa), which subsequently trigger a series of responses in the host cell. These responses include activation of signal transduction pathways, cell

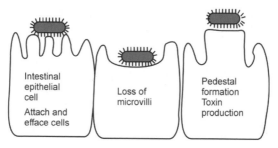

Fig. 3.16   EPEC infection of gut mucosa cells.

depolarisation and the binding of the outer membrane protein intimin (94 kDa). Intimin is encoded by the gene *eae*. Intimin probably binds to Tir (translocated intimin receptor), a bacterial protein (78 kDa) which is secreted by EPEC and binds to the host cell. The characteristic diarrhoea of EPEC infections is possibly due to an induced efflux of chloride ions due to activation of protein kinase C. The loss of microvilli would contribute due to the decreased absorption ability.

Enterohaemorrhagic *E. coli* (EHEC) strains bind tightly to cultured mammalian cells and produce the same type of attachment-effacement phenomenon as EPEC strains (Fig. 3.17). Although EHEC strains cause dysentery similar to that caused by *Shigella* species, they probably do not invade mucosal cells as readily as *Shigella* strains. The best studied strain is *E. coli* O157:H7 (Section 5.2.3).

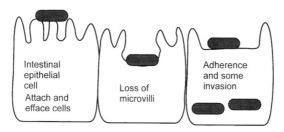

Fig. 3.17   EHEC infection of gut mucosa cells.

Enteroinvasive *E. coli* (EIEC) strains cause a disease that is indistinguishable at the symptomatic level from the dysentery caused by *Shigella* species. EIEC strains actively invade colonic cells and spread laterally to adjacent cells (Fig. 3.18). The steps in invasion and cell-to-cell spread appear to be virtually identical to *Shigella* species. However, EIEC do not produce shiga toxin, which may account for the lack of haemolytic uraemic syndrome as a complication of EIEC dysentery.

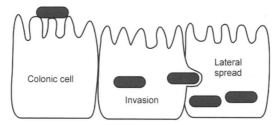

Fig. 3.18 EIEC infection of gut mucosa cells.

3.9.5 Salmonella *species*

Survival of *Salmonella* species through the acidic environment of the human stomach can be enhanced by the induction of acid tolerance due to heat treatment and exposure to short chain fatty acids (stress responses, Section 2.7.1). Like many other enteric pathogens *Salmonella* species invade mammalian cells of the lower intestinal tract by inducing actin rearrangements that result in the formation of pseudopods that engulf the bacteria. Invasion of epithelial cells by *Salmonella* species occurs following adhesion to microvilli via mannose-specific type 1 fimbriae. *Salmonella* species cause a change in the appearance of the surface of the host cell that resembles a droplet splash. This 'splash' effect results in internalisation of the bacteria inside an endocytic vesicle. The effect of the bacteria on host cells is called ruffling because of the appearance of the deformed host cell membrane (Fig. 3.19). Ruffling and internalisation of the bacteria are accompanied by extensive actin rearrangements in the vicinity of the invading bacteria. Invasion is enhanced under anaerobic conditions, when the cells are in the stationary phase and when osmolarity is high. A large number of gene loci are required for invasion and many are located on SPI-1 (*Salmonella* pathogenicity island-1; Table 3.11; Section 3.8.5). One of the genes (*inv*) encodes for the formation of surface structures produced when the bacterium adheres to the host cell. After the bacteria are engulfed in a vesicle, however, the host cell surface and

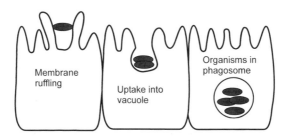

Fig. 3.19 *Salmonella* spp. infection of gut mucosa.

actin filament organisation return to normal. The vesicles may coalesce and this may be followed by bacterial multiplication. This may lead to cell death.

The organism can survive in the phagocytes due to its resistance to oxidative bursts through the production of catalase and superoxide dismutase and resistance to defensins (toxic peptides) due to products from the *pho*P/*pho*Q operon. The symptoms of typhoid fever are caused by LPS which induces a local inflammatory response during invasion of the mucosa. Unlike *S. typhi*, it is uncertain whether food-poisoning strains of *Salmonella* (i.e. *S. enteritidis* and *S. typhimurium*) are inside macrophages of the liver and spleen. The symptoms of gastroenteritis probably result from the invasion of mucosal cells. *S. typhi* causes a systemic infection whereas enteric pathogens such as *S. typhimurium* rarely penetrate beyond the submucosal tissues.

3.9.6   Listeria monocytogenes

Listeria monocytogenes attaches to the intestinal mucosa – possibly α-D-galactose residues on the bacterial surface bind to α-D-galactose receptors on the intestinal cells – and invades the mucosal cells (Fig. 3.20). The bacteria are taken up by induced phagocytosis, whereby the organism is engulfed by pseudopodia on the epithelial cells, resulting in vesicle formation. *L. monocytogenes* is encased in a vesicle membrane which it disrupts using listeriolysin O (responsible for the β haemolysis zone surrounding isolates on blood agar plates). The organism also produces catalase and superoxide dismutase which may protect it from oxidative burst in the phagosome. *L. monocytogenes* also produces two phospholipase C enzymes which disrupt host cell membranes by hydrolysing membrane lipids such as phosphatidylinositol and phosphatidylcholine. In the host cell cytoplasm the bacterium multiplies rapidly (possibly doubling every 50 minutes). The organism moves through the cytoplasm to invade adjacent cells by polymerising actin to form long tails. This form of motility enables it to move at about 1.5 μm/second. The bacterium produces

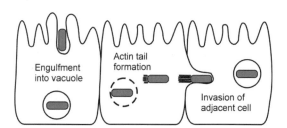

Fig. 3.20   *Listeria monocytogenes* infection of gut mucosa.

protrusions into adjacent cells, enters the cytoplasm and repeats the cycle of escaping the vacuole and multiplying.

3.9.7 Bacillus cereus

Bacillus cereus is a sporeforming facultative organism that grows well after cooking in the absence of competing microorganisms. It produces two types of toxins: one emetic toxin and three enterotoxins (Table 3.13). The emetic toxin is similar to the potassium ionophore valinomycin in that it is ring-shaped with three repeating units of the amino and/or oxy acids '-D-O-leucine-D-alanine-O-valine-L-valine-'. Two of the enterotoxins are involved in food poisoning and are composed of three different protein subunits. The enterotoxins are produced during growth in the human small intestine. One of these enterotoxins is a haemolysin. The third enterotoxin is a single protein unit and has not been linked to food poisoning.

Table 3.13 Characteristics of food poisoning caused by *B. cereus* (adapted from Granum & Lund 1997).

	Emetic syndrome	Diarrhoeal syndrome
Infective dose	10^5-10^8 cells/g	10^5-10^7 total
Toxin produced	Preformed	Produced in the small intestine
Type of toxin	Cyclic peptide (1.2 kDa)	Three protein subunits (L_1, L_2, B; 37-105 kDa)
Toxin stability	Very stable (126°C, 90 min; pH 2-11)	Inactivated at 56°C, 30 minutes
Incubation period	0.5-6 hours	8-24 hours
Duration of illness	6-24 hours	12-24 hours
Symptoms	Abdominal pain, watery diarrhoea, nausea	Nausea, vomiting and malaise, watery diarrhoea
Foods most frequently implicated	Fried and cooked rice, pasta, noodles and potatoes	Meat products, soups, vegetables, fish, puddings, sauces and milk and dairy products

4

THE MICROBIAL FLORA OF FOOD

4.1 Spoilage microorganisms

Spoilt food is food that tastes and smells off. It is due to the undesirable growth of microorganisms that produce volatile compounds during their metabolism, which the human nose and mouth detect. Spoilt food is not poisonous. The food is not of the quality expected by the consumer and therefore it is a quality, not safety, issue. The terms 'spoilt' and 'unspoilt' are subjective since acceptance is dependent upon consumer expectation and is not related to food safety. Soured milk is unacceptable as a drink, but can be used to make scones. The overgrowth of *Pseudomonas* spp. is undesirable in meat, yet desirable in hung game birds. Acetic acid production during wine storage is unacceptable, yet acetic acid production from wine (and beer) is necessary for the production of vinegar.

Food spoilage involves any change that renders the food unacceptable for human consumption. It may be due to a number of causes:

- Insect damage
- Physical injury due to bruising, pressure, freezing, drying and radiation
- The activity of indigenous enzymes in animal and plant tissues
- Chemical changes not induced by microbial or naturally occurring enzymes
- The activity of bacteria, yeasts and moulds

(adapted from Forsythe & Hayes 1998).

During harvesting, processing and handling operations food may become contaminated with a wide range of microorganisms. Subsequently, during distribution and storage the conditions will be favourable for certain organisms to multiply and cause spoilage. Which microorganisms will develop or what biochemical and chemical reactions occur is dependent upon the food and intrinsic and extrinsic parameters (Section 2.5.1). Spoilage can be delayed by lowering the storage temperature (Fig. 4.1).

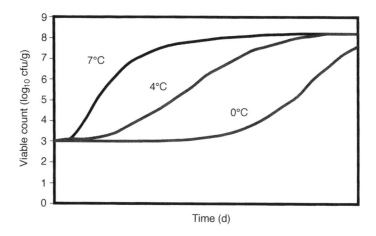

Fig. 4.1 Temperature effect on food spoilage.

For fresh foods the primary quality changes may be categorised as:

• Bacterial growth and metabolism resulting in possible pH changes and formation of toxic compounds, off odours, gas and slime formation.
• Oxidation of lipids and pigments in fat-containing foods resulting in undesirable flavours, formation of compounds with adverse biological effects or discoloration.

Spoilage is not only due to the visible growth of microorganisms but also to the production of end metabolites which result in off odours, gas and slime. The spoilage of milk (Section 4.1.2) illustrates these two aspects with the visible growth of thermoduric microorganisms (*B. cereus*) and the production of undesirable taste (bitterness due to thermostable proteases).

Although there is much progress in the characterisation of the total microbial flora and metabolites developing during spoilage, not much is known about the identification of specific microorganisms in relation to food composition. Despite the fact that food spoilage is a huge economical problem worldwide, the mechanisms and interaction leading to food spoilage are very poorly understood. Although the total microbial flora may increase during storage, it is specific spoilage organisms which cause the chemical changes and the production of off odours (Fig. 4.2). The shelf life is dependent upon the growth of the spoilage flora and can be reduced through cold storage to retard microbial growth and packaging (vacuum packaging and modified atmosphere packaging; Section 4.3).

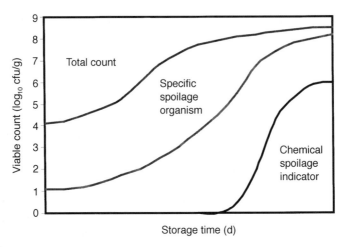

Fig. 4.2   Food spoilage indicators.

The higher the initial microbial load the shorter the shelf life due to the increased microbial activities (Fig. 4.3).

A wide variety of microorganisms may initially be present on food and grow if favourable conditions are present. On the basis of susceptibility to spoilage, foods may be classed as nonperishable (or stable), semi-perishable and perishable. The classification depends on the intrinsic factors of water activity, pH, presence of natural antimicrobial agents, etc. Flour is a stable product because of the low water activity. Apples are semi-perishable since poor handling and improper storage can result in slow

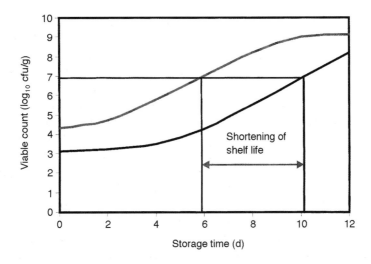

Fig. 4.3   Effect of initial microbial load on shelf life.

fungal rot. Raw meat is perishable since the intrinsic factors of pH and water activity favour microbial growth.

4.1.1 Spoilage microorganisms

Gram-negative spoilage microorganisms
Pseudomonas, Alteromonas, Shewanella putrefaciens and *Aeromonas* spp. spoil dairy products, red meat, fish, poultry and eggs during cold storage. These foods have high water activity and neutral pH and are stored without a modified atmosphere (i.e. normal levels of oxygen). There are a variety of spoilage mechanisms including production of heat-stable proteases and lipases which produce off flavours in milk after the organism has been killed by pasteurisation. They also produce coloured pigments in egg spoilage. *Erwinia carotovora* and various *Pseudomonas* spp. are responsible for approximately 35% of vegetable spoilage.

Gram-positive, non-sporeforming spoilage microorganisms
The lactic acid bacteria (Section 4.4.1) and *Brocothrix thermosphacta* are Gram-positive rods which typically cause spoilage of meats stored under modified atmosphere packaging or vacuum packaging. Acetobacters and *Pediococcus* spp. can produce a thick polysaccharide slime in beer ('rope'). The production of diacetyl (see later, Fig. 4.6) by lactic acid bacteria causes beer spoilage. The organisms can also produce lactic acid in wine, giving it a sour taste. The acetobacter produce acetic acid in beer, giving it a sour taste.

Gram-positive sporeforming spoilage microorganisms
The sporeforming *Bacillus* spp. and *Clostridium* spp. can be important spoilage organisms of heat-treated foods, since the spores may survive the process. *B. cereus* can grow in pasteurised milk at 5°C and cause 'sweet curdling' (renin coaggulation without acidification) and 'bitty cream'. *B. stearothermophilus* causes 'flat sour spoilage' of canned foods. It grows in the can, producing acids (hence the term 'sour') but without any gas production and subsequently the can does not swell. *Cl. thermosaccharolyticum* spoils canned foods and the can swells due to gas production. *Desulfotomaculum nigrificans* produces hydrogen sulphide which causes canned foods to swell and smell. This is known as 'sulphur stinker' spoilage. *B. stearothermophilus*, *D. nigrificans* and *Cl. thermosaccharolyticum* are thermophilic organisms. Therefore they are only able to grow at temperatures above ambient. Hence canned foods are processed to be 'commercially sterile' since under normal storage conditions no microbial growth is expected to occur.

Yeast and mould spoilage microorganisms

Yeast and moulds are more tolerant of low water activity and low pH than bacteria. Subsequently they typically spoil foods such as fruit and vegetables and bakery products (Table 4.1). They produce pectinolytic enzymes which soften the plant tissues causing 'rot'. As much as 30% of all fruit spoilage may be due to *Penicillium*. Fungi also produce large quantities of coloured sporangia which visibily colour the food. Bread is spoiled by *Rhizopus nigricans* ('bread mould', black spots), *Penicillium* (green mould), *Aspergillus* (green mould) and *Neurospora sitophila* ('red bread'). Osmophilic yeast (*Saccharomyces* and *Torulopsis* spp.) are able to grow in high sugar concentrations (65–70%) and spoil jams, syrups and honey. Fungi also cause various types of meat spoilage such as 'whiskers' due to *Mucor*, *Rhizopus* and *Thamnidium*.

Table 4.1 Food spoilage fungi (adapted from Brul & Klis 1999).

Organisms	Products typically affected
Aspergillus versicolor	Bread and dairy products
A. flavus	Cereals and nuts
A. niger	Spices
Byssochlamys fulva	Cereals in airtight packs
Fusarium oxysporum	Fruit
Mucor spp.	Meat
Neosartorya fischeri	Pasteurised foods
Neurospora sitophila	Bread
Penicillium roqueforti	Meat, eggs and cheese
P. expansum	Fruit and vegetables
P. commune	Margarines
P. discolor	Cheese
Penicillium spp.	Bread
Rhizopus nigricans	Bread
Rhizopus spp.	Meat
Saccharomyces spp.	Soft drinks, jams, syrups and honey
Thamnidium spp.	Meat
Torulopsis spp.	Jams, syrups and honey
Trichoderma harzianum	Margarines
Zygosaccharomyces bailii	Dressings

Standard preservation methods to control fungal growth are high temperatures to denature spores, weak acid preservatives, such as sorbic acid, and antibiotics such as natalycin. However, some fungi have become resistant to sorbic acid and some fungi degrade sorbic acid to the malodorous compound pentadiene.

4.1.2 Spoilage of dairy products

Milk is an ideal growth medium for bacteria and therefore needs to be kept refrigerated. The intrinsic flora (around 10^2-10^4 cfu/ml) is derived from the cow's milk ducts in the udder and the milking equipment, etc., during production. The flora includes *Pseudomonas* spp., *Alcaligenes* spp., *Aeromonas* spp., *Acinetobacter-Moraxella* spp., *Flavobacterium* spp., *Micrococcus* spp., *Streptococcus* spp., *Corynebacterium* spp., *Lactobacillus* spp. and coliforms. Milk spoilage is primarily due to the growth of the psychrophilic microorganisms which produce heat-stable lipases and proteases which are not denatured during pasteurisation. The pseudomonads, flavobacteria and *Alcaligenes* spp. produce lipases which produce medium- and short-chain fatty acids from milk triglycerides. These fatty acids spoil the milk due to their rancid off flavour. Proteases are produced by pseudomonads, aeromonads, *Serratia* and *Bacillus* spp. The enzyme hydrolyses milk proteins to produce bitter peptides. Therefore it is undesirable to have a high microbial count before pasteurisation as the action of the residual enzymes during storage will result in a shortening of the milk's shelf life.

Pasteurisation at 72°C for 15 seconds is designed to kill all pathogenic bacteria such as *Mycobacterium tuberculosis*, *Salmonella* spp. and *Brucella* spp. at levels expected in fresh milk. Thermoduric organisms are those that survive pasteurisation. These include *Streptococcus thermophilus*, *Enterococcus faecalis*, *Micrococcus luteus* and *Microbacterium lacticum*. The spores of *Bacillus cereus* and *B. subtilis* will survive the heat treatment. *B. cereus* growth causes pasteurised milk spoilage known as 'bitty cream'.

4.1.3 Spoilage of meat and poultry products

Meat (usually muscle) is a highly perishable food product with water activity (Section 2.5.2) suitable for the growth of most microorganisms. Since the meat is highly proteinaceous it is relatively strongly buffered and the growth of microorganisms does not significantly lower the pH. Because it is nutritious it can quickly become spoilt through the growth of microorganism and even poisonous if contaminated with foodborne pathogens. The meat itself is sterile within the animal's body. However, it can easily become contaminated during slaughter, abattoir practice, handling during processing and improper storage. If microorganisms such as *Pseudomonas* spp., *Brochothrix thermosphacta* and lactic acid bacteria grow on the meat, then it will become spoilt and unacceptable for eating. It is the growth of foodborne pathogens such as *Salmonella*, toxin-producing strains of *E. coli, L. monocytogenes, Cl. perfringens* and toxin production by *St. aureus* that are of concern with meat and poultry products.

Poultry skin can carry a range of spoilage organisms: *Pseudomonas* spp., *Acinetobacter-Moraxella* spp., *Enterobacter* spp., *Sh. putrefaciens, Br. thermosphacta* and *Lactobacillus* spp. The foodborne pathogen *C. jejuni* may also be present on the skin and subsequently transferred to work surfaces. Pathogens such as *S. enteritidis* can infect the ovaries and the oviducts of hens and subsequently the egg prior to shell formation. Additionally, egg shells become contaminated with intestinal bacteria during passage through the cloaca and the incubator surface. *Cl. perfringens* is isolated in small numbers from raw poultry. It is unable to grow due to the cold temperature of storage and the presence of competitive psychrotrophic organisms. Bacteria found on poultry include *St. aureus, L. monocytogenes* and *Cl. botulinum* type C (not toxic to healthy adults). The *St. aureus* strains found on poultry are not pathogenic to man, and staphylococcal food poisoning associated with poultry is normally due to contamination of cooked meat by an infected food handler.

Carcass meat at temperatures above 20°C will readily be spoilt by bacteria originating from the animal's intestines which have contaminated the meat during slaughtering. The spoilage flora is dominated by mesophilic organisms such as *Escherichia coli, Aeromonas* spp., *Proteus* spp. and *Micrococcus* spp. At temperatures below 20°C, the spoilage flora will be predominantly psychrotrophs such as *Pseudomonas* spp. (fluorescent and non-fluorescent types) and *Brochothrix thermosphacta*. Poultry meat will also contain small number of *Acinetobacter* spp. and *Shewanella putrefaciens*. At refrigeration temperatures of 5°C and below, the spoilage flora will be dominated by pseudomonads. The pseudomonads are aerobes and hence only grow on the food surface to a depth of 3–4 mm in the underlying tissues. The spoilage is due to the degradation of proteins producing volatile off flavours such as indole, dimethyl disulphide and ammonia. Chemical oxidation of unsaturated lipids results in a rancid off flavour.

Fungal growth occurs under prolonged storage periods (Forsythe & Hayes 1998):

- 'Whiskers' are due to *Mucor, Rhizopus* and *Thamnidium*.
- Black spot is due to *Cladosporium herbarum* and *Cl. cladosporoides*.
- *Penicillium* spp. and *Cladosporium* spp. cause coloured spoilage due to their yellow and green coloured spores.
- 'White spot' is caused by *Sporotrichum carnis* growth.

4.1.4 Spoilage of fish

Bacteria can be detected in fresh fish from the slime coat on the skin (10^3-10^5 cfu/cm^2), gills (10^3-10^4 cfu/g) and the intestines (10^2-10^9 cfu/g). Fish from the marine environments of the North Seas will be dominated by psychrotrophs whereas the fish from warmer waters will be colonised from a higher proportion of mesotrophs. The psychrophilic flora will be composed of *Pseudomonas* spp., *Alteromonas* spp., *Shewanella putrefaciens*, *Acinetobacter* spp. and *Moraxella* spp. The mesophilic flora will contain micrococci and coryneforms as well as the acinetobacters. The flora of fish landed at ports will also include organisms from the ice used to preserve the fish and the flora of the boat holds. Due to the high psychrotrophic flora, spoilage occurs very rapidly since the low temperature storage favours psychrotrophs. At 7°C for 5 days the microbial load can be as high as 10^8/g. This is faster than meat spoilage which would take about 10 days to obtain this microbial load. The off odour of spoiled fish is probably produced by certain strains of *Pseudomonas* producing volatile esters (ethyl acetate, etc.) and volatile sulphide compounds (methyl mercaptan, dimethyl sulphide, etc.). Trimethylamine oxide is reduced by certain fish spoilage organisms, such as *Shewanella putrefaciens* to produce trimethylamine which gives a smell characteristic of spoiled fish. *Photobacterium phosphoreum* (a luminescent bacterium) causes the spoilage of vacuum packed cod (Dalgaard *et al.* 1993).

4.1.5 Egg spoilage

Eggs contain a barrage of antimicrobial factors including iron chelating agents (conalbumin) and lysozyme in the albumen (egg white). The shell is covered with a water-repellent cuticle and two inner membranes. In contrast, the yolk does not contain any antimicrobial factors. Egg spoilage is principally due to *Pseudomonas* spp. as well as *Proteus vulgaris*, *Alteromonas* spp. and *Serratia marcescens*. These produce a variety of coloured rots (Table 4.2).

Table 4.2 Bacterial causes of egg spoilage.

Spoilage organism	Type of rot
Pseudomonas spp.	Green
Pseudomonas, Proteus, Aeromonas, Alcaligenes and *Enterobacter* spp.	Black
Pseudomonas spp.	Pink
Pseudomonas and *Serratia* spp.	Red
Acinetobacter-Moraxella spp.	Colourless

4.2   Shelf life indicators

The shelf life is an important attribute of all foods (Ellis 1994). It can be defined as the time between the production and packaging of the product and the point at which it becomes unacceptable to the consumer. It is therefore related to the total quality of the food and linked to production design, ingredient specifications, manufacturing process, transportation and storage (at retail and in the home). The shelf life depends upon the food itself (Table 4.3) and it is essential for the food manufacturer to identify the intrinsic and extrinsic parameters which limit the shelf life.

Table 4.3   Food products and associated shelf lives.

Food product	Typical shelf life
Bread	Up to 1 week at ambient temperature
Sauces, dressings	1 to 2 years at ambient
Pickles	2 to 3 years at ambient
Chilled foods	Up to 4 months at 0–8°C
Frozen foods	12 to 18 months in freezer cabinets
Canned foods	Unlacquered cans, 12 to 18 months Lacquered cans, 2 to 4 years

Shelf life can be determined by combined microbiological and chemical analysis of food samples taken over the product's anticipated shelf life. There are two approaches to shelf-life determination:

(1) *Direct determination and monitoring.* This requires batches of samples to be taken at specified stages in the development of the product. Typically these samples are taken at intervals equal to 20% of the expected shelf life to give samples of six different ages. The samples are stored under controlled conditions until their quality becomes unacceptable. Attributes tested include smell, texture, flavour, colour and viscosity. This method is not ideal for products with shelf lives in the order of 1 year.

(2) *Accelerated estimation.* The need to meet product launch dates may require the use of accelerated shelf life estimations by raising the storage temperature to increase any ageing processes. This method has to be used with care since different microbial floras may develop

which differ in off flavour formation from the unaccelerated spoilage flora.

Suggested microbiological shelf life limits (index of spoilage) are given in Table 4.4.

Table 4.4 Suggested microbiological limits for end of shelf life.

Product	Microbial count	Comments
Raw meat	1×10^6 cfu/g aerobic plate count	Visible deterioration and/or slime at 1×10^7 cfu/g
Ground beef	1×10^7 cfu/g	End of shelf life, colour begins to fade and slime forms
Vacuum packaged cooked products	1×10^6 cfu/g aerobic plate count	Represents the point where a pH shift > 0.25 occurs
Fully cooked products	1×10^6 cfu/g aerobic plate count 1×10^3 cfu/g Enterobacteriaceae 1×10^3 cfu/g lactic acid bacteria 500 cfu/g yeast and moulds	

There is a wide range of compounds, principally end products of microbial growth, which have potential for use in the determination and prediction of shelf life (Dainty 1996). Chemical indicators of food spoilage are:

Glucose
The main substrate for microbial growth of red meats, including modified atmosphere and vacuum packaged. Spoilage is associated with post-glucose utilisation of amino acids by pseudomonads and hence monitoring glucose depletion can indicate onset of spoilage. The technique is of limited value, however, due to an ill-defined lag phase before spoilage.

Gluconic and 2-oxogluconic acid
Pseudomonas metabolism of glucose results in the accumulation of gluconic and 2-oxogluconic acids in beef.

L- *and* D-*Lactic acids, acetic acid and ethanol*

There is an ill-defined lag between maximum microbial cell number and sensory detection of spoilage in vacuum packed and low oxygen modified atmosphere packed red meats. Therefore the production of L- and D-lactic acids, acetic acid and ethanol from glucose may be good indicators of the onset of spoilage in certain foods, such as pork and beef. Dainty (1996) reported that acetate levels greater than 8 mg/100 g meat indicate a microbial flora $> 10^6$ cfu/g.

Biologically active amines

Tyramine is produced by certain lactic acid bacteria and has no sensory property of relevance to spoilage, but does have vasoactive properties. It is detectable (0.1–1.0 mg/100 g meat) in vacuum packed beef with high microbial loads ($> 10^6$ cfu/g).

Volatile compounds

The major advantages of volatile compound determination are that no food extraction method is required and it enables the simultaneous determination of microbial and chemical activities. Vallejo-Cordoba & Nakai (1994) demonstrated the presence of volatile compounds during milk spoilage which could be attributed to spoilage bacteria: 2- and 3-methylbutanals, 2-propanol, ethyl hexanoate, ethyl butanoate, 1-propanol, 2-methylpropanol and 1-butanol. Other volatile compounds could be attributed to chemical oxidation of lipids. Stutz *et al.* (1991) proposed acetone, methyl ethyl ketone, dimethyl sulphide and dimethyl disulphide as indicators of ground beef spoilage. Other volatiles include acetoin and diacetyl for pork spoilage and trimethylamine for fish (Dalgaard *et al.* 1993; Rehbein *et al.* 1994).

Shelf life of foods can be determined microbiologically:

Storage trials

As described above, samples are taken at timed intervals and analysed for total microbial load and specific spoilage organisms such as pseudomonads, *B. thermosphacta* and lactic acid bacteria. The viable counts are compared with chemical and sensory evaluation of the product and correlations between the variables determined to identify key indicators of early food spoilage.

Challenge tests

Samples of the food are incubated under conditions which reproduce the large-scale food production and storage period. The food may be inoculated with specified target organisms of interest, such as sporeformers.

Predictive modelling

This technique is described in more detail in Section 2.8. Its advantages are that the method can simultaneously predict the growth of micro-organisms over a broader range of conditions than is feasible in a microbiology laboratory (Walker 1994). The essential aspect is the need to validate the model using published and in-house laboratory data. A major limitation at present is that the majority of predictive models are concerned with food pathogens, whereas it is the food spoilage organisms which primarily limit a product's shelf life.

4.3 Methods of preservation and shelf life extension

All foods can be spoilt between harvesting, processing and storage before consumption (Gould 1996). Spoilage can be due to physical, chemical and microbial factors. Most preservation methods are designed to inhibit the growth of microorganisms (Table 4.5).

Methods which prevent or inhibit microbial growth are chilling, freezing, drying, curing, conserving, vacuum packaging, modified atmosphere packaging, acidifying, fermenting and adding preservatives (Tables 2.4 and 2.12). Other methods inactivate the microorganisms, for example pasteurisation, sterilisation and irradiation. Novel methods include the use of high pressure. The major preservation methods are based upon the reduction of microbial growth due to unfavourable environmental conditions (see intrinsic and extrinsic parameters, Section 2.5.1) such as temperature reduction, lowering of pH and water activity and denaturation due to heat treatment. Due to consumer pressure, the trend in recent years has been to use less severe, milder preservation methods including combined methods such as cooked-chill foods (prolonged shelf life) and modified atmosphere packaging (prolonged quality). The effect on microbial growth with regard to stress response is considered in Section 2.7.

4.3.1 Preservation by heat treatment

For the kinetics of microbial death refer to Section 2.4.

Pasteurisation

Milk is often pasteurised using the high temperature short time (HTST) 72°C, 15 seconds process. This is designed to kill all pathogenic bacteria such as *Mycobacterium tuberculosis*, *Salmonella* spp. and *Brucella* spp. at levels expected in fresh milk. Thermoduric organisms are those that survive pasteurisation. These include *Streptococcus thermophilus*,

Table 4.5 Preservation methods and effect on microorganisms (adapted from Gould, and reprinted with permission from Elsevier Science from *International Journal of Food Microbiology*, 1996, **33**, 51–64).

Effect on microorganisms	Preservation factor	Method of achievement
Reduction or inhibition of growth	Low temperature	Chill and frozen storage
	Low water activity	Drying, curing and conserving
	Restriction of nutrient availability	Compartmentalisation in water-in-oil emulsions
	Lowered oxygen	Vacuum and nitrogen packaging
	Raised carbon dioxide	Modified atmosphere packaging
	Acidification	Addition of acids: fermentation
	Alcoholic fermentation	Brewing; vinification; fortification
	Use of preservatives	Addition of preservatives: inorganic (sulphite, nitrite); organic (propionate, sorbate, benzoate, parabens); antibiotic (nisin, natamycin)
Inactivation of microorganisms	Heating	Pasteurisation and sterilisation
	Irradiating	Ionising irradiation
	Pressurising	Application of high hydrostatic pressure

Enterococcus faecalis, *Micrococcus luteus* and *Microbacterium lacticum*. The spores of *Bacillus cereus* and *B. subtilis* will survive the heat treatment. *B. cereus* growth causes pasteurised milk spoilage known as 'bitty cream'. Other pasteurisation time and temperature regimes are given in Table 4.6.

Sterilisation

Milk sterilisation (130°C, at least 1 second) is a well established method for prolonged milk storage. The heat treatment is severe enough to kill all microorganisms present, both spoilage and foodborne pathogens, to an acceptable level. There is a statistical chance of an organism surviving the process, but this is acceptable in the normal sense of safe food production. The 12D botulinum cook of canned foods is considered in Section 2.4.2.

Sous-vide

Sous-vide products are vacuum packed and undergo a mild heat treatment and have very carefully controlled chilled storage to prevent the outgrowth of sporeforming pathogens, notably *Cl. botulinum*. The heat treatment should be equivalent to 90°C for 10 minutes to kill the spores of psychrotrophic *Cl. botulinum*. Spoilage organisms, especially psychrotrophs, are also reduced during the heat treatment and this prolongs the shelf life (for example 3 weeks, 3°C). Mesophiles and thermophiles tend to survive mild heat treatments to a greater extent than psychrophiles, but are unable to multiply at the chill storage temperatures.

High hydrostatic pressure

High hydrostatic pressures of 300 to 500 MPa inactivate vegetative bacteria, whereas higher pressures (> 1000 MPa) are required to kill bacterial spores. The process can be combined (synergistically) with heat treatment such that lower pressures of 100 to 200 MPa are required. This is a novel method of food preservation and is not yet in large-scale usage. Nevertheless, the microbial stress response to pressure is covered in Section 2.7.5.

4.3.2   Food irradiation

Food irradiation is the exposure of food, either packaged or in bulk, to controlled amounts of ionising radiation (usually gamma rays from ^{60}Co) for a specific period of time to achieve certain desirable objectives. The process does not increase the background radioactivity of the food. It does prevent the growth of bacteria by damaging their DNA. It can also delay ripening and maturation by causing biochemical reactions in the plant tissues. The same irradiation process is used to sterilise medical products

Table 4.6 Pasteurisation time and temperature regimes.

Food	Pasteurisation process	Main objective	Secondary effects
Milk	63°C, 30 min 71.5°C, 15 sec	Kill pathogens: *Br. abortis, My. tuberculosis, C. burnetti*	Kill spoilage microorganisms
Liquid egg	64.4°C, 2.5 min 60°C, 3.5 min	Kill pathogens	Kill spoilage microorganisms
Ice cream	65°C, 30 min 71°C, 10 min 80°C, 15 sec	Kill pathogens	Kill spoilage microorganisms
Fruit juice	65°C, 30 min 77°C, 1 min 88°C, 15 sec	Enzyme inactivation: pectin esterase and polygalacturonase	Kill spoilage yeast and fungi
Beer	65-68°C, 20 min (in bottle) 72-75°C, 1-4 min (900-1000 kPa)	Kill spoilage microorganisms: yeast and lactic acid bacteria	

such as bandages, contact lens solutions and hospital supplies such as gloves, sutures and gowns.

Irradiation technology

The type of radiation used in processsing materials is limited to radiations from high energy gamma rays, X-rays and accelerated electrons. These radiations are also known as ionising radiations because their energy is high enough to dislodge electrons from atoms and molecules and to convert them to electrically charged particles called ions. Gamma and X-rays, like radiowaves, microwaves, ultraviolet and visible light rays, form part of the electromagnetic spectrum, occuring in the short wavelength, high energy region of the spectrum. Only certain radiation sources can be used in food irradiation. These are the radionuclides ^{60}Co or ^{137}Cs, X-ray machines having a maximum energy of 5 million electron volts (MeV) or electron machines having a maximum energy of 10 MeV. Energies from these radiation sources are too low to induce radioactivity in food. ^{60}Co is used considerably more than ^{137}Cs. The latter can only be obtained from nuclear waste reprocessing plants whereas ^{60}Co is produced by neutron bombardment of ^{59}Co in a nuclear reactor. High energy electron beams can be produced from machines capable of accelerating electrons, but electrons cannot penetrate very far into food, compared with gamma radiation or X-rays. X-rays of various energies are produced when a beam of accelerated electrons bombards a metallic target.

Radiation dose is the quantity of radiation energy absorbed by the food as it passes through the radiation field during processing. It is measured in the unit called the gray (Gy). One gray equals one joule of energy absorbed per kilogram of food being irradiated. International health and safety authorities have endorsed the safety of irradiation for all foods up to a dose level of 10 kGy.

Reasons for the application of food irradiation

The reasons for using food irradiation are:

- It decreases the considerable food losses due to infestation, contamination and spoilage.
- Concerns about foodborne diseases.
- The increase in the international trade of food products that must meet strict import standards of quality.

The Food and Agriculture Organisation of the United Nations (FAO) has estimated that, worldwide, about 25% of all food products is lost after harvesting to insects, bacteria and rodents. Irradiation could reduce losses and dependence upon chemical pesticides. Many countries experience considerable losses of grain due to insect infestation, moulds and pre-mature germination. Sprouting is the major cause of losses for roots and tubers. Countries, including Belgium, France, Hungary, Japan, the Netherlands and Russia, irradiate grains, potatoes and onions on an industrial scale.

As described in Chapter 3 even countries such as the United States experience considerable numbers of food poisoning cases – about 76 million cases per year and 5000 deaths. The economical aspect is a loss of $5 to $17 billion, estimated by the US Food and Drug Administration (FDA). Belgium and the Netherlands irradiate frozen seafoods and dry food ingredients to control foodborne diseases. France uses electron beam irradiation on blocks of mechanically deboned, frozen poultry products. Spices are irradiated in Argentina, Brazil, Denmark, Finland, France, Hungary, Israel, Norway and the USA. In December 1997 the FDA approved irradiation for beef, pork and lamb. The irradiation of poultry was approved by the FDA in 1990 and poultry processing guidelines were approved by the US Department of Agriculture (USDA) in 1992. Annually, about 500 000 tonnes of food products and ingredients are irradiated worldwide. In most countries, irradiated foods must be labelled with the international symbol for irradiation (the radura): simple green petals in a broken circle. This symbol must be accompanied by the words 'Treated by irradiation' or 'Treated with radiation'. Processors may add information explaining why irradiation is used; for example, 'Treated with irradiation to inhibit spoilage' or 'Treated with irradiation instead of chemicals to control insect infestation'. When used as ingredients in other foods, however, the label of the other food does not need to describe these ingredients as irradiated. Irradiation labelling also does not apply to restaurant foods.

The world trade in food has increased and subsequently the health and safety requirements of importing countries must be met (Chapter 10). There are differences in requirements, however. For example, not all countries allow the importation of chemically treated fruit; some countries (USA and Japan) have banned the use of certain fumigants. Ethylene oxide is extremely toxic. Methyl bromide, which is the primary fumigant that allows fruits, vegetables and grain to be exported, will be banned on 1 January 2001. Food irradiation technology may be of particular usefulness to developing countries whose economies rely on food and agricultural production as it is an alternative to fumigation.

Food irradiation standard
The technology of food irradiation has been accepted in about 37 countries for 40 different foods. Twenty-four countries are using food irradiation at a commercial level. The standard covering irradiated food has been adopted by the Codex Alimentarius Commission (Section 10.2). It

was based on the findings of a Joint Expert Committee on Food Irradiation (JECFI), which concluded that the irradiation of any food commodity up to an overall average dose of 10 kGy presented no toxicological hazard and required no further testing. It stated that irradiation up to 10 kGy introduced no special nutritional or microbiological problems in foods. The accepted dosage (10 kGy) is the equivalent of pasteurisation and cannot sterilise food. The method does not inactivate foodborne viruses such as Norwalk or hepatitis A. Irradiation does not denature or inactivate enzymes. Therefore, despite the lack of microbial growth, residual enzymes could limit a product's shelf life. It is very important that foods intended for processing (as for any other preservation method) are of good quality and handled and prepared according to good manufacturing practice established by national or international authorities.

The FDA approved the first use of irradiation on a food product (wheat and wheat flour) in 1963. Table 4.7 gives the levels of approved doses for different products. The microbial survival following irradiation for different organisms is given in Fig. 2.6.

Testing for irradiated food

Irradiated food can be detected to a limited extent (Haire *et al.* 1997). Tests include thermoluminescence measurement for detection of irradiated spices and electron spin resonance spectroscopy for determining irradiation of meats, poultry and seafoods containing any bone or shells and some specific chemical tests. However, no single method has been developed that reliably detects irradiation of all types of foods or the radiation dose level that was used. This is partly because the irradiation process does not physically change the appearance, shape or temperature of products and causes negligible chemical changes in foods.

> Irradiation should be regarded as complementing good hygienic practice and not a replacement for it.

Acceptability of food irradiation by the consumer

One major factor influencing the large scale application of food irradiation is public understanding and acceptance of the technology. Consumer reservation is based upon a number of issues:

(1) Fears linked to other nuclear-related technologies. However, irradiated food cannot become radioactive itself.

(2) Ingestion of radiolytic product. The maximum dose (10 kGy) produces 3–30 ppb radiolytic products. This would account for 5% of the total diet. The probability of harm occuring from radiolytic product formation from food additives is extremely low.

Table 4.7 Approved food irradiation levels.

Food	Purpose	Average dose (kGy)
Chicken	Prolong storage life	7
	Reduce the number of certain pathogenic microorganisms, such as *Salmonella*, from eviscerated chicken	7
Cocoa beans	Control insect infestation in storage	1
	Reduce microbial load of fermented beans with or without heat treatment	1
Dates	Control insect infestation during storage	1
Mangoes	Control insect infestation	1
	Improve keeping quality by delaying ripening	1
	Reduce microbial load by combining irradiation and heat treatment	1
Onions	Inhibit sprouting during storage	0.15
Papaya	Control insect infestation and improve its keeping quality by delaying ripening	1
Potatoes	Inhibit sprouting during storage	0.15
Pulses	Control insect infestation in storage	1
Rice	Control insect infestation in storage	1
Spices and condiments, dehydrated onions, onion powder	Control insect infestation	1
	Reduce microbial load	10
	Reduce the number of pathogenic microorganisms	10
Strawberry	Prolong shelf life by partial elimination of spoilage organisms	3
Teleost fish and fish products	Control insect infestation of dried fish during storage and marketing	1
	Reduce microbial load of the packed or unpacked fish and fish products[a]	2.2
	Reduce the number of certain pathogenic microorganisms in packaged or unpackaged fish and fish products[a]	2.2
Wheat and ground wheat products	Control insect infestation in the stored product	1

[a] During irradiation and storage the fish and fish products should be kept at the temperature of melting ice.

(3) Food irradiation would mask poor quality (i.e. spoilt) food. However, although spoilage bacteria may not be detectable, the off odours and poor appearance would still be evident.

(4) The lack of a suitable detection method is also of concern. However, organically grown produce cannot be identified analytically, nor can unacceptable temperature fluctuations of chilled and frozen foods be determined.

(5) Vitamin losses, especially B_1 in pork and C in fruit. The vitamin B_1 loss is estimated at 2.3% of the total B_1 in an American diet, and vitamin C is converted into an equally usable form.

(6) Induction of chromosomal abnormalities. The publicised polyploidy from the consumption of irradiated wheat in India has never been reproduced by subsequent investigations and the validity of the original data has been disputed by international scientific committees.

Current consumer pressure has hindered the large scale expansion of irradiation technology despite its promise of safe food production.

4.3.3 Reduced oxygen packaging

Reduced oxygen packaging (ROP) results in the food being surrounded by an atmosphere which contains little or no oxygen. The term ROP is defined as any packing procedure that results in a reduced oxygen level in a sealed package. It covers a range of packaging processes:

- *Cook-chill*, whereby a plastic bag is filled with hot cooked food from which the air has been expelled and which is closed with a plastic or metal crimp.
- *Controlled atmosphere packaging (CAP)* is an active system in which a desired atmosphere is maintained for the duration of the shelf life by the use of agents to bind or scavenge oxygen or a gas generating kit. CAP is defined as packaging of a product in a modified atmosphere followed by maintaining subsequent control of that atmosphere.
- *Modified atmosphere packaging (MAP)* uses a gas flushing and sealing process or reduction of oxygen through respiration of vegetables or microbial action. MAP is defined as packaging of a product in an atmosphere which has had a one-time modification of gaseous composition so that it is different from that of air. MAP is defined as the enclosing of food products in an atmosphere inside gas-barrier materials, in which the gaseous environment has been modified to slow down respiration rate, microbial growth and reduce enzymatic degradation, with the intention of extending the shelf life (Parry 1993). Carbon

dioxide is the most important headspace gas. A concentration of 20–60% retards the spoilage due to pseudomonads (Section 4.1.1). Carbon dioxide is a non-combustible, colourless gas that is sometimes favoured as a preservative since it leaves no toxic residues in the food. Hence it can also be used in modified atmosphere packaging. Solid CO_2 (dry ice) controls microbial growth by acting as a refrigerant during transport and storage. Then, as the dry ice sublimes, the CO_2 gas further inhibits bacterial growth by displacing the oxygen required by aerobic organisms, as well as by forming carbonic acid and thus possibly lowering the pH of the food to bacteriostatic levels (Foegeding & Busta 1991).

- *Sous vide* is a specialised process for partially cooked ingredients alone or combined with raw foods that require refrigeration (<3°C) or frozen storage until the package is throughly heated immediately before service. The sous vide process is a pasteurisation step that reduces bacterial load but is not sufficient to make the food shelf stable.
- *Vacuum packaging* reduces the amount of air from a package and hermetically seals it so that a near-perfect vacuum remains inside.

ROP can prevent the growth of aerobic spoilage organisms (Section 4.1) and hence extend the shelf life for foods in the distribution chain. However, the foods may be subject to temperature abuse during distribution and therefore at least one more barrier (or hurdle, Section 2.6.5) needs to be incorporated into the ROP process. ROP products commonly do not contain preservatives and do not have intrinsic factors (pH, a_w, etc.) to prevent microbial growth and subsequently need to be modified to ensure safety. The lack of growth of spoilage organisms enables pathogenic organisms such as *Cl. botulinum* and *L. monocytogenes* to grow and the food to appear normal or 'safe' to consume (Palumbo 1986). Hence the storage temperature must be maintained below 3.3°C (cf Table 2.7) for an extended shelf life, otherwise 5°C for the duration of the shelf life, and inhibitory intrinsic parameters achieved. The temperature treatment for ready-to-eat, including cook-chill, products should achieve a 4D kill of *L. monocytogenes* (US recommendation; Brown 1991; Rhodehamel 1992). In Europe, sous vide products should be subjected to a 12–13D *Ent. faecalis* heat treatment. This organism is used on the assumption that thermal inactivation of *Ent. faecalis* would ensure the destruction of all other vegetative pathogens.

4.3.4 *Preservation due to weak acids and low pH*

Foods can be divided into two groups: low-acid food with pH values above 4.5 and high-acid foods with pH values less than 4.5. The importance of the pH value of food is that *Clostridium botulinum* is unable to grow at pH values lower than 4.5. Therefore the processing of high-acid foods does not need to take into account the possible growth of *Cl. botulinum*.

Most food preservatives are effective at acid pH values. Organic acids (Section 2.6.1) include acetic, propionic, sorbic and benzoic acids. These compounds in their undissociated form are membrane soluble. In the cytoplasm the acid dissociates, with the resultant release of a proton. The expulsion of the proton is energy demanding and therefore restricts cell growth. Carbon dioxide forms a weak acid, carbonic acid, in solution. It prevents the growth of pseudomonads at levels as low as 5%, whereas the lactic acid bacteria are unaffected. Therefore modified atmosphere packaging changes the spoilage flora. The area of fermented foods is covered in Section 4.4.

4.3.5 Non-acidic preservatives

Nitrate and nitrite are traditional curing salts used to prevent microbial growth, especially of *Cl. botulinum*. Many organisms in the microbial flora on the food are able to reduce nitrate to nitrite, which is subsequently reduced to nitric oxide. Nitric oxide in *Cl. botulinum* binds to the pyruvate phosphotransferase enzyme, preventing ATP generation. Nitric oxide also binds to the haem ring of myoglobin in meat tissue forming nitrosomyoglobin, which produces nitrosylhaemochrome which is the characteristic pink coloration of cooked cured meats.

Nisin and pediocin PA-1 are examples of class IIa bacteriocins from lactic acid bacteria that are used as food preservatives, primarily in dairy products, and have an established safety record. They have a broad inhibitory spectrum and even prevent the outgrowth of bacterial spores. Nisin is an antibacterial peptide composed of 34 amino acids. It is produced on an industrial scale from certain strains of *Lactococcus lactis* belonging to serological group N. It is a lantibiotic which is a novel class of antimicrobial peptide that contains atypical amino acids and single sulphur lanthionine rings (Fig. 4.4). It has two natural variants, 'nisin A' and 'nisin Z', which differ in a single amino acid residue at position 27 (histidine in nisin A and asparagine in nisin Z). Their mode of action is primarily at the cytoplasmic membrane of susceptible microorganisms. The bacteriocins dissipate the proton motive force which stops the generation of ATP via oxidative phosphorylation. Models for nisin interaction with membranes propose that the peptide forms poration complexes in the membrane though a multistep process of binding, insertion and pore formation which requires a trans-negative membrane potential of 50–100 mV. Intermediate moisture foods have water activity values between 0.65 and 0.85 (Section 2.5.2). This is equivalent to a moisture content of

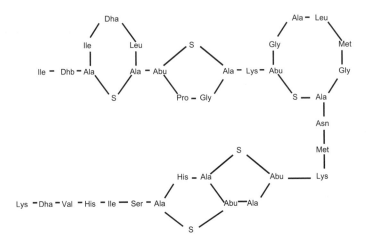

Fig. 4.4   The structure of nisin.

about 20–40%, but this conversion must be used with care as free water is not synonymous with available water.

Heat treatment can be used synergistically with water activity. Food with pH values greater than 4.5 would normally require a botulinum cook (Section 2.4.2) to ensure safety. However, in the presence of curing salt (nitrate and nitrite) the water activity is reduced and a shelf stable product with a long, safe ambient stability can be achieved that has had a milder (pasteurisation) heat treatment.

4.3.6   Ultrasound

Ultrasound waves inactivate microorganisms by introducing alternating compression and expansion cycles in a liquid medium. The bactericidal effect of high intensity ultrasound waves is caused during the expansion phase of ultrasound, when small bubbles grow until they implode. The temperature and pressure reached inside these bubbles can become extremely high. Ultrasound has not been adapted for food preservation, probably because of its adverse effect on the quality of the treated food when applied at the intensity required to kill bacteria. However, by using ultrasound in conjunction with pressure and heat, it may be possible to enhance the lethality of this technology for food applications (Raso *et al.* 1998).

4.4   Fermented foods

Fermentation is one of the oldest means of food processing. It is considered that early civilisations in the fertile crescent of the Middle East ate fermented milk, meats and vegetables. This established traditions of production methods and the incidental selection of microbial strains. In 1910 Metchnikoff (Stanley 1998) proposed that lactic acid bacteria could be used to aid human health, a topic which has resurfaced recently in the area of 'probiotics' (see Section 4.5 for details). Nowadays the production of fermented foods is practised across the world with many diverse national products (Table 4.8). The important microbial strains have been identified through conventional techniques and their metabolic pathways studied (Caplice & Fitzgerald 1999). Such fermented foods are consequently heavily colonised by bacteria and yeast, depending upon the specific product, yet there is no associated health risk. The reason is that the organisms are nonpathogenic, having been selected during the passage of time through trial and error. In fact the organisms inhibit the growth of pathogens and relatively recently a new range of food products termed functional foods, in particular probiotics (for example 'Yakult', Section 4.5), has arisen from the ancient practice of fermented food production. The lactic acid bacteria used in the production of most fermented foods produce a range of antimicrobial factors including organic acids, hydrogen peroxide, nisin and bacteriocins (see Sections 2.6.1 and 4.3.5). Because lactic acid bacteria have such a proven history of being nonpathogenic they are termed 'Generally Regarded As Safe' (GRAS) organisms. This title, however, may come under closer scrutiny with the advent of genetically engineered strains and the area of probiotics (Section 4.5). Only the essential aspects of fermented foods pertaining to their safety will be covered here. For a detailed study of fermented foods the reader is directed to Wood's (1998) double volume book entitled *Microbiology of Fermented Foods*.

The main factors resulting in the safety of fermented foods are:

- Their acidity generated through lactic acid production
- The presence of bacteriocins
- High salt concentration
- The anaerobic environment

There are many sources quoting the pH growth range of bacterial pathogens (see Table 2.7). However, it is very important to appreciate that the acid tolerance of bacterial pathogens can be enhanced during processing. The stress response (heat shock proteins, etc.) of bacterial pathogens is currently under investigation and is summarised in Section 2.7. Also any preformed bacterial or fungal toxins and viruses will persist during storage before ingestion. Examples of food poisoning related to fermented products are dealt with under individual pathogens in Chapter 5.

Table 4.8 Fermented foods from around the world.

Product	Substrate	Microorganism(s)
Beer, Ale	Grain	*Sac. cerevisiae*
Lager	Grain	*Sac. carlsbergensis*
	Millet	*Sac. fibulger*
Bread	Grain	*Sac. cerevisiae*, other yeasts, lactic acid bacteria
Bongkrek	Coconut	*Rh. oligosporus*
Cheese		
Cheddar	Milk	*Strep. cremoris, Strep. lactis, Strep. diacetylactis*, lactobacilli
Cottage	Milk	*Step. diacetylactis*
Mould ripened	Milk	*Strep. cremoris, Strep. lactis, Pen. caseicolum*
Swiss	Milk	As per cheddar, plus *Prop. sbermanii*
Coffee	Bean	*Leuconostoc, Lactobacillus, Bacillus, Erwinia, Aspergillus* and *Fusarium* spp.
Cocoa	Bean	*Lb. plantarum, Lb. mali, Lb. fermentum, Lb. collinoides, Ac. rancens, Ac. aceti, Ac. oxydans, Sac. cerevisiae* var. *ellipsoideus, Sac. apiculata*
Fish, fermented	Fish	*B. pumilus, B. licheniformis*
Gari	Cassava	*Corynebacterium manibot, Geotrichum* spp., *Lb. plantarum*, streptococci
Idii	Rice	*Leu. mesenteroides, Ent. faecalis, Torulopsis, Candida, Tricbosporon pullulans*
Kimchi	Cabbage, nuts, vegetables, seafoods	Lactic acid bacteria

(Contd)

Table 4.8 *(Contd)*

Product	Substrate	Microorganism(s)
Mahewu	Maize	*Corynebacterium* spp., *Aerobacter* spp., *Sac. cerevisiae*, *Lactobacillus* spp. *Candida mycoderma*
Nan	Wheat	*Sac. cerevisiae*, lactic acid bacteria
Ogi	Maize	Lactic acid bacteria, *Cephalosporium*, *Fusarium*, *Aspergillus*, *Penicillium* spp., *Sac. cerevisiae*, *Can. mycoderma*, *Can. valida*, *Can. vini*
Olives	Green olives	*Ln. mesenteroides*, *Lb. plantarum*
Oncom	Peanuts	*Neur. intermedia*, *Rh. oligosporus*
Pickles	Vegetables	*Lac. mesenteroides*, *Lac. brevis*, *Pen. cerevisiae*, *Lac. plantarum*
Sauerkraut	Cabbage	*Lac. mesenteroides*, *Lac. brevis*, *Pen. cerevisiae*, *Lac. plantarum*
Sausages, fermented	Meat	*Ped. cerevisiae*, *Lac. plantarum*, *Lac. curvatus*
Soy sauce	Soybeans, wheat	*Asp. oryzae*, *Asp. soya*, *Mucor* spp., *Rhizopus* spp., *Ped. soyae*, *Zygosac. rouxii*, *Lactobacillus* spp.
Tempeh	Soybeans	*Rh. oligosporus*
Vinegar	Wine	*Ac. aceti*
Wine		
Agave	Cactus	*Sac. carbajaoli*
Common	Grape	*Sac. cerevisiae var. ellipsoideus*
Palm		*Zym. mobilis*
Sake	Rice	*Sac. sake*
Yoghurt	Milk	*Strep. thermophilus*, *Lac. bulgaricus*

4.4.1   Lactic acid bacteria

Lactic acid bacteria (LAB) are primarily mesophiles (with a few thermophilic strains) and are able to grow in the range 5–45°C. They are able to grow at pHs as low as 3.8 and are proteolytic with fastidious amino acid growth requirements. The organisms are so called due to their ability to produce lactic acid. The lactic acid may be either or both the L(+) and D(−) isomers. Lactic acid bacteria produce a range of antimicrobial factors including organic acids, hydrogen peroxide, nisin and bacteriocins. Organic acids such as lactic acid, acetic acid and propionic acid interfere with the proton motive force and active transport mechanisms of the bacterial cytoplasmic membrane (Davidson 1997; Section 2.6.1). The production of hydrogen peroxide is due to the lactic acid bacteria lacking the enzyme catalase. H_2O_2 can consequently cause membrane oxidation (Lindgren & Dobrogosz 1990). Additionally, H_2O_2 activates the lactoperoxidase system of fresh milk, causing the formation of antimicrobial hypothiocyanate (Section 2.6.2). Lactic acid bacteria produce four groups of bacteriocins (Klaenhammer 1993; Nes *et al.* 1996). The best documented and exploited bacteriocin is nisin (class 1 bacteriocin) which is a post-translationally modified amino acid (also known as a lantibiotic). Nisin is produced by *Lactococcus lactis* subsp. *lactis* and has a broad inhibitory spectrum against Gram-positive bacteria (Section 4.3.5). It is formulated (concentration range 2.5–100 ppm) into a range of food products including cheeses, canned foods and baby foods and is particularly stable in high acid foods.

4.4.2   Glucose and lactose metabolism of lactic acid bacteria

The lactic acid bacteria organisms are divided into two groups, the homofermentative and heterofermentative lactic acid bacteria (Table 4.9). The homofermentative lactic acid bacteria produce two molecules of lactic acid for each molecule of glucose fermented, whereas the heterofermentative bacteria produce one molecule of lactic acid and one molecule of ethanol (Fig. 4.5). The homofermentative lactic acid bacteria are *Pediococcus*, *Streptococcus*, *Lactococcus* and some lactobacilli. The heterofermentative lactic acid bacteria are *Weisella*, *Leuconostoc* and some lactobacilli. The uptake of lactose (a disaccharide) is facilitated either by a carrier, permease or via the phosphoenolpyruvate-dependent phosphotransferase (PTS) system (Fig. 4.6). Lactose is cleaved to yield galactose (or galactose-6-phosphate) and glucose. Since lactose is the major sugar in milk, its metabolism has been well studied. Subsequently it has been demonstrated that the dairy starter culture *Lc. lactis* has the PTS

Table 4.9 Lactic acid bacteria groups.

Fermentation type	Main products	Lactate isomer	Organism
Homofermentative	Lactate	L(+)	*Lactobacillus bulgaricus*, *Lactobacillus*, *Enterococcus faecalis*
Homofermentative	Lactate	DL, L(+)	*Pediococcus pentosaceus*
Homofermentative	Lactate	D(−), L(+), DL	*Lactobacillus plantarum*
Heterofermentative	Lactate, ethanol, CO_2	DL	*Lactobacillus brevis*
Heterofermentative	Lactate, ethanol, CO_2	D(−)	*Leuconostoc mesenteroides*

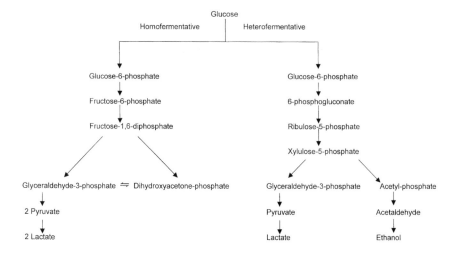

Fig. 4.5 Production of lactic acid by homo- and heterofermentative lactic acid bacteria.

system encoded on a plasmid, which may partially explain the instability of certain starter cultures.

4.4.3 Citrate metabolism and diacetyl production of lactic acid bacteria

Citrate metabolism is of particular importance in *Lc. lactis* subsp. *lactis* (biovar *diacetylactis*) and *Lc. mesenteroides* subsp. *cremoris*. These organisms metabolise excess pyruvate via an unstable intermediate, α-acetolactate, to produce acetoin via the enzyme α-acetolactate, decarboxylase. However in the presence of oxygen α-acetolactate is chemically converted into diacetyl (Fig. 4.7). It is diacetyl which gives the characteristic aroma of butter and certain yoghurts.

4.4.4 Proteolytic activity of lactic acid bacteria

The proteolytic activity of lactic acid bacteria, especially *Lactococcus,* has been studied in detail (Fig. 4.8). Proteolysis is a prerequisite for the growth of lactic acid bacteria in milk and the subsequent degradation of milk proteins (casein) is of central importance in the production of cheese. Hence a fundamental understanding of the process is necessary. The growth of *Lc. lactis* is enhanced by the overproduction of the membrane-anchored serine proteinase (PrtP) which can degrade casein. The resultant oligopeptides enter the cell via an oligopeptide transport system

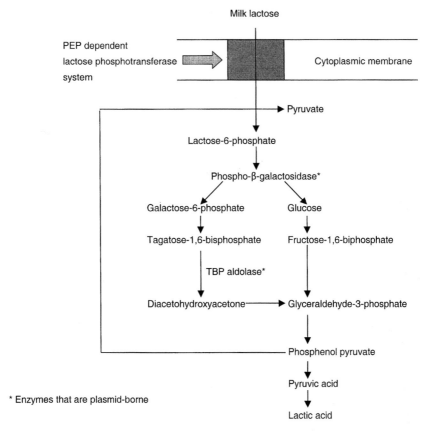

Fig. 4.6   Lactose metabolism.

(OPP) where they are further degraded by intracellular peptidases. The complete degradation of peptides is achieved by a wide range of intracellular peptidases with overlapping specificities (Mierau *et al.* 1996). The metabolism of proteins and amino acids by the lactic acid bacteria is part of the ripening process of cheese and flavour development.

4.4.5   Fermented milk products

There are many regional fermented milk products produced around the world (Table 4.10). The production of most fermented milk products (such as cheese and yoghurt) requires the addition of starter cultures. These are laboratory maintained cultures of lactic acid bacteria. The cultures may be well defined or composed of different strains depending upon the scale of production and historical experience (Table 4.11). The

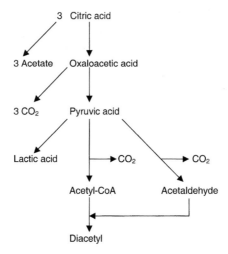

Fig. 4.7 Production of diacetyl by lactic acid bacteria.

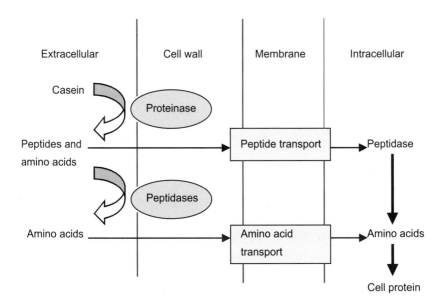

Fig. 4.8 Protein metabolism in lactic acid bacteria.

cultures will be rotated at regular intervals to avoid the possible accumulation of lactic acid bacteria specific bacteriophages in the production plant. Bacteriophages would slow the fermentation activity of the start cultures and even enable pathogenic organisms to multiply in the fermented food. Typical starter cultures include the mesophiles *Lactococcus*

Table 4.10 Fermented milk products around the world.

Product	Type	Microorganisms
Yoghurt	Moderate acid	*Strep. thermophilus, Lb. bulgaricus*
Cheddar cheese (UK)	Moderate acid	*Strep. cremoris, Strep. lactis, Strep. diacetylactis*
Cultured buttermilk (USA)	Moderate acid	*Lc. cremoris, Lc. lactis. citrovorum*
Acidophlus milk (USA)	High acid	*Lb. acidophilus*
Bulgarican milk (Europe)	High acid	*Lb. bulgaricus*
Dahi (India)	Moderate acid	*Lc. lactis, Lc. cremoris, Lc. diacetylactis, Leuconostoc* spp.
Leben (Egypt)	High acid	Streptococci, lactobacilli, yeast
Surati cheese (India)	Mild acid	Streptococci
Kefir (Russia)	High alcohol	Streptococci, *Leuconostoc* spp., yeast
Koumiss (Russia)	High alcohol	*Lb. acidophilus, Lb. bulgaricus, Sacharomyces lactis*

Table 4.11 Categories of lactic starter strains.

Type	Species	Application method
Single strain	*Strep. cremoris* *Strep. lactis* *Strep. lactis* subsp. *diacetylactis*	Single or paired
Multiple strains	As above plus *Leuconostoc* spp.	Defined mixture of two or more strains
Mixed strain starters	*Strep. cremoris*	Unknown proportions of different strains, which may vary on subculturing

lactis subsp. *lactis* and *Lc. lactis* subsp. *cremoris* or thermophilic *Strep. thermophilus, Lb. helveticus* and *Lb. delbrueckii* subsp. *bulgaricus*.

In cheese manufacture (Fig. 4.9a) the addition of the protease rennin, either from calf stomachs or from the fungus *Mucor pusillis*, and acidification from microbial lactic acid production results in the precipitation

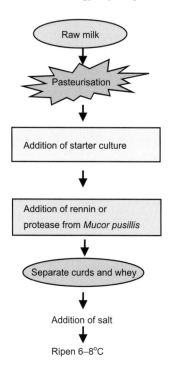

Fig. 4.9a   Generalised production of cheese.

of casein as 'curds' which are separated from the remaining liquid fraction ('whey'). The curds are salted and pressed into shape and left to ripen at 6–8°C. The characteristics of many regional products are dependent upon secondary bacterial or fungal colonisation (Table 4.12). For example, *Propionibacterium shermanii* causes the production of CO_2 gas and the characteristic eyes in Swiss cheese. *Penicillium roqueforti* (a fungus) produces the blue veining in certain cheeses. *Candida utilis* and *C. kefir* are involved in the production of blue cheese and kefir, respectively.

Pathogenic bacteria are killed by pasteurisation and therefore cheese made from pasteurised milk should be safe. However, inadequate pasteurisation and post-pasteurisation contamination can occur. In addition, some cheeses are made from nonpasteurised milk. Nevertheless, the rapid acidification of the milk due to lactic acid production should inhibit the multiplication of any pathogens present. However, if the starter culture is not fully active, possibly due to bacteriophage infection, then pathogens can multiply to infectious levels. A number of food poisoning incidences linked to cheese have been reported.

Yoghurt manufacture uses the complementary metabolic activities of *Strep. thermophilus* and *Lb. bulgaricus* (1:1 inoculation ratio, Fig. 4.9b).

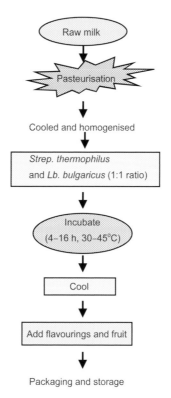

Fig. 4.9b   Production of yoghurt.

The lactobacillus requires the initial growth of the streptococci to produce folic acid, which is essential for the lactobacillus growth. Subsequently *Lb. bulgaricus* produces diacetyl and acetaldehyde which add flavour and aroma to the product (Fig. 4.7). A slow fermentation by the starter cultures can enable *St. aureus* to multiply to numbers sufficient to produce enterotoxins at levels high enough to induce vomiting (Section 3.2.7).

There are many health benefits attributed to the ingestion of fermented milk products:

- Increased digestibility and the nutritive value of milk itself.
- Reduced lactose content, important to a lactose intolerant population.
- Increased aborption of calcium and iron.
- Increased content of some B-type vitamins.
- Control of the intestinal microbial flora composition.
- Inhibition of the multiplication of pathogenic microorganism in the intestinal tract.
- Decreased cholesterol level in blood.

Table 4.12 Secondary flora of fermented dairy products (adapted from Wood 1998).

Type	Microorganism	Cheese types
Bacteria	Cornebacteria *Micrococcus* spp. *Lactobacillus* spp. *Pediococcus* spp.	} Red smear-ripened, washed rind } Hard and semi-hard
	Propionibacteria	Swiss types
Yeasts	*Kluyveromyces lactis* *Saccharomyces cerevisiae* *Candida utilis* *Debaryomyces hansensii* *Rhodosporidium* *infirmominiatum*	Blue, soft mould-ripened, red smear-ripened, washed rind
	Candida kefir	Kefir
Mould	*Penicillium camemberti* *Geotrichum candidum*	} Soft mould-ripened (Brie, Camembert, Coulommiers)
	Penicillium roqueforti	Blue cheese (Danish Blue, Roquefort, Stilton)
	Penicillium nalgiovensis	Tomme
	Verticillium lecanii	Tomme

(after Oberman & Libudzisz 1998.) There are a number of commercially available fermented milk products which claim to promote health (see Table 4.16 in Section 4.5.3).

4.4.6   Fermented meat products

Fermented sausages are produced from the bacterial metabolism of meat. The meat is first comminuted and mixed with fat, salt, curing agents (nitrate and nitrite), sugar and spices before being fermented. The sausages have a low water activity (Section 2.5.2) and are generally classified as either dry ($a_w < 0.9$) or semi-dry (a_w 0.9–0.95). The dry sausage is normally eaten without any cooking or other form of processing. In contrast, the semi-dry sausages are smoked and hence receive a heat treatment of between 60 and 68°C. Sausages produced without the addition of a starter culture have a pH of 4.6–5.0, whereas sausages with starter cultures usually have a lower final pH of 4.0–4.5. These low pH values are inhibitory to the growth of most food pathogens (Table 2.7). The starter cultures are composed of lactobacilli (such as *Lb. sake* and *Lb. curvatus*), pediococci, *St. carnosus* and *Micrococcus varians*. The fungi *Debaryomyces hansensii*, *Candida famata*, *Penicillium nalgiovense* and *P. chrysogenum* may also form part of the starter culture (Jessen 1995).

The bacterial pathogens which are potential hazards in fermented sausages are *Salmonella*, *St. aureus*, *L. monocytogenes* and the spore-forming bacteria *Bacillus* and *Clostridium*. The high level of salt (around 2.5%) inhibits pathogen growth and during the drying process *Salmonella* (and other Enterobacteriaceae) die off. Controlling *Salmonella* contamination also controls the growth of *L. monocytogenes*. Although *St. aureus* is resistant to nitrite and salt, it is a poor competitor of the remnant microflora under the acidic anaerobic conditions.

4.4.7   Fermented vegetables

According to Bückenhuskes (1997) there are 21 different commercial vegetable fermentations in Europe. Additionally there are a large variety of fermented vegetable juices, olives, cucumbers and cabbage. Vegetables have a high microbial load. However, if they were pasteurised this would affect their final properties and would be detrimental to the product's quality. Therefore the majority of vegetable fermentations use lactic acid production and no heat treatment (smoking, etc.). Lactic acid bacteria are present only in low numbers on vegetables and therefore starter cultures are usually used: *Lb. plantarium*, *Lb. casei*, *Lb. acidophilus*, *Lc. lactis* and *Leuc. mesenteroides*.

4.4.8   Fermented protein foods, shoyu and miso

In the Orient the fungi *Aspergillus* (*A. oryzae, A. sojae* and *A. niger*), *Mucor* and *Rhizopus* have been used to ferment grain, soybeans and rice (Table 4.13). This is called the koji process (Fig. 4.10). The main microbial activity is amylolytic degradation. The presence of mycotoxins has not been reported in koji products despite extensive investigations. Mycotoxins can potentially be produced after prolonged incubation by commercial strains of *Aspergillus* under specific environmental stress conditions (Table 4.14; Trucksess *et al.* 1987). In contrast, *Aspergillus* fermentations in the koji process seldom exceed 48–72 hours.

Table 4.13   Yeast and moulds from koji starter cakes (adapted from Lotong 1999).

Starter cake	Moulds	Yeasts
Chinese yeast	*Rhizopus javanicus R. chinensis, Amylomyces rouxii*	*Endomycopsis* spp.
Ragi	*Amy. rouxii, Mucor dubius, M. javanicus, R. oryzae, Aspergillus niger R. stolonifer, M. rouxii R. cohnii, Zygorrhynchus moelleri A. oryzae, A. flavus, R. oligosporus, R. arrhizus Fusarium* spp.	*Torula indica, Hansenula anomala, Saccharomyces cerevisiae Endomycopsis chodati, E. fibuligera H. subpelliculosa, H. malanga, Candida guilliermondii, C. humicola, C. intermedia, C. japonica, C. pelliculosa*
Loog-pang	*Amy. rouxii, Mucor* spp., *Rhizopus* spp., *A. oryzae*	*Endomycopsis fibuligera Saccharomyces cerevisiae*
Murcha	*M. fragilis, M. rouxii, R. arrhizus*	*H. anomala*
Bubod	*Rhizopus* spp., *Mucor* spp.	*Endomycopsis* spp. *Saccharomyces* spp.

4.4.9   The future of the lactic acid bacteria

The advent of DNA microarrays and the application of genomic and proteomic studies will extend the application of lactic acid bacteria in food microbiology (Schena *et al.* 1998; Blackstock & Weir 1999; Section 6.4.3). Genomic science has recently emerged as a powerful means to compare genomes and examine differentially expressed genes (see Fig. 6.10; Kuipers 1999). Microarray analysis enables differential gene

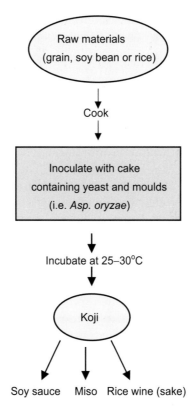

Fig. 4.10 Production of koji.

Table 4.14 Mycotoxins produced by koji moulds (adapted from Rowan *et al.* 1998).

Organism	Mycotoxin
A. oryzae	Cyclopiazonic acid, kojic acid, maltoryzine, β-nitropropionic acid
A. sojae	Aspergillic acid, kojic acid
A. tamarii	Kojic acid

expression of the total genome to be examined (Klaenhammer & Kullen 1999). Initial genetic studies of lactic acid bacteria demonstrated that many important traits were plasmid borne these included lactose metabolism, bacteriophage resistance and proteolytic activity. Subsequently stable starter cultures were developed for large scale production and have already been applied in the dairy industry. The genomic sequences of *Lc. lactis, Lb. acidophilus* and *Strep. thermophilus* have been published or are in progress. This increase in genetic knowledge will also improve the use of lactic acid bacteria as cell factories for the production of food-grade polysaccharides to add texture to food (Kuipers *et al.* 1997; de Vos 1999). It is anticipated that genome sequencing of probiotic cultures will identify gene systems responsible for cell survival and activity and secondly responsive gene systems that enable the organism to adapt to the changing environment within the intestinal tract (Klaenhammer & Kullen 1999).

In the future, molecular engineering of starter cultures will be possible in order to optimise production of end products such as diacetyl, acetate, acetaldehyde and flavour compounds from proteolytic activity (Hugen-holtz & Kleerebezem 1999). Other health promoting products produced from lactic acid bacteria will be antioxidants and vitamins. The use of lactic acid bacteria to deliver vaccines is currently under investigation (Wells *et al.* 1996). The microbial response to stress (such as heat and pH shock) is an increasing area of genetic and physiological research (Section 2.7). The ability to survive in a low pH environment (such as the stomach) may be increased due to previous exposure to a heat treatment due to the production of heat shock proteins.

There is obviously a considerable potential for the scientific exploitation of genetically engineered lactic acid bacteria. However, it must be ensured that the general public regard these organisms, previously 'Generally Regarded As Safe', as beneficial to their health and not an example of 'Frankenstein Food' which has recently caused such public reluctance in certain European countries. These factors are of considerable importance in the controversial area of pre- and probiotics.

4.5 Prebiotics, probiotics and synbiotics

4.5.1 Background

Probiotics, prebiotics and synbiotics are controversial topics in the general area of 'functional foods' (Berg 1998; Rowland 1999; Atlas 1999). Prebiotics are 'non-digestible food components that beneficially affect the host by selectively stimulating the growth and/or activity of one or a limited number of bacteria in the colon, that have the potential to improve host health' (Zoppi 1998). A probiotic can be defined as 'a live microbial food supplement which beneficially affects the host animal by improving its intestinal microbial balance' (Fuller 1989). This definition was broadened by Havenaar & Huis in't Veld (1992) to a 'mono- or mixed-culture of live microorganisms which benefits man or animals by improving the properties of the indigenous microflora'. The Lactic Acid Bacteria Industrial Platform workshop proposed a different definition for probiotics as 'oral probiotics are living microorganisms, which upon ingestion in certain numbers, exert health benefits beyond inherent basic nutrition'. This definition means that probiotics may be consumed either as a food component or as a nonfood preparation (Guarner & Schaafsma 1998). Synbiotics can be described as 'a mixture of probiotics and prebiotics that beneficially affects the host by improving the survival and implantation of live microbial dietary supplements in the gastrointestinal tract'. The reader is referred to Tannock (1999a) for a fuller coverage of the probiotic topic.

The idea of using lactic acid bacteria as an aid to improved health is not new (see Section 4.4 on fermented foods for related stories). However, the current trend is largely driven by consumer demands rather than through scientific investigation. For example, Green *et al.* (1999) demonstrated that two *B. subtilis* oral probiotics did not even contain *B. subtilis*.

The gastrointestinal tract of vertebrate animals is the most densely colonised region of the human body (Tannock 1995, Section 3.7). There are aproximately 10^{12} bacteria per gram (dry weight) of contents of the large intestine, which is estimated to contain several hundred bacterial species (Savage 1977). It is widely accepted that this collection of microbes has a powerful influence on the host in which it resides. The production of bacterial fermentation products (acetate, butyrate and propionate) commonly known as short-chain fatty acids (SCFA) for colonic health has been documented. For example, butyrate has a trophic effect on the mucosa, is an energy source for the colonic epithelium and regulates cell growth and differentiation. The human intestinal gut-associated lymphoid tissue represents the largest mass of lymphoid tissue in the body, and about 60% of the total immunoglobulin produced daily is secreted into the gastrointestinal tract. The colonic flora is the major antigenic stimulus for specific immune responses at local and systemic levels. Abnormal intestinal response to foreign antigen, as well as local immunoinflammatory reactions, might, as a secondary event, induce impairment of the intestine's function because of breakdown of the intestinal barrier (Diplock *et al.* 1999).

4.5.2 Claims of probiotics

It is implicit in the definition of probiotics that consumption of probiotic cultures positively affects the composition of this microflora and extends a range of host benefits (Sanders 1998; Tannock 1999b; Klaenhammer & Kullen 1999; German *et al.* 1999), including:

(1) Pathogen interference, exclusion and antagonism (Mack *et al.* 1999)
(2) Immunostimulation and immunomodulation (Schiffrin *et al.* 1995; Marteau *et al.* 1997)
(3) Anticarcinogenic and antimutagenic activities (Rowland 1990; Reddy & Riverson 1993; Matsuzaki 1998)
(4) Alleviation of symptoms of lactose intolerence (Marteau *et al.* 1990; Sanders 1993)
(5) Reduction in serum cholesterol
(6) Reduction in blood pressure
(7) Decreased incidence and duration of diarrhoea (antibiotic-associated diarrhoea, *Clostridium difficile*, travellers and rotaviral) (Isolauri *et al.* 1991; Kaila *et al.* 1992; Biller *et al.* 1995)
(8) Prevention of vaginitis
(9) Maintenance of mucosal integrity

It has been proposed that prebiotics help the colonic microbial flora to maintain a composition in which bifidobacteria and lactobacilli become predominant in number. This change in flora composition is regarded by believers in probiotics as optimal for health promotion. Nondigestible carbohydrates (for example inulin and fructo-oligosaccharides) increase faecal mass both directly by increasing nonfermented material and indirectly by increasing bacterial biomass; they also improve stool consistency and stool frequency. Promising food components for functional foods will be those, such as specific nondigestible carbohydrates, that can provide optimal amounts and proportions of fermentation products at relevant sites in the colon, particularly the distal colon, if they are to help reduce the risk of colon cancer.

4.5.3 Probiotic studies

Many of the specific effects attributed to the ingestion of probiotics, however, remain controversial and it is rare that specific health claims can be made (Sanders 1993; Ouwehand *et al.* 1999). In order to test the claims of pre- and probiotics there needs to be advances in the detailed characterisation of the intestinal microflora, especially the organisms currently regarded as unculturable. Molecular approaches are required such as 16S ribosomal RNA analysis, denaturing gradient gel electrophoresis (DGGE), temperature gradient gel electrophoresis (TGGE) and fluorescent *in situ* hybridisation (FISH; Muyzer 1999; Rondon *et al.* 1999). Molecular biology has been used to establish the phylogenetic relationship among members of the *Lactobacillus acidophilus* complex (Schleifer *et al.* 1995a,b). This may in part explain the variation in effects from earlier studies due to the use of unrelated organisms with identical names due to inadequate (classical) identification techniques.

The lactobacilli and the bifidobacteria constitute the most important probiotic organisms under investigation (Tannock 1998, 1999b; Reid 1999; Vaughan *et al.* 1999). The reason for this is that they are recognised as part of the human intestinal tract flora (especially in the newborn infant) and they have been safely ingested for many centuries in fermented foods (Section 4.4; Adams & Marteau 1995). Particular lactic acid bacteria strains which have been investigated as probiotics include *Lb. rhamnosus* GG, *Lb. johnsonii* IJ1, *Lb. reuteri* MM53 and *Bifidobacterium lactis* Bb12 (Table 4.15; Reid 1999; Vaughan *et al.* 1999). Important attributes of probiotic strains are colonisation factors and conjugated bile salt hydrolysis which aid persistence in the intestinal tract. Additional characteristics include effects on immunocompetent cells and anti-mutagenicity.

Table 4.15 Lactic acid bacteria used as probiotics (adapted from Tannock 1995 and Klaenhammer & Kullen 1999).

Genus	Species
Lactobacillus	*Lb. acidophilus, Lb. amylovorus, Lb. bulgaricus, Lb. casei, Lb. crispatus, Lb. gallinarum, Lb. gasseri, Lb. johnsonii* (strain Lal), *Lb. plantarum, Lb. reuteri* (strain MM53), *Lb. rhamnosus* (strain GG), *Lb. salivarius*
Bifidobacterium	*Bif. adolescentis, Bif. animalis, Bif. bifidum, Bif. breve, Bif. infantis, Bif. longum, Bif. lactis* (strain Bb12)
Streptococcus	*Strep. thermophilus, Strep. salivarius*
Enterococcus	*Ent. faecium*

Lactobacillus rhamnosus GG is probably the most studied probiotic prepartion. There is some evidence that *Lb. rhamnosus* GG colonises the intestine and reduces diarrhoea (Kaila *et al.* 1995). The oral dose required is greater than 10^9 cfu/day. There are many fermented milk products commercially available which claim to promote health (Table 4.16). *Lb. casei* (Shirota strain) fermented milk is sold as a product called 'Yakult' and is estimated to be consumed daily by 10% of the Japanese population (Reid 1999). On a worldwide basis it has been predicted that 30 million people ingest this product each day. It is prepared from skimmed milk with the addition of glucose and a *Chlorella* (algae) extract followed by inoculation with *Lb. casei* (Shirota). The fermentation takes 4 days at 37°C. The lactobacillus strain was initially isolated in 1930 and appears to

Table 4.16 Microbial composition of fermented milk products for which health benefits are claimed.

Product	Country	Microorganism
Yakult	Japan	*Lb. casei* (Shirota)
Miru-Miru	Japan	*Lb. acidophilus, Lb. casei, Bif. breve*
Liquid yoghurt	Korea	*Lb. bulgaricus, Lb. casei, Lb. helveticus*
AB-fermented milk	Denmark	*Lb. acidophilus, Bif. bifidum*
A-38 fermented milk	Denmark	*Lb. acidophilus* and mesophilic lactic acid bacteria
Real Active	UK	Yoghurt culture, *Bif. bifidum*
Acidophilus milk	USA	*Lb. acidophilus*

colonise the intestine and improve infant recovery time from rotaviral gastroenteritis (Sugita & Togawa 1994). The strain appears to have anti-tumour and antimetastasis effects in mice (Matsuzaki 1998). In Canada the application of probiotics via the urogenital tract has been investigated. Trials with *Lb. rhamnosus* GR-1 and *Lb. fermentum* B-54 (applied as a milk-based suspension or freeze-dried preparation) have indicated that the vaginal microflora can be re-established and the incidence of urinary tract infections reduced (Reid *et al.* 1994).

Although probiotics are promoted as aids to health, it is plausible that there may be some risks associated with the ingestion of probiotic bacteria. The major risk would be the unrestricted stimulation of the immune system in people suffering from autoimmune diseases (Guarner & Schaafsma 1998).

Lactic acid bacteria are also being studied for vaccine delivery in the intestinal tract (Fischetti *et al.* 1993). A strain of *Lb. lactis* containing the luciferase gene has been used as an experimental model to study gene promoter activities in the intestinal tract as a first step to predicting the expression of heterogenous proteins *in vivo* (Corthier *et al.* 1998).

4.6   Microbial biofilms

In nature and food systems, microorganisms attach to surfaces and grow as a microbial community (Denyer *et al.* 1993; Zottola & Sasahara 1994; Kumar & Anand 1998; Stickler 1999). The resulting build-up of micro-organisms is called a 'biofilm', or less attractively 'slime'. When biofilms are a nuisance the term 'biofouling' or 'fouling' is used. This commonly occurs in heat exchanges.

The cells are embedded in a polymeric matrix and are phenotypically different from when they are growing in suspension ('planktonic'). One major difference is the increased resistance (orders of 10–100-fold) to antimicrobial agents (Le Chevallier *et al.* 1988; Holmes & Evans 1989; Krysinski *et al.* 1992; Druggan *et al.* 1993). Food can become contaminated with undesirable spoilage and pathogenic bacteria from sloughed portions of biofilms. Hence biofilm formation leads to serious hygiene problems and economic losses due to food spoilage and the persistence of food pathogens.

Biofilms form on any submerged surface where bacteria are present. They form in a sequence of events (Fig. 4.11):

(1) Nutrients from the food are adsorbed to the surface to form a conditioning film. This leads to a higher concentration of nutrients compared to the fluid phase and favours biofilm formation. The conditioning also affects the physicochemical properties of the surface, i.e. surface free energy, changes in hydrophobicity and electrostatic charges which influence the microbial colonisation.

(2) Microorganisms attach to the conditioned surface. The adhesion of bacteria is initially reversible (van der Walls attraction forces, electrostatic forces and hydrophobic interactions) and subsequently irreversible (dipole–dipole interactions, hydrogen, ionic and covalent bonding and hydrophobic interactions). The bacterial appendages flagella, fimbriae and exopolysaccharide fibrils (also known as glycocalyx, slime or capsule) are involved in contact with the conditioning film. The exopolysaccharide is important in cell–cell and cell–surface adhesion and also protects the cells from dehydration.

(3) The irreversibly attached bacteria grow and divide to form microcolonies which enlarge and subsequently coalesce to form a layer of cells covering the surface. During this phase the cells produce more additional polymer which increases the cells' anchorage and stabilises the colony against fluctuations of the environment.

(4) The continuous attachment and growth of bacterial cells, and exopolysaccharide formation, lead to biofilm formation. The biofilm layer may be several millimeters thick within a few days.

(5) As the biofilm ages, sloughing occurs whereby detachment of relatively large particles of biomass from the biofilm occurs. The sloughed bacteria can either contaminate the food (note the subsequent nonhomogenous presence of bacteria in food) or initiate a new biofilm further along the production line.

(a) Conditioning film of food residues on work surface

Food residues (nutrient source)

Work surface

(b) Microorganisms attached to conditioned surface

(c) Microorganisms divide and form microcolonies. Polysaccharide formation stabilises the biofilm.

Secreted polysaccharide layer

(d) Fragments of biofilm shed periodically

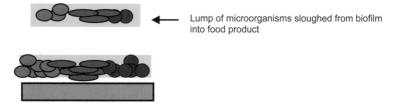

Lump of microorganisms sloughed from biofilm into food product

Fig. 4.11   Biofilm formation.

Equipment design and plant layout are also important aspects of prevention and control strategies. Obviously good design of equipment such as tanks, pipelines and joints facilitates cleaning of the production line. The microtopography of a surface can complicate cleaning procedures when crevices and other surface imperfections shield attached cells from the rigours of cleaning. Stainless steel resists impact damage but is vulnerable to corrosion, while rubber surfaces are prone to deterioration and may develop surface cracks where bacteria can accumulate (LeClercq-Perlat & Lalande 1994).

Generally, an effective cleaning and sanitation programme will inhibit biofilm formation. Mechanical treatment and chemical breakage of the polysaccharide matrix are very necessary for removal of biofilms. As can

(a) Mature biofilm

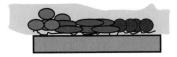

(b) Detergent removal of surface (polysaccharide, food residues, etc.) material

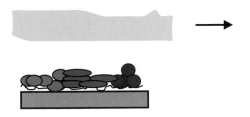

(c) Disinfectant killing of microorganisms

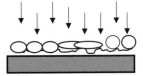

(d) Rinse to remove dead microbial cells which otherwise could act as a nutrient source for fresh biofilm

Fig. 4.12   Biofilm removal.

be seen in Fig. 4.12, the adhered bacteria are covered with organic material (polysaccharide and food residues) which may inhibit the penetration of the disinfectant due to its lack of wetting properties. Therefore a detergent activity is required to remove this outer layer prior to disinfection. The dead microbial biomass must be removed otherwise it may act as a conditioning film and nutrient source for further biofilm formation. New cleaning agents and enzyme treatments are being formulated for the effective removal of biofilms.

5

FOOD POISONING
MICROORGANISMS

Food poisoning is caused by food which:

(1) Looks normal
(2) Smells normal
(3) Tastes normal

Because the consumer is unaware that there is a potential problem with the food, a significant amount is ingested which exceeds the infectious dose (Table 7.7, see also Table 9.1) and hence they become ill. Consequently it is hard to trace which food was the original cause of food poisoning since the consumer will not recall noticing anything appropriate in their recent meals. Food poisoning is caused by a variety of organisms and the incubation period and duration of illness varies considerably (Table 5.1).

Food poisoning organisms are normally divided into two groups:

• Infections e.g. salmonella, campylobacter and pathogenic *E. coli*
• Intoxications e.g. *B. cereus, St. aureus, Cl. botulinum*

The first group comprises organisms which multiply in the human intestinal tract, whereas the second group comprises organisms that produce toxins either in the food or during passage in the intestinal tract. This division is very useful to help recognise the routes of food poisoning. Vegetative organisms are killed by heat treatment, whereas spores may survive and hence germinate if the food is not kept sufficiently hot or cold.

An alternative grouping would be according to severity of illness. This approach is useful in setting microbiological criteria (sampling plans) and risk analysis. The ICMSF (proposed revised version 2000) divided the common foodborne pathogens into such groups in order to aid decision

Table 5.1   Common food poisoning organisms.

Microorganism	Incubation Period	Duration of Illness
Aeromonas species	Unknown	1–7 days
Campylobacter jejuni	3–5 days	2–10 days
Escherichia coli		
ETEC	16–72 hours	3–5 days
EPEC	16–48 hours	5–15 days
EIEC	16–48 hours	2–7 days
EHEC	72–120 hours	2–12 days
Hepatitis A	30–60 days	2–4 weeks
Listeria monocytogenes	3–70 days	Variable
Norwalk-like virus	24–48 hours	1–2 days
Rotavirus	24–72 hours	4–6 days
Salmonellae	16–72 hours	2–7 days
Shigellae	16–72 hours	2–7 days
Yersinia enterocolitica	3–7 days	1–3 weeks

making of sampling plans (Section 8.3). The ICMSF groupings are given in Table 5.2. Detailed descriptions of foodborne pathogens are given in the following sections of this chapter. Initially, however, it is useful to appreciate that some foodborne pathogens are very difficult to detect and hence 'indicator' organisms, whose presence may indicate the possible presence of a pathogen, may be used which are easier to detect.

5.1   Indicator organisms

The term 'indicator organisms' can be applied to any taxonomic, physiological or ecological group of organisms whose presence or absence provides indirect evidence concerning a particular feature in the past history of the sample. It is often associated with organisms of intestinal origin, but other groups may act as indicators of other situations. For example, the presence of members of 'all Gram-negative bacteria' in heat-treated foodstuffs is indicative of inadequate heat treatment (relative to the initial numbers of these organisms) or of contamination subsequent to heating. Coliform counts, since coliforms represent only a subset of 'all Gram-negative bacteria', provide a much less sensitive indicator of problems associated with heat treatment, but are still frequently used in the examination of heat-treated foodstuffs. The term 'index organism' was suggested by Ingram in 1977 for a marker whose presence indicated the possible presence of an ecologically similar pathogen.

Microbial indicators are more often employed to assess food safety and

Table 5.2 Microbiological hazards categorisation.

Effects of hazards	Pathogen	Cases (see Table 8.1)
Categorisation of common foodborne pathogens (ICMSF 1986)		
(1) Moderate, direct, limited spread, death rarely occurs	*B. cereus, C. jejuni, Cl. perfringens, St. aureus, Y. enterocolitica, Taenia saginata, Toxoplasma gondii*	7, 8, 9
(2) Moderate, direct, potentially extensive spread, death or serious sequelae can occur. Considered severe	Pathogenic *E. coli, S. enteritidis* and other salmonellae other than *S. typhi* and *S. paratyphi*, shigellae other than *Sh. dysenteriae, L. monocytogenes*	10, 11, 12
(3) Severe, direct	*Cl. botulinum* types A, B, E and F, hepatitis A virus, *Sh. dysenteriae, S. typhi* and *S. paratyphi* A, B, and C, *T. spiralis*	13, 14, 15
Proposed up-dated categorisation (ICMSF 2000)		
(1) Food poisoning organisms causing moderate, not life-threatening, no sequelae, normally short duration, self-limiting	*B. cereus* (including emetic toxin), *Cl. perfringens* type A, Norwalk-like viruses, *E. coli* (EPEC, ETEC), *St. aureus, V. cholerae* non-O1 and non-O139, *V. parahaemolyticus*	
(2) Serious hazard, incapacitating but not life-threatening, sequelae rare, moderate duration	*C. jejuni, C. coli, S. enteritidis, S. typhimurium*, shigellae, hepatitis A, *L. monocytogenes, Cryptosporidium parvum*, pathogenic *Y. enterocolitica, Cyclospora cayetanensis*	
(3) Severe hazard for general population, life-threatening, chronic sequelae, long duration	Brucellosis, botulism, EHEC (HUS), *S. typhi, S. paratyphi*, tuberculosis, *Sh. dysenteriae*, aflatoxins, *V. cholerae* O1 and O139	
(4) Severe hazard for restricted populations, life-threatening, chronic sequelae, long duration	*C. jejuni* O:19 (GBS), *C. perfringens* type C, hepatitis A, *Cryptosporidium parvum, V. vulnificus, L. monocytogenes*, EPEC (infant mortality), infant botulism, *Ent. sakazakii*	

hygiene than quality. Ideally, a food safety indicator should meet certain important criteria.

It should:

- Be easily and rapidly detectable.
- Be easily distinguishable from other members of the food flora.
- Have a history of constant association with the pathogen whose presence it is to indicate.
- Always be present when the pathogen of concern is present.
- Be an organism whose numbers ideally should correlate with those of the pathogen of concern.
- Possess growth requirements and a growth rate equalling those of the pathogen.
- Have a die-off rate that at least parallels that of the pathogen and ideally persists slightly longer than the pathogen of concern.
- Be absent from foods that are free of the pathogen, except perhaps at certain minimum numbers.

Common indicator organisms are:

- Coliforms
- *E. coli*
- Enterobacteriaceae
- Faecal streptococci

Coliforms are Gram-negative, rod-shaped, facultatively anaerobic bacteria. They are also known as the 'coli-aerogenes' group. Identification criteria used are the production of gas from glucose (and other sugars) and fermentation of lactose to acid and gas within 48 hours at 35°C (Hitchins *et al.* 1998). The coliform group includes species from the genera *Escherichia*, *Klebsiella*, *Enterobacter* and *Citrobacter*, and includes *E. coli*. Coliforms were historically used as indicator microorganisms to serve as a measure of faecal contamination, and therefore potentially the presence of enteric pathogens in fresh water. However, since most coliforms are found throughout the environment they have little direct hygiene significance. Since coliforms are easily killed by heat, coliform counts can be useful when testing for post-processing contamination.

In order to differentiate between faecal and non-faecal coliforms the faecal coliform test was developed. Faecal coliforms are defined as coliforms that ferment lactose in EC medium with gas production within 48 hours at 45.5°C (except shellfish isolates, 44.5°C).

5.2   Foodborne pathogens: bacteria

5.2.1   Campylobacter jejuni, C. coli, *Guillain-Barré syndrome*

The campylobacters were initially recognised as animal pathogens and have only in the past 15 to 20 years been recognised as human pathogens (Butzler *et al.* 1973; Skirrow 1977). Nowadays the recorded incidence of campylobacter is greater than that of any other food poisoning organisms. The annual cost of campylobacteriosis in the USA is estimated at $1.3 to $6.2 billion.

There are two major species of *Campylobacter* causing food poisoning. *C. jejuni* causes the majority of outbreaks (89–93%), secondly *C. coli* (7–10%) and finally *C. upsaliensis* and *C. lari* have occasionally been implicated. The reservoirs of this organism include poultry, cattle, swine, sheep, rodents and birds (Skirrow 1991). The routes of infection are via contaminated water, milk and meat. Poultry is the largest potential source of infectious campylobacter. Subsequently most sporadic infections are associated with improper preparation or consumption of mishandled poultry products. Most *C. jejuni* outbreaks, which are far less common than sporadic illnesses, are associated with the consumption of raw milk or unchlorinated water. Campylobacteriosis may lead to Guillain-Barré syndrome, a cause of flacid paralysis (Allos 1998).

Campylobacters are Gram-negative thin rods (0.5–0.8 µm × 0.2–0.5 µm). They are microaerophiles (requiring 3–5% oxygen and 2–10% carbon dioxide) with an optimum growth at 42–43°C and no growth at 25°C (room temperature). These species are commonly known as 'thermophilic campylobacters' since they grow at higher temperatures than other campylobacters. The organism's morphology can change between vibroid (curved), spiral, doughnut, S-shaped and coccoid. The short vibroid morphology is associated with the 'viable but nonculturable' (VNC) state (Rollins & Colwell 1986). By definition, it is not recoverable using conventional methods from the VNC condition (Section 6.2.2). The organism is very sensitive to drying and is readily destroyed by cooking at 55–60°C for several minutes (D_{50} 0.88–1.63 min; Table 2.3). Although the organism does not multiply at room temperature, its low infectious dose (500 cells) means it can easily cause cross-contamination from raw meats to processed meats. This may be why cases of campylobacter gastroenteritis outnumber salmonella in many countries. There is a notable seasonality to campylobacter enteritis, with peak incidences in the summer months. Temperature, pH and water activity growth ranges and D values for campylobacter are given in Tables 2.3 and 2.7.

On blood agar plates campylobacter colonies are nonhaemolytic, flat, about 1–2 mm in diameter and either spreading with an irregular edge or discrete, circular-convex. Most campylobacter isolates are not speciated in routine analytical laboratories. In contrast, public health laboratories investigating a food poisoning outbreak will 'fingerprint' isolates using a range of procedures: biotyping, phage typing and serotyping. The biotyping (biochemical activity profiling) scheme of Preston is more extensive than that of Lior. Two serotyping procedures are used: the Penner and Hennessy scheme for the heat-stable antigens (lipopolysaccharide) and the Lior scheme for the heat-labile antigens (flagella). Phage typing can differentiate strains within a serotype and is a simple enough technique to apply to a large number of strains simultaneously. An increase in ciprofloxacin resistance (a medically important antibiotic) has been reported possibly due to the veterinary use of the structurally related (fluoroquinolone) antibiotic enrofloxacin used in poultry husbandry. Direct enumeration of *Campylobacter* species is uncommon. Usually an enrichment step is used to recover low numbers of the organism from processed foods. For detailed detection methods see Section 6.5.3.

The characteristics of campylobacter enteritis are:

- Flu-like illness
- Abdominal pain
- Fever
- Diarrhoea, which may be profuse, watery and frequent or bloody

The incubation period is 2–10 days, it lasts for about one week and is usually self-limiting. The organism is excreted in the faeces for several weeks after the symptoms have ceased. Relapses occur in about 25% of cases.

There is no concensus of opinion concerning the virulence factors of campylobacter. The organism is reported to produce at least three toxins (Wassenaar 1997):

(1) Heat-labile enterotoxin (60 to 70 kDa) which increases cyclic AMP levels of intestinal cells and cross-reacts with anticholera toxin antibodies. This is also referred to as the 'cholera-like' toxin. It is iron-regulated and causes elongation of Chinese hamster ovary cells and binds to GM_1 ganglioside. It has an AB_5 domain structure and there is homology between the B subunit and cholera toxin and *E. coli* heat-labile toxin (LT, Section 3.9.1; Baig *et al.* 1986).

(2) Cytoskeletal altering toxin that may also cause diarrhoea (Section 3.8). This is also known as the cytolethal distending toxin (CLDT) and has been the most studied toxin from campylobacter. This toxin causes characteristic elongation of cultured mammalian cells followed by distension. It differs from other toxins produced by enteric

bacteria in that it is destroyed by heating at 70°C for 15 minutes and by trypsin.

(3) Heat-labile protein cytotoxin which is not neutralised by anti-cholera toxin.

Campylobacter can easily cross-contaminate processed food. A contaminated piece of raw meat can leave 10 000 campylobacter cells per cm^2 on a work surface. Since the infectious dose is only around 1000 cells the residual microbial load must be reduced to < 2 per cm^2. *C. jejuni* is quickly killed during cooking at 55–60°C for several minutes and is not a sporeformer. Hence the main control mechanisms are adequate cooking regimes and prevention of cross-contamination for raw meats and poultry.

Since the eradication of polio in most parts of the world, Guillain-Barré syndrome (GBS) has become the most common cause of acute flaccid paralysis (Allos 1998). GBS is an autoimmune disorder of the peripheral nervous system characterised by weakness which is usually symmetrical, evolving over a period of several days or more. Since laboratories began to isolate *Campylobacter* species from stool specimens some 20 years ago, there have been many reports of GBS following campylobacter infection. Only during the past few years has strong evidence supporting this conjecture developed. Campylobacter infection is now known as the single most identifiable antecedent infection associated with the development of GBS. Of an estimated annual number of predicted 2628–9575 US cases, 526–3830 are triggered by campylobacter infection. *Campylobacter* serotype O:19 is thought to cause this autoimmune disease through a mechanism called molecular mimicry, whereby campylobacter contains ganglioside-like epitopes in the lipopolysaccharide moiety that elicit autoantibodies reacting with peripheral nerve targets. Campylobacter is associated with several pathologic forms of GBS, including the demyelinating (acute inflammatory demyelinating polyneuropathy) and axonal (acute motel axonal neuropathy) forms. Different strains of campylobacter as well as host factors probably play an important role in determining who develops GBS as well as the nerve targets for the host immune attack of peripheral nerves. Estimated total costs (in US$) of campylobacter-associated GBS are $0.2 to $1.8 billion.

5.2.2 Salmonella *spp.*

Salmonella is a genus of the Enterobacteriaceae family. They are Gram-negative, facultatively anaerobic, non-sporeforming short (1–2 μm) rods. The majority of species are motile with peritrichous flagella; *S. gallinarum* and *S. pullorum* are nonmotile. Salmonellae ferment glucose with the production of acid and gas but are unable to metabolise lactose and sucrose. Their optimum growth temperature is about 38°C and their minimum is about 5°C (Table 2.7). Since they are non-sporeformers, they are relatively heat sensitive, being killed at 60°C in 15–20 minutes ($D_{62.8}$ = 0.06 minutes, Table 2.3). There are only two species of *Salmonella* (*S. enterica* and *S. bongori*) which are divided into eight groups (Boyd *et al.* 1996). This is useless with regard to epidemiological investigations and therefore detailed characterisation is required.

The genus *Salmonella* contains over 2324 different 'strains' which are also called serovars or serotypes. These are distinguished by their O, H and Vi antigens (Sections 2.2.3–5) using the Kaufmann-White scheme (Brenner 1984; Ewing 1986; Le Minor 1988). The serotypes are then put into serogroups according to common antigenic factors (Table 2.2). The salmonellae have a complex lipopolysaccharide (LPS) structure (Section 2.2.3) which gives rise to the O antigen. The number of repeating units and sugar composition varies considerably in *Salmonella* LPS and is of vital importance with regard to epidemiological studies. The sugars are antigenic and therefore can be used immunologically to identify *Salmonella* isolates. It is the serotype of the *Salmonella* isolate that aids epidemiological studies tracing the vector of *Salmonella* infections. Further characterisation is required for epidemiological studies and this includes biochemical profiles and phage typing. Some serotypes were initially named after the place where they were first isolated, e.g. *S. dublin* and *S. heidelberg*. Others were named after the disease and affected animal (e.g. *S. typhimurium*, which causes typhoid in mice). *S. typhi* and the paratyphoid bacteria are normally septicaemic and produce typhoid or typhoid-like fever in humans. Other forms of salmonellosis generally produce milder symptoms.

Characteristic symptoms of salmonella food poisoning included:

- Diarrhoea
- Nausea
- Abdominal pain
- Mild fever and chills
- Sometimes vomiting, headache and malaise

The incubation period before the illness is between 16 and 72 hours. The illness is usually self-limiting, lasting 2–7 days. The infected person will be shedding large numbers of salmonellae in the faeces during the period of illness. The numbers of salmonellae in the faeces will decrease, but may in exceptional cases (symptomless carriers) continue for up to 3 months. Chronic consequences such as postenteritis reactive arthritis and Reiter's syndrome may follow 3–4 weeks after onset of acute symptoms. Reactive arthritis may occur in about 2% of cases.

The infective dose varies according to the age and health of the victim, the food and also the salmonella strain. The infectious dose (see Tables 7.7 and 9.1) varies from 20 cells to 10^6 according to serotype, food and vulnerability of the host. It should be noted that the first 50 ml of liquid passes straight through the stomach into the small intestines and is therefore protected from the stomach's hostile acidic environment. Likewise it is believed that chocolate can protect salmonella while transient in the stomach, hence reducing the infectious dose. The disease is caused by the penetration and passage of *Salmonella* organisms from the gut lumen into the epithelium of the small intestine, where they multiply. Subsequently the bacteria invade the ileum and even occasionally the colon (Section 3.9.5). The infection illicits an inflammatory response. The number of salmonellosis cases shows a marked seasonal trend, with peak incidences in the summer.

A wide range of contaminated foods is associated with salmonella food poisoning, including raw meats, poultry, eggs, milk and dairy products, fish, shrimp, frog legs, yeast, coconut, sauces and salad dressing, cake mixes, cream-filled desserts and toppings, dried gelatin, peanut butter, cocoa and chocolate. Contamination of the food is through poor temperature control or handling practices, or cross-contamination of processed foods from raw ingredients. The organism multiplies on the food to an infectious dose.

In addition to contaminating egg shells, *S. enteritidis* can be isolated from the egg yolk due to transovarian infection. The organism travels up the anus from the environment and colonises the ovaries. Subsequently *S. enteritidis* infects the egg before the protective shell is formed. An infected unfertilised egg will result in contaminated egg products, whereas an unfertilised egg results in a chronically ill chick with systemic infection and hence a contaminated carcass.

The dominant serotype causing food poisoning has changed over recent decades from *S. agona*, *S. hadar* and *S. typhimurium* to the current *S. enteritidis* (D'Aoust 1994). In fact, one phage type (PT4) *S. enteritidis* is the predominant cause of salmonellosis in many countries. The changes in serotypes probably reflect changes in animal husbandry and the dissemination in the food chain of new serotypes from increased world trade. Current concern is the increase in multiple antibiotic resistant serotypes.

S. typhi and *S. paratyphi* A, B and C produce typhoid and typhoid-like fever in humans. Typhoid fever is a life-threatening illness. The organism multiplies in the submucosal tissue of the ileal epithelium and then spreads throughout the body via macrophages. Subsequently, various internal organs such as the spleen and liver become infected. The bacteria infect the gallbladder from the liver and finally infect the intestines using bile as the transportation medium. If the organism does not progress past the gallbladder then no typhoid fever develops. Nevertheless, the person may continue to shed the organism in their faeces.

Typical symptoms of typhoid fever are:

- Sustained fever as high as 39–40°C
- Lethargy
- Abdominal cramps
- Headache
- Loss of appatite
- A rash of flat, rose-coloured spots may appear

The fatality rate of typhoid fever is 10% compared to less than 1% for most forms of salmonellosis. A small number of people recover from typhoid fever but continue to shed the bacterium in their faeces. *S. typhi* and *S. paratyphi* enter the body through food and drinks that may have been contaminated by a person who is shedding the organism in their faeces.

5.2.3 *Pathogenic* E. coli

Pathogenic strains of *E. coli* are divided according to clinical symptoms and mechanisms of pathogenesis into the following groups (see also Section 3.9.4 and Figs 3.12–17):

- Enterotoxigenic *E. coli* (ETEC) are commonly known as traveller's diarrhoea. ETEC cause watery diarrhoea, rice water-like, and a low grade fever. The organism colonises the proximal small intestine.
- Enteropathogenic *E. coli* (EPEC) cause watery diarrhoea of infants. EPEC cause vomiting, fever and diarrhoea which is watery with mucus but no blood. The organism colonises the microvilli over the entire intestine to produce the characteristic 'attaching and effacing' lesion in the brush border microvillus membrane.
- Enterohaemorrhagic *E. coli* (EHEC) cause bloody diarrhoea, haemorrhagic colitis, haemolytic uraemic syndrome and thrombic thrombocytopenic purpura. This group includes the verotoxigenic *E. coli* (VTEC, also known as shiga toxin-producing *E. coli* or STEC) serotypes O157, O26 and O111.
- Enteroaggregative *E. coli* (EAggEC) cause persistent watery diarrhoea, especially in children, lasting more than 14 days. The EAggEC align themselves in parallel rows on either tissue cells or glass. This aggregation has been described as 'stacked brick-like'. They produce a heat-labile toxin, antigenically related to haemolysin but not haemolytic, and

a plasmid-encoded heat-stable toxin (EAST1) unrelated to the heat-stable enterotoxin of ETEC. It is thought that EAggEC adhere to the intestinal mucosa and elaborate the enterotoxins and cytotoxins, which result in secretory diarrhoea and mucosal damage. Recent studies support the association of EAggEC with malnutrition and growth retardation in the absence of diarrhoea.

- Enteroinvasive *E. coli* (EIEC) cause a fever and profuse diarrhoea containing mucus and streaks of blood. The organism colonises the colon and carries a 120–140 mD plasmid known as the invasiveness plasmid which carries all the genes necessary for virulence.
- Diffusely adherent *E. coli* (DAEC) have been associated with diarrhoea in some studies, but not consistently.

EHEC were first described in 1977 and recognised as a disease of animals and humans in 1982. They harbour plasmids of various sizes, the most common being 75–100 kbp. The EHEC belong to many serogroups. The serotype O157:H7 is the most important in the UK and USA; other serogroups are dealt with later. The 'O157' and 'H7' refer to the serotyping of the strain's O and H antigens respectively (Sections 2.2.3 and 2.2.4). It can cause very severe forms of food poisoning resulting in death. It has been postulated that *E. coli* O157:H7 evolved from EPEC and acquired the toxin genes from *Sh. dysenteriae* via a bacteriophage, and that the newly emerging pathogen arrived in Europe from South America (Coghlan 1998). The reported incidence and serotype vary from country to country (Table 5.3). In the UK the only EHEC serotype recognised is O157:H7, whereas in France it is serotype O111.

E. coli O157:H7 differs from the majority of *E. coli* strains in that it does not grow or grows poorly at 44°C and does not ferment sorbitol or produce β-glucuronidase. It should be noted that there are other *E. coli* serotypes producing verocytotoxins, however routine examination for them is not yet feasible. The growth of this strain in the human intestine produces a large quantity of toxin(s) that cause severe damage to the lining of the intestine and other organs of the body. These toxins, referred to as verotoxin (VT) or Shiga-like toxin (SLT), are very similar, if not identical, to the toxins produced by *Sh. dysenteriae* (Sections 3.8.4, 3.9.3 and 3.9.4). There are four subgroups VT1 or SLT1, VT2 or SLT2, VT2c or SLT IIc and VT2e or SLT IIe. Strains producing shiga toxins were initially recognised by the cytotoxicity towards Vero cells (green monkey kidney cell line) and subsequently the term 'verotoxigenic *E. coli*' or VTEC arose. However, since the purified and sequenced verotoxin has been found to be nearly identical to shiga toxin, the organisms are referred to as 'shiga toxin-producing *E. coli*' or STEC. The phage-encoded toxin consists of one A subunit and five identical B subunits. The 32 kDa A subunit is an RNA

Table 5.3 Incidence of EHEC in Europe (1996) (adapted from Anon. 1997c).

Country	EHEC infections	Millions of inhabitants	Per million inhabitants
Spain	4	39.6	0.1
Italy	9	57.1	0.2
Netherlands	10	15.4	0.6
Finland	5	5.1	1.0
Denmark	6	5.2	1.2
Austria	11	8.0	1.4
Germany	314	81.5	3.9
Belgium	52	10.0	5.2
Sweden	118	8.7	13.6
United Kingdom	1180	58.1	20.3
Northern Ireland	14	1.6	8.8
Wales	36	2.9	9.2
England	624	48.5	12.4
Scotland	506	5.1	99.2

N-glycosidase that removes a specific adenine residue from 28S rRNA. The subunit is noncovalently associated with a pentamer of 7.7 kDa B subunit. The pentamer binds the toxin to plasma membrane globotriosylceramide (Gb_3) present on susceptible mammalian cells. The toxins destroy the intestinal cells of the human colon and may cause additional damage to the kidneys, pancreas and brain. The incubation period for EHEC diarrhoea is usually 3 to 4 days, although incubation times can be as long as 5 to 8 days or as short as 1 to 2 days.

Healthy adults suffer from thrombic thrombocytopenic purpura (TTP), where blood platelets surround internal organs leading to kidney damage and central nervous system damage. The most vulnerable of the population – children and the elderly – develop haemorrhagic colitis (HC), which may lead to haemolytic uraemic syndrome (HUS).

HC is a less severe form of *E. coli* O157:H7 infection than HUS. The first symptom of HC is the sudden onset of severe crampy abdominal pains. About 24 hours later nonbloody (watery) diarrhoea starts. Some victims have a fever of short duration. Vomiting occurs in about half of the patients during the period of nonbloody diarrhoea and/or other times in the illness. After 1 or 2 days, the diarrhoea becomes bloody and the patient experiences increased abdominal pain. This usually lasts between 4 to 10 days. In severe cases, faecal specimens are described as 'all blood and no stool'. In most patients, the bloody diarrhoea resolves with no long-term impairment. Unfortunately 2–7% of patients (up to

30% in certain outbreaks) will progress to HUS and subsequent complications.

In HUS the patient suffers from bloody diarrhoea, haemolytic anaemia, kidney disorder and renal failure, and requires dialysis and blood transfusions. Central nervous disease may develop which can lead to seizures, coma and death. The mortality rate is 3–17%.

HUS is a leading cause of kidney failure in children, which often requires dialysis and may ultimately be fatal. Other systemic manifestations of illness due to *E. coli* O157:H7 include a central nervous system involvement, hypertension, myocarditis, and other cardiovascular complications that may result in death or severe disability. In some cases, the illness is indicative of some forms of heart disease and has been responsible for strokes in small children. These complications are attributed to direct or indirect actions of verotoxins absorbed from the intestinal tract.

Cattle appear to be the main reservoir of *E. coli* O157:H7. Transmission to humans is principally through the consumption of contaminated foods, such as raw or undercooked meat products and raw milk. Fresh-pressed apple juice or cider, yoghurt, cheese, salad vegetables and cooked corn have also been implicated (Fig. 5.1). Faecal contamination of water and other foods, as well as cross-contamination during food preparation may also be responsible. There is evidence of transmission of this pathogen

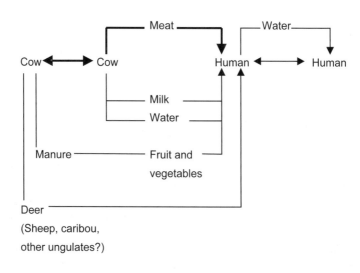

Fig. 5.1   *E. coli* O157:H7 transmission.

through direct contact between people. *E. coli* O157:H7 can be shed in faeces for a median period of 21 days with a range of 5 to 124 days.

According to the Food and Drug Administration of the United States, the infectious dose for *E. coli* O157:H7 is unknown. However, a compilation of outbreak data indicates that it may be as low as 10 organisms. The data show that it takes a very low number of microorganisms to cause illness in young children, the elderly and immune-compromised people.

Most reported outbreaks of enterohaemorrhagic *E. coli* infection have been caused by O157:H7 strains. This suggests that this serotype is more virulent or more transmissible than other serotypes. However, other serotypes of enterohaemorrhagic *E. coli* have been implicated in outbreaks, and the incidence of disease due to non-O157:H7 serotypes seems to be rising. More than 50 of these serotypes have been associated with bloody diarrhoea or HUS in humans. The most common non-O157:H7 serotypes associated with human disease include: O26:H11, O103:H2, O111:NM and O113:H21. At least 10 of the outbreaks due to these organisms have been reported in Japan, Germany, Italy, Australia, the Czech Republic and the United States. These outbreaks have involved 5 to 234 people and, for most of them, the source of infection could not be determined. In many countries such as Chile, Argentina and Australia, non-O157:H7 serotypes have been found to be responsible for the majority of HUS cases. Non-bloody diarrhoea has also been associated with some of these non-O157:H7 serotypes.

Infants are particularly prone to EPEC infections. The EPEC cause vomiting, fever and diarrhoea which is watery with mucus but no blood. The organism colonises the microvilli over the entire intestine to produce the characteristic 'attaching and effacing' lesion in the brush border microvillus membrane.

5.2.4 Shigella *species*

Shigella is a highly contagious bacterium that infects the intestinal tract (Section 3.9.3). It is closely related to *E. coli* but can be differentiated from *E. coli* by the lack of gas production from carbohydrates (anaerogenic) and it is lactose negative. The genus *Shigella* consists of four species: *Sh. dysenteriae* (serotype A), *Sh. flexneri* (serotype B), *Sh. boydii* (serotype C) and *Sh. sonnei* (serotype D). In general, *Sh. dysenteriae*, *Sh. flexneri* and *Sh. boydii* predominate in developing countries. In contrast, *Sh. sonnei* is most common and *Sh. dysenteriae* is least common in developed countries. *Shigella* is spread by direct and indirect contact with infected individuals. Food or water may be contaminated by direct or direct contact with faecal material from infected people. *Shigella* often causes outbreaks in daycare centres.

The main symptoms of shigellosis are:

• Diarrhoea, mild or severe, be watery or bloody
• Fever and nausea
• Vomiting and abdominal cramping may also occur

The symptoms appear within 12 to 96 hours after exposure to *Shigella*; the incubation period is typically 1 week for *Sh. dysenteriae*. The symptoms from *Sh. sonnei* are generally less severe than for the other *Shigella* species. *Sh. dysenteriae* may be associated with serious disease, including toxic megacolon and the haemolytic uraemic syndrome. *Shigella* cells are shed in the faeces for 1 to 2 weeks of infection.

5.2.5　Listeria monocytogenes

Listeria are Gram-positive, non-sporeforming bacteria. They are motile by means of flagella and grow between 0 and 42°C. They are less sensitive to heat compared with *Salmonella* and hence pasteurisation is sufficient to kill the organism. The genus is divided into eight species of which *L. monocytogenes* is the species of primary concern with regard to food poisoning. The species is further subdivided using serotyping and the epidemiologically imporant serotypes are 1/2a, 1/2b and 4b.

　L. monocytogenes has been found in at least 37 mammalian species, both domestic and feral, as well as at least 17 species of birds and possibly some species of fish and shellfish. It is plausible that 1–10% of humans may be intestinal carriers of *L. monocytogenes*. This ubiquitous bacterium has been isolated from various environments, including decaying vegetation, soil, animal feed, sewage and water. It is resistant to diverse environmental conditions and can grow at temperatures as low as 3°C. It is found in a wide variety of foods, both raw and processed, where it can survive and multiply rapidly during storage. These foods include supposedly pasteurised milk and cheese (particularly soft-ripened varieties), meat (including poultry) and meat products, raw vegetables, fermented raw-meat sausages as well as seafood and fish products. *L. monocytogenes* is quite hardy and resists the deleterious effects of freezing, drying, and heat remarkably well for a bacterium that does not form spores. Its ability to grow at temperatures as low as 3°C permits multiplication in refrigerated foods.

　L. monocytogenes is responsible for opportunistic infections, preferentially affecting individuals whose immune system is perturbed, including pregnant women, newborns and the elderly (Section 3.9.6). Listeriosis is clinically defined when the organism is isolated from blood, cerebrospinal fluid or an otherwise sterile site, for example placenta and fetus.

Symptoms of listeriosis are:

- Meningitis, encephalitis or septicaemia
- It can lead to abortion, stillbirth or premature birth when pregnant women are infected in the second and third trimesters

The infective dose of *L. monocytogenes* is unknown but is believed to vary with the strain and susceptibility of the victim. From cases contracted through raw or supposedly pasteurised milk it is evident that fewer than 1000 total organisms may cause disease. The incubation period is extremely wide at 1–90 days. *L. monocytogenes* may invade the gastrointestinal epithelium. Once the bacterium enters the host's monocytes, macrophages or polymorphonuclear leukocytes it is bloodborne (septicaemic) and can grow (Fig. 3.19). Its intracellular presence in phagocytic cells also permits access to the brain and probably transplacental migration to the fetus in pregnant women. The pathogenesis of *L. monocytogenes* centres on its ability to survive and multiply in phagocytic host cells. Listeriosis has a very high mortality rate. When listeric meningitis occurs, the overall mortality may be as high as 70%. Cases of septicaemia have a 50% fatality rate whereas in perinatal-neonatal infections the rate is greater than 80%. In infections during pregnancy, the mother usually survives. Infection can be symptomless, resulting in faecal excretors of infectious listeria. Consequently approximately 1% of faecal samples are positive for *L. monocytogenes* and about 94% of sewage samples.

5.2.6   Yersinia enterocolitica

There are three pathogenic species in the genus *Yersinia*, but only *Y. enterocolitica* and *Y. pseudotuberculosis* cause gastroenteritis. *Y. pestis*, the causative agent of 'the plague', is genetically very similar to *Y. pseudotuberculosis* but infects humans by routes other than food.

Y. enterocolitica, a small (1–3.5 μm × 0.5–1.3 μm) rod-shaped, Gram-negative bacterium, is often isolated from clinical specimens such as wounds, faeces, sputum and mesenteric lymph nodes.Young cultures are oval or coccoid cells. The organism produces peritrichous flagella when it is grown at 25°C, but not when grown at 35°C. The organism has an optimal growth temperature of 30–37°C, however it is also able to grow on food at refrigeration temperatures (8°C). *Y. pseudotuberculosis* has been isolated from the diseased appendix of humans. Both organisms have often been isolated from such animals as pigs, birds, beavers, cats and dogs. Only *Y. enterocolitica* has been detected in environmental and food sources, such as ponds, lakes, meats, ice cream, and milk. Most isolates have been found not to be pathogenic.

Typical symptoms of foodborne yersiniosis are:

- Abdominal pain
- Fever
- Diarrhoea (lasting several weeks)
- Other sypmtoms may include sore throat, bloody stools, rash, nausea, headache, malaise, joint pain and vomiting.

Yersiniosis is frequently characterised by such symptoms as gastro-enteritis with diarrhoea and/or vomiting; however, fever and abdominal pain are the hallmark symptoms. Yersinia infections mimic appendicitis and mesenteric lymphadenitis, but the bacteria may also cause infections of other sites such as wounds, joints and the urinary tract. The minimum infectious dose is unknown. Illness onset is usually between 24 and 48 hours after ingestion, although the maximum incubation period can be as long as 11 days. Yersiniosis has been misdiagnosed as Crohn's disease (regional enteritis) as well as appendicitis.

There are four serotypes of *Y. enterocolitica* associated with pathogenicity: O:3, O:5, O:8 and O:9. The genes (*inv* and *ail*) encoding for invasion of mammalian cells are located on the chromosome while a 40–50 MDa plasmid encodes most of the other virulence associated phenotypes. The 40–50 MDa plasmid is present in almost all the pathogenic *Yersinia* species and the plasmids appear to be homologous. A heat-stable enterotoxin has been isolated from most clinical isolates, however it is uncertain if this toxin has a role in the pathogenicity of the organism.

Y. enterocolitica is ubiquitous. It can be found in meat (pork, beef, lamb, etc.), oysters, fish and raw milk. The exact cause of the food contamination is unknown. However, the prevalence of this organism in the soil and water and in animals such as beavers, pigs and squirrels offers ample opportunities for it to enter our food supply. The principal host recognised for *Y. enterocolitica* is the pig.

Yersiniosis does not occur frequently. It is rare unless a breakdown occurs in food processing techniques. Yersiniosis is a far more common disease in Northern Europe, Scandinavia and Japan than in the USA. The major 'complication' with the disease is the performance of unnecessary appendectomies, since one of the main symptoms of infections is abdominal pain of the lower right quadrant. Both *Y. enterocolitica* and *Y. pseudotuberculosis* have been associated with reactive arthritis, which may occur even in the absence of obvious symptoms. The frequency of such postenteritis arthritic conditions is about 2–3%. Another complication is bacteraemia (entrance of organisms into the blood stream), in which case the possibility of a disseminating disease may occur. This is rare, however, and fatalities are also extremely rare.

The organism is resistant to adverse storage conditions (such as freezing for 16 months). Poor sanitation and improper sterilisation techniques by food handlers, including improper storage, cannot be overlooked as contributing to contamination. Therefore primary control of the organism would require changes to current slaughtering practice (Büllte *et al.* 1992). Since the organism is able to grow at refrigeration temperatures this is not an effective means of control unless combined with the addition of preservatives.

5.2.7 Staphylococcus aureus

St. aureus is a Gram-positive, spherical bacterium (coccus) which occurs in pairs, short chains or bunched, grape-like clusters. The organism was first described in 1879. It is a facultative anaerobe and is divided into a number of biotypes on the basis of biochemical tests and resistance patterns. The biotypes are then further divided according to phage typing, serotyping, plasmid analysis and ribotyping. *St. aureus* produces a wide range of pathogenicity and virulence factors: staphylokinase, hyaluronidases, phophatases, coagulases and haemolysins. Food poisoning is specifically caused by enterotoxins. The enterotoxins are low molecular weight proteins (26 000–34 000 Da) which can be differentiated by serology into seven antigenic types: SEA, SEB, SEC_1, SEC_2, SEC_3, SED, SEE). A toxin previously designated enterotoxin F is now recognised as responsible for toxic shock syndrome and not enteritis. These toxins are highly heat-stable ($D_{98.9} \geq 2$ h) and resistant to cooking and proteolytic enzymes. They are superantigenic and stimulate monocytes and macrophages to produce cytokines (Section 3.8.2). A toxin dose of less than 1.0 µg/kg (300–500 ng) in contaminated food will produce symptoms of staphylococcal intoxication. This amount of toxin is produced by 10^5 organisms per gram. The resistance to heat and proteolysis in the intestinal tract means that it is important that foods which may have supported the growth of *St. aureus* are tested for toxin after heat processing (and subsequent bacterial cell death). A comparison of characteristics for *St. aureus* growth and toxin production is given in Table 5.4.

Staphylococci exist in air, dust, sewage, water, milk, and food or on food equipment, environmental surfaces, humans and animals. Humans and animals are the primary reservoirs. Staphylococci are present in the nasal passages and throats and on the hair and skin of 50% or more of healthy individuals. This incidence is even higher for those who associate with or who come in contact with sick individuals and hospital environments. Although food handlers are usually the main source of food contamination in food poisoning outbreaks, equipment and environmental surfaces can also be sources of contamination with *St. aureus*. Human

Table 5.4   Conditions for *St. aureus* growth and toxin production.

Parameter	Growth	Toxin production
Temperature (°C)	7–48	10–48
pH	4–10	4.5–9.6
Water activity	0.83–0.99	0.87–0.99

Note: To convert to °F use the equation °F = (9/5)°C + 32. As a guidance: 0°C = 32°F, 4.4°C = 40°F, 60°C = 140°F.

intoxication is caused by ingesting enterotoxins produced in food by some strains of *St. aureus*, usually because the food has not been kept hot enough (60°C (140°F) or above) or cold enough (7.2°C (45°F) or below). Foods that are frequently incriminated in staphylococcal food poisoning include meat and meat products, poultry and egg products, salads such as egg, tuna, chicken, potato and macaroni, bakery products such as cream-filled pastries, cream pies and chocolate eclairs, sandwich fillings and milk and dairy products. Foods that require considerable handling during preparation and that are kept at slightly elevated temperatures after preparation are frequently involved in staphylococcal food poisoning. The organism competes poorly with other bacteria and therefore rarely causes food poisoning in a raw product. The organism is quickly killed by heat ($D_{65.5}$ = 0.2–2.0 min) but is resistant to drying and is salt-tolerant (see Tables 2.3 and 2.7).

Symptoms of staphylococcal food poisoning are a rapid onset of:

- Nausea
- Vomiting
- Abdominal cramps

The onset of symptoms in staphylococcal food poisoning is usually rapid, within hours of ingestion. The symptoms can be very acute, depending on individual susceptibility to the toxin, the amount of contaminated food eaten, the amount of toxin in the food ingested and the general health of the victim. The most common symptoms are nausea, vomiting and abdominal cramping. Some individuals may not always demonstrate all the symptoms associated with the illness. In severe cases, headache, muscle cramping, and transient changes in blood pressure and pulse rate may occur. The illness is usually self-limiting and generally takes 2–3 days. Severe cases will take longer.

Since the staphylococcal toxin is very heat stable it cannot be inactivated by standard cooking regimes. Therefore avoiding contamination of food by the organism and maintaining low temperatures are necessary to eliminate the microbial load.

5.2.8 Clostridium perfringens

Clostridium perfringens is an anaerobic, Gram-positive, sporeforming rod. 'Anaerobic' means the organism is unable to grow in the presence of free oxygen. It was first associated with diarrhoea in 1895, but the first reports of the organism causing food poisoning were in 1943. It is widely distributed in the environment and frequently occurs in the intestines of humans and animals. Spores of the organism persist in soil, sediments and areas subject to human or animal faecal pollution.

There are five types (A–E) of *Cl. perfringens* which are divided according to the presence of exotoxins (Section 3.8.1). *Cl. perfringens* types A, C and D are human pathogens, whereas types B, C, D and E are animal pathogens. The acute diarrhoea caused by *Cl. perfringens* is due to the production of α toxin, an enterotoxin (Titbull *et al.* 1999). A more serious but rare illness is also caused by ingesting food contaminated with type C strains. The latter illness is known as enteritis necroticans jejunitis, or 'pig-bel disease', and is due to the β exotoxin.

Characteristics of perfringens food poisoning are:

• Abdominal pain
• Nausea
• Acute diarrhoea
• Symptoms appear 8–12 hours after ingestion of the organism

The common form of perfringens poisoning is characterised by intense abdominal cramps and diarrhoea which begin 8–12 hours after consumption of foods containing large numbers of those *Cl. perfringens* bacteria capable of producing the food poisoning toxin. The illness is usually over within 24 hours but less severe symptoms may persist in some individuals for 1 or 2 weeks. A few deaths have been reported as a result of dehydration and other complications. In most instances, the actual cause of poisoning by *Cl. perfringens* is temperature abuse of prepared foods. Meats, meat products, and gravy are the foods most frequently implicated. Small numbers of the organisms are often present as spores after cooking. The spores germinate and the clostridia multiply to food poisoning levels during the cooling period and storage. The cooking process drives off oxygen, hence creating an anaerobic environment favourable to clostridial growth. After ingestion the enterotoxin is produced in the intestine, after the organism has passed through the stomach. The enterotoxin is associated with sporulation, possibly induced by the acidic environment of the stomach. The enterotoxin is a heat-sensitive protein, 36 000 Da in size. It is destroyed by heat (the D_{90} value is 4 minutes) and a few cases of food poisoning due to ingestion of the toxin have been reported.

Necrotic enteritis (pig-bel) caused by *Cl. perfringens* is often fatal. This disease also begins as a result of ingesting large numbers (greater than 10^8) of *Cl. perfringens* type C in contaminated foods. Deaths from necrotic enteritis are caused by infection and necrosis of the intestines and from resulting septicaemia.

The isolation of low numbers of *Cl. perfringens* from food does not necessarily mean that a food poisoning danger exists. Only when large numbers are present is there a definite hazard and therefore enumeration techniques are essential for this organism. Although the *Cl. perfringens* vegetative cell is killed by chilling and freezing, the spore form may survive. Further details on the detection of clostridia and growth characteristics can be found in Section 6.5.9; see also Tables 2.3 and 2.7. Control of the organism is mainly achieved through the cooking and cooling steps. Rapid cooling through the temperature range 55 to 15°C reduces the opportunity for any surviving clostridia spores to germinate. Thorough reheating of food to 70°C immediately before consumption will kill any vegetative cells present.

5.2.9   Cl. botulinum *types A, B, E and F*

Cl. botulinum is a Gram-positive, rod-shaped, strict anaerobe (0.3–0.7 × 3.4–7.5 μm) with peritrichous flagella. It causes the foodborne illness called 'botulism'. This is a food intoxication due to the ingestion of preformed neurotoxins (Fig. 3.5). The organism is ubiquitous in nature. There are four clinically important species: *Cl. botulinum*, *Cl. perfringens*, *Cl. difficile* and *Cl. tetani*. However, only *Cl. botulinum* and *Cl. perfringens* cause food poisoning.

There are seven types of *Cl. botulinum*: A, B, C, D, E, F and G. These are differentiated on the basis of toxin antigenicity (Table 3.10). Types A, B, E and F are the main causes of human botulism (types C and D in animals). A number of serologically distinct neurotoxins (BoNTA to BoNTG) have been identified and are among the most potent toxins known to man. The toxin is made of two proteins, fragment A (light chain (LC), 50 kDa) and fragment B (heavy chain (HC), 100 kDa), which are linked by a disulphide bridge (Fig. 3.6). The LC is responsible for the toxin's effect on nerve cells. The HC contains the membrane translocation domain and the toxin receptor binding moiety (Boquet *et al.* 1998). The organism forms spores which may become airborne and contaminate open jars or cans. Once sealed, the anaerobic conditions favour spore outgrowth and the production of toxins (Figs 3.7 and 3.8).

Symptoms of botulism are:

- Double vision
- Nausea
- Vomiting
- Fatigue
- Dizziness
- Headache
- Dryness in the throat and nose
- Respiratory failure

The onset of the symptoms is from 12 to 36 hours after ingestion of the bacterial toxins. Botulinum toxins block the release of the neurotransmitter acetylcholine, which results in muscle weakness and subsequently paralysis. The illness may last from 2 hours to 14 days, depending upon dose and vulnerability of host. The fatality rate is about 10%.

Botulism is associated with canned (especially home canned) low acid foods, vegetables, fish and meat products. It is also associated with honey and hence honey should not be given to children under one year old. Infant botulism is milder than the adult version. The spores germinate in the intestinal tract and the bacterium produces the toxins causing 'floppy baby syndrome'.

Heat treating low acid canned foods to 121°C for 3 minutes or equivalent will eliminate the *Cl. botulinum* spores. *Cl. botulinum* cannot grow in acid or acidified foods with a pH less than 4.6.

5.2.10 Bacillus cereus

B. cereus is a sporeforming food pathogen that was first isolated in 1887. The spores can survive many cooking processes. The organism grows well in cooked food because of the lack of a competing microflora. *B. cereus* is a Gram-positive, facultatively aerobic sporeformer whose cells are large rods and whose spores do not swell the sporangium. These and other characteristics, including biochemical features, are used to differentiate and confirm the presence of *B. cereus*. However, these characteristics are shared with *B. cereus* var. *mycoides*, *B. thuringiensis* and *B. anthracis*. Differentiation of these organisms depends upon determination of motility (most *B. cereus* are motile), presence of toxin crystals (*B. thuringiensis*), haemolytic activity (*B. cereus* and others are β-haemolytic whereas *B. anthracis* is usually nonhaemolytic) and rhizoid growth which is characteristic of *B. cereus* var. *mycoides*. *B. subtilis* and *B. licheniformis* have been proposed as food poisoning organisms, but the evidence to date is not complete. Nevertheless, the Public Health Laboratory Service (UK) ready-to-eat guidelines do include them in the sampling plan (Section 8.10).

B. cereus is ubiquitous in nature, being isolated from soil, vegetation, fresh water and animal hair. It is commonly found at low levels in food ($< 10^2$ cfu/g), which is considered acceptable. Food poisoning outbreaks usually occur when the food has been subjected to time-temperature abuse which was sufficient for the low level of organisms to multiply to a significant (intoxication) level ($> 10^5$ cfu/g). Characteristics of temperature and water activity growth range and D values are given in Tables 2.3 and 2.7. Isolation methods are covered in Section 6.5.10.

There are two recognised types of *B. cereus* food poisoning: diarrhoeal and emetic (Section 3.9.7). Both types of food poisoning are self-limiting and recovery is usually within 24 hours. *B. cereus* produces the diarrhoeal toxins during growth in the human small intestines, whereas the emetic toxins are preformed on the food. A wide variety of toxins have been identified (Table 3.10; Kramer & Gilbert 1989; Granum & Lund 1997):

- Diarrhoeal enterotoxin
- Emetic factor
- Haemolysin I
- Haemolysin II
- Phosphatase C

The diarrhoeal type of illness is caused by a large molecular weight protein (diarrhoeal enterotoxin), of 38 to 46 kDa. This toxin is inactivated at 56°C for 30 minutes. The vomiting (emetic) type of illness is caused by a low molecular weight (1.2 kDa), heat-stable peptide. It is very resistant to heat (126°C for 90 minutes) and pH values of 2 to 11 (Turnbull *et al.* 1979). It is produced in the stationary phase of growth, but it is uncertain whether its production is linked with spore formation.

Symptoms of *B. cereus* diarrhoeal food poisoning are:

- Watery diarrhoea
- Abdominal cramps and pain
- Nausea, rarely vomiting

The symptoms of *B. cereus* diarrhoeal type food poisoning mimic those of *Cl. perfringens* food poisoning. The onset of watery diarrhoea, abdominal cramps and pain occurs 8–24 hours after consumption of contaminated food. Nausea may accompany diarrhoea, but vomiting (emesis) rarely occurs. Symptoms persist for 24 hours in most instances, during which time the organism is excreted in large numbers.

Symptoms of *B. cereus* emetic food poisoning are:

- Nausea
- Vomiting
- Abdominal cramps and diarrhoea may occur

The emetic type of food poisoning is characterised by nausea and vomiting within 0.5 to 6 hours after consumption of contaminated foods. Occasionally, abdominal cramps and/or diarrhoea may also occur. Duration of symptoms is generally less than 24 h. The symptoms of this type of food poisoning parallel those caused by *St. aureus* foodborne intoxication. Some strains of *B. subtilis* and *B. licheniformis* have been isolated from lamb and chicken incriminated in food poisoning episodes. These organisms demonstrate the production of a highly heat-stable toxin, which may be similar to the vomiting-type toxin produced by *B. cereus*.

A wide variety of foods including meats, milk, vegetables and fish have been associated with the diarrhoeal-type food poisoning. The vomiting-type outbreaks have generally been associated with rice products; however, other starchy foods such as potato, pasta and cheese products have also been implicated. Food mixtures such as sauces, puddings, soups, casseroles, pastries and salads have frequently been incriminated in food poisoning outbreaks. The presence of large numbers of *B. cereus* ($> 10^6$ organisms/g) in a food is indicative of active growth and proliferation of the organism and is consistent with a potential hazard to health.

Since the organism is ubiquitous in the environment low numbers commonly occur in food, so the main control mechanism is to prevent spore germination and multiplication in cooked, ready-to-eat foods. Storage of foods below 10°C will inhibit *B. cereus* growth.

5.2.11   Vibrio parahaemolyticus

V. parahaemolyticus means 'dissolving blood' and the organism was first isolated in 1951. The organism is not isolated in the absence of additional NaCl (2–3%) and was therefore not cultivated in early studies of gastroenteritis. *V. parahaemolyticus* is now recognised as a major cause of foodborne gastroenteritis in Japan. This is because the organism is associated with the consumption of seafood, which is a significant part of the average diet in Japan.

Typical symptoms of *V. parahaemolyticus* food poisoning are:

- Diarrhoea
- Abdominal cramps
- Nausea
- Vomiting
- Headaches
- Fever and chills

The incubation period is 4–96 hours after the ingestion of the organism, with a mean of 15 hours. The illness is usually mild or moderate, although some cases may require hospitalisation. On average the illness lasts about 3 days. The disease is caused when the organism attaches itself via cell-associated adhesins to the small intestine and excretes an as yet uncharacterised enterotoxin. The enterotoxin is possibly Shiga-like. The infective dose may be greater than one million organisms. Although the demonstration of the Kanagawa haemolysin was initially regarded as indicative of pathogenicity, this is now uncertain.

The organism is usually less than 10^3 cfu/g on fish and shellfish, except in warm waters where the count may increase to 10^6 cfu/g. Infections with this organism have been associated with the consumption of raw, improperly cooked or cooked, recontaminated, fish and shellfish. A correlation exists between the probability of infection and the warmer months of the year. Improper refrigeration of seafoods contaminated with this organism will allow its proliferation, which increases the possibility of infection. The organism is very sensitive to heat and outbreaks are therefore frequently due to improper handling procedures and temperature abuse. Control of the organism is therefore through prevention of multiplication of the organism after harvesting by chilling ($<5°C$) and cooking to an internal temperature of $>65°C$. Isolation of any *Vibrio* species from cooked food indicates poor hygiene practice, since the organism is rapidly killed by heat.

5.2.12   V. vulnificus

V. vulnificus was first reported in 1976 as 'lactose-positive vibrios'. 'Vulnificus' means 'wound inflicting', which reflects the organism's ability to invade and destroy tissue. The organism is therefore associated with wound infections and fatal septicaemia (Linkous & Oliver 1999). It has the highest death rate of any foodborne disease agent (Todd 1989) and causes 95% of all seafood-related deaths in the United States.

Typical symptoms of *V. vulnificus* food poisoning are:

- Fever
- Chills
- Nausea
- Skin lesions

The onset of symptoms takes about 24 hours (range 12 hours to several days) after ingestion of contaminated raw shellfish (especially oysters) by vulnerable people. Individuals most susceptible to infection include the elderly, the immunocompromised and those suffering from chronic liver disease and chronic alcoholism. The organism differs from other pathogenic vibrios in that it invades and multiplies in the blood stream. The mortality is between 40 and 60% of cases. The organism is highly invasive and produces various factors which protect it from the host immune system, including a serum resistance factor, capsular polysaccharide and the ability to acquire iron from iron-saturated transferrin. It produces a range of exoenzymes including a heat-labile haemolysin/cytolysin and an elastolytic protease which probably causes the cellular damage. The lipopolysaccharide is endotoxic. Since a large amount of shellfish is consumed each year and only relatively small numbers of *V. vulnificus* food poisoning cases are reported, it seems reasonable to predict that not all strains are pathogenic.

V. vulnificus is isolated from shellfish and coastal waters. The organism is rarely isolated from sea water at < 10–15°C, but numbers rise when the water temperature is > 21°C. The principal routes of infection are through wounds and septicaemia after ingestion. It is not a significant cause of food poisoning among healthy adults. Therefore the main means of prevention is for immunocompromised individuals to avoid eating raw shellfish, in particular oysters. Isolation of any *Vibrio* species from cooked food indicates poor hygiene practice since the organism is rapidly killed by heat and has not been reported in processed foods. Control of the organism is primarily by ceasing oyster harvesting if the water temperatures exceed 25°C, and by chilling and holding oysters to < 15°C post-harvesting.

5.2.13   Brucella melitensis, Br. abortus *and* Br. suis

Brucella is a strictly aerobic, Gram-negative coccobacillus which causes brucellosis. This organism is carried by animals and causes incidental infections in humans. The four species that infect humans are named after the animal they are commonly isolated from: *Br. abortus* (cattle), *Br. suis* (swine), *Br. melitensis* (goats) and *Br. canis* (dogs). The cattle and dairy industries are the prime source of infection. *Brucella* can enter the body via the skin, respiratory tract or the digestive tract. Once there the intracellular organism can enter the blood and the lymphatics where it multiplies inside phagocytes and eventually cause bacteraemia (bacterial infection of blood). Symptoms vary from patient to patient, but can include high fever, chills and sweating.

Worldwide brucellosis remains a major source of disease in humans and domesticated animals. In humans *Br. melitensis* is the most important clinically apparent disease. Pathogenicity is related to the production of lipopolysaccharides containing poly *N*-formyl perosamine O chain, Cu-Zn superoxide dismutase, erythrulose phosphate dehydrogenase, stress-induced proteins related to intracellular survival and adenine and guanine monophosphate inhibitors of phagocyte functions.

Prevention of brucellosis depends on its eradication or control in animal hosts, hygiene precautions to limit exposure to infection through occupational activities and the heat treatment of dairy and other potentially contaminated foods.

5.2.14   Aeromonas hydrophila, A. caviae *and* A. sobria

The genus *Aeromonas* was proposed in 1936 for rod-shaped Enterobacteriaceae possessing a polar flagellum. *A. hydrophila* is present in all freshwater environments and in brackish water. It has frequently been found in fish and shellfish and in market samples of red meats (beef, pork, lamb) and poultry. The organism is able to grow slowly at $0°C$. It is presumed that, given the ubiquity of the organism, not all strains are pathogenic. Some strains of *A. hydrophila* are capable of causing illness in fish and amphibians as well as in humans, who may acquire infections through open wounds or by ingestion of a sufficient number of the organisms in food or water. *A. hydrophila* may cause gastroenteritis in healthy individuals or septicaemia in individuals with impaired immune systems or various malignancies. *A. caviae* and *A. sobria* may also cause enteritis in anyone or septicaemia in immunocompromised persons or those with malignancies.

There is controversy as to whether *A. hydrophila* is a cause of human gastroenteritis. The uncertainty is because volunteer human feeding studies (10^{11} cell dose) have failed to demonstrate any associated human illness. However, its presence in the stools of individuals with diarrhoea, in the absence of other known enteric pathogens, suggests that it has some role in disease and thus it has been included in this survey of foodborne pathogens. Similarly, *A. caviae* and *A. sobria* are putative pathogens associated with diarrhoeal disease, but as of yet they are unproven causative agents.

General symptoms of *A. hydrophila* gastroenteritis are:

- Diarrhoea
- Abdominal pain
- Nausea
- Chills and headache
- Dysentery-like illness
- Colitis
- Additional symptoms include septicaemia, meningitis, endocarditis and corneal ulcers

Two distinguishable types of gastroenteritis have been associated with *A. hydrophila*: a cholera-like illness with a watery ('rice and water') diarrhoea and a dysenteric illness characterised by loose stools containing blood and mucus. The infectious dose of this organism is unknown. *A. hydrophila* may spread throughout the body in the bloodstream and cause a general infection in persons with impaired immune systems. Those at risk are individuals suffering from leukaemia, carcinoma and cirrhosis and those treated with immunosuppressive drugs or who are undergoing cancer chemotherapy. *A. hydrophila* produces a cytotonic enterotoxin(s), haemolysins, acetyl transferase and a phospholipase. Together these virulence factors may account for the organism's pathogenicity.

5.2.15 Pleisiomonas shigelloides

Pl. shigelloides is a facultatively anaerobic, rod-shaped, Gram-negative bacterium related to the *Aeromonas* genus and classified in the Vibrionaceae family. Sometimes the organism has the apprearance of long or filamentous cells. It has been isolated from fresh water, fresh water fish and shellfish, and from many types of animals including cattle, goats, swine, cats, dogs, monkeys, vultures, snakes and toads. Most human *Pl. shigelloides* infections are suspected to be waterborne. The organism may be present in contaminated or untreated water which has been used as drinking water or recreational water. *Pl. shigelloides* does not always cause illness after ingestion but may reside temporarily as a transient, noninfectious member of the intestinal flora. It has been isolated from the stools of patients with diarrhoea, but is also sometimes isolated from healthy individuals (0.2–3.2% of the population). As with *A. hydrophila*, *Pl. shigelloides* is not a proven foodborne pathogen but has nevertheless been included in this survey as a few studies have linked water and food contamination with *Pl. shigelloides* gastroenteritis outbreaks (Lieb 1983; Claesson *et al.* 1994).

Typical symptoms of *Pl. shigelloides* gastroenteritis include:

- Diarrhoea
- Abdominal pain
- Nausea
- Chills
- To a lesser extent fever, headaches and vomiting

The symptoms may begin 20 to 24 hours after consumption of contaminated food or water and last from 1 to 9 days. Diarrhoea is watery, non-mucoid and non-bloody. In severe cases, however, diarrhoea may be greenish-yellow, foamy and blood-tinged.

The infectious dose is presumed to be quite high, at least greater than one million organisms. The pathogenesis of *Pl. shigelloides* infection is not known. The organism produces a heat-stable cytotonic enterotoxin and possibly haemolysins, proteases and endotoxins. Its significance as an enteric (intestinal) pathogen is presumed because of its predominant isolation from stools of patients with diarrhoea.

The organism is ubiquitous in the environment. Most *Pl. shigelloides* infections occur in the summer months and correlate with environmental contamination of fresh water (rivers, streams, ponds, etc.). The usual route of transmission of the organism in sporadic or epidemic cases is by ingestion of contaminated water or raw shellfish. Most *Pl. shigelloides* strains associated with human gastrointestinal disease have been isolated from stools of diarrhoeic patients living in tropical and subtropical areas. Such infections are rarely reported in the USA or Europe because of the self-limiting nature of the disease. Subsequently it may be included in the group of diarrhoeal diseases 'of unknown aetiology' which are treated with and respond to broad-spectrum antibiotics. Since the organism is heat sensitive, the main means of control is the proper cooking of shellfish before ingestion.

5.2.16 Streptococcus *spp.*

The genus *Streptococcus* comprises Gram-positive, microaerophilic cocci (round), which are nonmotile and occur in chains or pairs. The genus is defined by a combination of antigenic, haemolytic, and physiological characteristics into groups A, B, C, D, F and G. Groups A and D can be transmitted to humans via food. Group A contains one species (*Strep. pyogenes*) with 40 antigenic types. Group D contains five species: *Enterococcus (Streptococcus) faecalis*, *Ent. faecium*, *Strep. durans*, *Strep. avium*, and *Strep. bovis*. The link between scarlet fever and septic sore throat was linked to the consumption of contaminated milk almost 100 years ago. Outbreaks of septic sore throat and scarlet fever were numerous before the advent of milk pasteurisation.

Symptoms of group A streptococcal infection are:

- Sore and red throat
- Pain on swallowing
- Tonsillitis
- High fever and headache
- Nausea and vomiting
- Malaise
- Runny nose
- Rash

Group A streptococci cause septic sore throat and scarlet fever as well as other pyogenic and septicaemic infections. The organism may also cause toxic shock syndrome (Cone *et al.* 1987). The onset of illness is 1 to 3 days. The infectious dose is probably quite low, at less than 1000 organisms. Food sources include milk, ice cream, eggs, steamed lobster, ground ham, potato salad, egg salad, custard, rice pudding and shrimp salad. In almost all food poisoning cases, the foodstuffs had been subjected to temperature abuse between preparation and consumption. Contamination of the food is often due to poor hygiene, ill food handlers or the use of unpasteurised milk. Streptococcal sore throat is very common, especially in children. Usually it is successfully treated with antibiotics. Complications are rare and the fatality rate is low. Most current outbreaks have involved complex foods (for example salads) which were infected by a food handler with a septic sore throat. One ill food handler may subsequently infect hundreds of individuals.

The pathogenicity of group A streptococci is partially due to the presence of M proteins in the fibrillae on the cell surface. These adherence factors enable the organism to adhere to epithelial cells and evade phagocytosis. The organism produces haemolysin(s) as evidenced by the β type haemolysis on blood agar.

Group D streptococci may produce a clinical syndrome similar to staphylococcal intoxication.

General symptoms of group D streptococcal infection are:

- Diarrhoea
- Abdominal cramps
- Nausea and vomiting
- Fever and chills
- Dizziness

Symptoms occur 2 to 36 hours after ingestion of suspect food. The infectious dose is probably greater than 10^7 organisms. The diarrhoeal illness is poorly characterised, but is acute and self-limiting. Food sources include sausage, evaporated milk, cheese, meat croquettes, meat pie, pudding, raw milk and pasteurised milk. Entrance into the food chain is due to underprocessing and/or unhygienic practices during food preparation.

Control of foodborne streptococcal infections is by strict control of personal hygiene and excluding staff with sore throats from the production area.

Strep. parasanguinis is an emerging foodborne pathogen. This organism was isolated from two sheep in Spain during a bacteriological survey determining the prevalence of subclinical mastitis. Since the organism has been associated with the development of experimental endocarditis, its presence at relatively high concentrations in apparently healthy sheep milk may pose a health risk in persons with predisposing heart lesions.

5.3   Foodborne pathogens: viruses

5.3.1   Norwalk and Norwalk-like viruses (SRSV)

Norwalk and Norwalk-like viruses (also known as Small Round Structured Viruses or SRSV) cause diarrhoea and vomiting. This virus is classified in the Caliciviridae. The Caliciviridae are divided into five groups on the basis of sequence relatedness and genomic organisation (Berke *et al.* 1997). Four are known human pathogens: Norwalk-like small round structured viruses, hepatitis E, Sapporo and the marine (animal) calicivirus. The fifth group, rabbit haemorrhagic disease, is not yet proven to be a human pathogen. The virus was the first one to be associated with acute gastroenteritis outbreaks and has been named after the location of the outbreak at Norwalk (Snow Maontain, Hawaii; Kapikian *et al.* 1972). The virus is similar to caliciviruses in appearance, but cannot be cultivated in the laboratory. It has a positive strand RNA genome of 7.5 kb and a single structural protein of about 60 kDa. The 27–32 nm viral particles have a buoyant density of 1.39–1.40 g/ml in CsCl. The family consists of several serologically distinct groups of viruses that have been named after the places where the outbreaks occurred.

   Viral gastroenteritis due to the Norwalk virus is self-limiting and mild. Characteristics of Norwalk gastroenteritis are:

• Nausea
• Vomiting
• Diarrhoea
• Abdominal pain
• Headache, fever, chills, mucle aches and weakness may also occur

The infectious dose is unknown but is presumed to be low. A mild and brief illness usually develops 24 to 48 hours after contaminated food or water is consumed and lasts for 24 to 60 hours. The virus invades and damages the gastrointestinal tract, resulting in mucosal lesions of the small intestinal tract. Severe illness or hospitalisation is very rare.

   Norwalk virus gastroenteritis is transmitted by the faecal-oral route via contaminated water and foods. Secondary person-to-person transmission has been documented. Water is the most common source of outbreaks and may include water from municipal supplies, wells, recreational lakes and swimming pools and water stored aboard cruise ships. Shellfish and salad ingredients are the foods most often implicated in Norwalk virus outbreaks. Ingestion of raw or insufficiently steamed clams and oysters poses a high risk for infection with Norwalk virus. Foods other than shellfish are contaminated by ill food handlers. The virus does not multiply on food or in water.

The inability to grow the virus in the laboratory severely hinders studies of inactivation and detection. Nevertheless, control of the Norwalk and Norwalk-like viruses in food is the same as for hepatitis A virus. It is achieved through avoidance of food contamination from contaminated water and infectious personnel.

5.3.2 Hepatitis A

Hepatitis A virus (HAV) is classified with the enterovirus group of the Picornaviridae family. HAV has a single molecule of RNA surrounded by a small (27 nm diameter) protein capsid and a buoyant density in CsCl of 1.33 g/ml. Many other picornaviruses cause human disease, including polio viruses, coxsackie viruses, echo viruses and rhinoviruses (cold viruses). Only one serotype has been observed among HAV isolates collected from various parts of the world.

The incubation period for HAV varies from 10 to 50 days (mean 30 days). It is dependent upon the number of infectious particles consumed. The infectious dose is unknown but is presumed to be 10 to 100 virus particles. The period of communicability extends from early in the incubation period to about a week after the development of jaundice. The greatest danger of spreading the disease to others occurs during the middle of the incubation period, well before the first presentation of symptoms. Many infections with HAV do not result in clinical disease, especially in children. When disease does occur, it is usually mild and recovery is complete in 1 to 2 weeks. Occasionally the symptoms are severe and convalescence can take up to 6 months. Patients suffer from feeling chronically tired during convalescence, and their inability to work can cause financial loss. Up to 20% may have a relapse and be impaired for as long as 15 months. Approximately 15% of patients require hospitalisation.

Typical symptoms of hepatitis A are:

* Fever
* Chills
* Malaise
* Loss of appetite
* Nausea

- Jaundice
- Dark urine
- Light-coloured stools
- Abdominal pain in the liver area

HAV is usually a mild illness characterised by sudden onset of fever, malaise, nausea, loss of appetite and abdominal discomfort. This is followed in several days by jaundice. HAV is excreted in faeces of infected people and can produce clinical disease when susceptible individuals consume contaminated water or foods. Cold cuts and sandwiches, fruits and fruit juices, milk and milk products, vegetables, salads, shellfish and iced drinks are commonly implicated in outbreaks. Water, shellfish and salads are the most frequent sources. Contamination of foods by infected workers in food processing plants and restaurants is common. Viruses do not multiply on food, it is simply the vector.

The virus has not been isolated from any food associated with an outbreak. Because of the long incubation period, the suspected food is often no longer available for analysis. No satisfactory method is presently available for routine analysis of food, but sensitive molecular methods used to detect HAV in water and clinical specimens should prove useful to detect virus in foods. Among those, the PCR amplification method seems particularly promising.

HAV has a worldwide distribution occurring in both epidemic and sporadic fashions. HAV is primarily transmitted by person-to-person contact through faecal contamination, but common source epidemics from contaminated food and water also occur. Poor sanitation and crowding facilitate transmission. Outbreaks of HAV are common in institutions, crowded house projects and prisons, and in military forces in adverse situations. In developing countries, the incidence of disease in adults is relatively low because of exposure to the virus in childhood. Most individuals 18 and older demonstrate an immunity that provides lifelong protection against reinfection.

5.3.3 Hepatitis E

Hepatitis E virus (HEV) is the major aetiological agent of enterically transmitted non-A, non-B hepatitis world wide. It is a spherical, nonenveloped, single stranded RNA that is approximately 32–34 nm in diameter. HEV has provisionally been classifed in the Caliciviridae family. This virus enters the body through water or food, especially raw shellfish that has been contaminated by sewage. Anti-HEV activity has been determined in the serum of a number of domestic animals in areas with a high endemicity of human infection, indicating that this may be an emerging zoonosis.

5.3.4 Rotaviruses

Rotaviruses are classified with the Reoviridae family. They have a genome consisting of 11 double-stranded RNA segments surrounded by a distinctive two-layered protein capsid. Particles are 70 nm in diameter and have a buoyant density of 1.36 g/ml in CsCl. Six serological groups have been identified, three of which (groups A, B and C) infect humans. Rotavirus gastroenteritis is a self-limiting, mild to severe disease characterised by vomiting, watery diarrhoea and low-grade fever. The infective dose is presumed to be 10–100 infectious viral particles. Because a person with rotavirus diarrhoea often excretes large numbers of virus (10^8–10^{10} infectious particles/ml of faeces) infectious doses can be readily acquired through contaminated hands, objects or utensils. Asymptomatic rotavirus excretion has been well documented and may play a role in perpetuating endemic disease.

Rotaviruses are transmitted by the faecal-oral route. Person-to-person spread through contaminated hands is probably the most important means by which rotaviruses are transmitted in close communities such as paediatric and geriatric wards, day care centres and family homes. Infected food handlers may contaminate foods that require handling and no further cooking, such as salads, fruits and hors d'oeuvres. Rotaviruses are quite stable in the environment and have been found in estuary samples at levels as high as 0.2–1 infectious particles/litre. Sanitary measures adequate for bacteria and parasites seem to be ineffective in endemic control of rotaviruses, as a similar incidence of rotavirus infection is observed in countries with both high and low health standards.

Group A rotavirus is endemic worldwide. It is the leading cause of severe diarrhoea among infants and children, and accounts for about half of the cases requiring hospitalisation. Over 3 million cases of rotavirus gastroenteritis occur annually in the USA. In temperate areas it occurs primarily in the winter, but in the tropics it occurs throughout the year. The number of cases attributable to food contamination is unknown. Group B rotavirus, also called adult diarrhoea rotavirus or ADRV, has caused major epidemics of severe diarrhoea affecting thousands of persons of all ages in China. Group C rotavirus has been associated with rare and sporadic cases of diarrhoea in children in many countries. However, the first outbreaks were reported from Japan and England. The incubation period ranges from 1 to 3 days. Symptoms often start with vomiting followed by 4 to 8 days of diarrhoea. Temporary lactose intolerance may occur. Recovery is usually complete. However, severe diarrhoea without fluid and electrolyte replacement may result in death. Childhood mortality caused by rotaviruses is relatively low in the USA, with an estimated 100 cases/year, but reaches almost 1 million cases/year worldwide. Association with other enteric pathogens may play a role in the severity of the disease.

The virus has not been isolated from any food associated with an outbreak, and no satisfactory method is available for routine analysis of food. Control of the organism is the same as for hepatitis A and Norwalk virus, that is the prevention of food contamination by polluted water and infected food handlers.

5.3.5   Small round viruses, astroviruses, caliciviruses, adenoviruses and parvoviruses

Although the rotavirus and the Norwalk family of viruses are the major causes of viral gastroenteritis, a number of other viruses have been implicated in outbreaks, including astroviruses, caliciviruses, enteric adenoviruses and parvovirus. Astroviruses, caliciviruses and the Norwalk family of viruses possess well-defined surface structures and are sometimes identified as 'small round structured viruses' or SRSVs. Viruses with a smooth edge and no discernible surface structure are designated 'featureless viruses' or 'small round viruses' (SRVs). These agents resemble enterovirus or parvovirus and may be related to them.

Viral gastroenteritis is usually a mild illness characterised by nausea, vomiting, diarrhoea, malaise, abdominal pain, headache and fever. The infectious dose is not known but is presumed to be low. Viruses are transmitted by the faecal-oral route via person-to-person contact or ingestion of contaminated foods and water. Ill food handlers may contaminate foods that are not further cooked before consumption. Enteric adenovirus may also be transmitted by the respiratory route.

Astroviruses cause sporadic gastroenteritis in children under 4 years of age and account for about 4% of the cases hospitalised for diarrhoea. Most American and British children over 10 years of age have antibodies to the virus. Astroviruses are unclassified viruses which contain a single positive strand of RNA of about 7.5 kb surrounded by a protein capsid of 28–30 nm diameter. A five- or six-pointed star shape can be observed on the particles under the electron microscope. Mature virions contain two major coat proteins of about 33 kDa each and have a buoyant density in CsCl of 1.38–1.40 g/ml. At least five human serotypes been identified in England. The Marin County agent found in the USA is serologically related to astrovirus type 5.

Caliciviruses infect children between 6 and 24 months of age and account for about 3% of hospital admissions for diarrhoea. By 6 years of age, more than 90% of all children have developed immunity to the illness. Caliciviruses are classified in the family Caliciviridae. They contain a single strand of RNA surrounded by a protein capsid of 31–40 nm diameter.

Mature virions have cup-shaped indentations which give them a 'Star of David' appearance in the electron microscope. The particle contains a single major coat protein of 60 kDa and has a buoyant density in CsCl of 1.36–1.39 g/ml. Four serotypes have been identified in England.

The enteric adenovirus causes 5–20% of the gastroenteritis in young children, and is the second most common cause of gastroenteritis in this age group. By 4 years of age, 85% of all children have developed immunity to the disease. Enteric adenoviruses represent serotypes 40 and 41 of the family Adenoviridae. These viruses contain a double-stranded DNA surrounded by a distinctive protein capsid of about 70 nm diameter. Mature virions have a buoyant density in CsCl of about 1.345 g/ml.

Parvoviruses belong to the family Parvoviridae, the only group of animal viruses to contain linear single-stranded DNA. The DNA genome is surrounded by a protein capsid of about 22 nm diameter. The buoyant density of the particle in CsCl is 1.39–1.42 g/ml. The Ditchling, Wollan, Paramatta and cockle agents are candidate parvoviruses associated with human gastroenteritis. Shellfish have been implicated in illness caused by a parvo-like virus. Parvo-like viruses have been implicated in a number of shellfish-associated outbreaks, but the frequency of disease is unknown. A mild, self-limiting illness usually develops 10 to 70 hours after contaminated food or water is consumed and lasts for 2 to 9 days. The clinical features are milder but otherwise indistinguishable from rotavirus gastroenteritis. Co-infections with other enteric agents may result in more severe illness lasting a longer period of time. Only a parvovirus-like agent (cockle) has been isolated from seafood associated with an outbreak.

5.4   Seafood and shellfish poisoning

There are a number of causes of food poisoning originating from seafoods and shellfish (Table 5.5). Seafood poisoning can be caused by ciguatera poisoning, a toxin from microalgae which has been accumulated in fish flesh. It used to be believed that scrombroid poisoning was due to the consumption of fish flesh containing high levels of histamine from bacterial histidine dehydrogenase activity in mackerel and similar fish. Incriminated bacteria are *Morganella morganii*, *Proteus* spp., *Hafnia alvei* and *Klebsiella pneumoniae*. However, this has not been conclusively proven since human volunteers ingesting histamine do not always produce the characteristic scrombroid poisoning symptoms. Possibly there are other biogenic amines present.

Shellfish poisoning is caused by a group of toxins produced by planktonic algae (dinoflagellates, in most cases) upon which the shellfish feed. The toxins are accumulated and sometimes metabolised by the shellfish.

Table 5.5 Microorganisms and toxins associated with seafood and shellfish poisoning (adapted from ICMSF 1998).

Disease	Microorganism	Toxin	Incriminated seafood
Paralytic shellfish poisoning	*Alexandrium catenella* *Alexandrium tamarensis* Other *Alexandrium* spp. *Pyrodinium bahamense* *Gymnodinium catenatum*	Saxitoxin Neosaxitoxin Gonyautoxins Other saxitoxin derivatives	Mussels, oysters, clams, planktonivorous fish
Diarrhoeic shellfish poisoning	*Dinophysis fortii* *Dinophysis acuminata* *Dinophysis acuta* *Dinophysis mitra* *Dinophysis norvegica* *Dinophysis sacculus* *Prorocentrum lima* Other *Prorocentrum* spp.	Okadaic acid Dinophysis toxin Pectenotoxin Yessotoxin	Mussels, scallops, clams, oysters
Neurotoxic shellfish poisoning	*Gymnodinium breve*	Brevetoxins	Oysters, mussels, clams, scallops
Amnesic shellfish poisoning	*Pseudonitzschia pungens*	Domoic acid	Mussels
Ciguatera	*Gambierdiscus toxicus* *Ostreopsis lenticularis*	Ciguatoxin Maitotoxin Scaritoxin	Reef-associated fish
Scombroid poisoning	*Morganella morganii, Proteus* spp., *Hafnia alvei, Klebsiella pneumoniae,* and other bacteria capable of decarboxylating amino acids to biogenic amines	Histamine and other biogenic amines	Scombroid fish species, Mahi mahi, bluefish, tuna, sardines

Ingestion of contaminated shellfish results in a wide variety of symptoms, depending upon the toxins present, their concentrations in the shellfish and the amount of contaminated shellfish consumed. Paralytic shellfish poisoning is better characterised than the symptoms associated with diarrhoeic shellfish poisoning, neurotoxic shellfish poisoning and amnesic shellfish poisoning. All shellfish (filter-feeding molluscs) are potentially toxic. Paralytic shellfish poisoning is generally associated with a range of shellfish: mussels, clams, cockles and scallops. Neurotoxic shellfish poisoning is associated with shellfish harvested along the Florida coast and the Gulf of Mexico. Diarrhoeic shellfish poisoning is associated with mussels, clams, oysters and scallops, whereas amnesic shellfish poisoning is only associated with mussels.

5.4.1 Ciguatera poisoning

Ciguatera poisoning is characterised by:

- Prickling of the lips, tongue and throat
- Headache
- Severe pain in arms, legs and eyes
- Impaired vision
- Skin disorders: blisters, stinging sensation and erythema

The majority of cases recover within days to weeks and mortality is low. A lipid-soluble ciguatera toxin has been identified (Fig. 5.2; Murata *et al.* 1990). This polyether is similar in structure to the brevetoxins and is known to affect thermoregulation and sensory, motor, autonomic and muscular activities. The scaritoxin, which is less potent than ciguatera toxin, may actually be a derivative of ciguatera toxin. A third toxin, called maitotoxin, has been implicated in scombroid poisoning. It activates Ca^{2+} channels, releases neurotransmitters and increases the contraction of smooth, cardiac and skeletal muscle. Since there is a variety of symptoms it is possible that a range of toxins is involved in the illness.

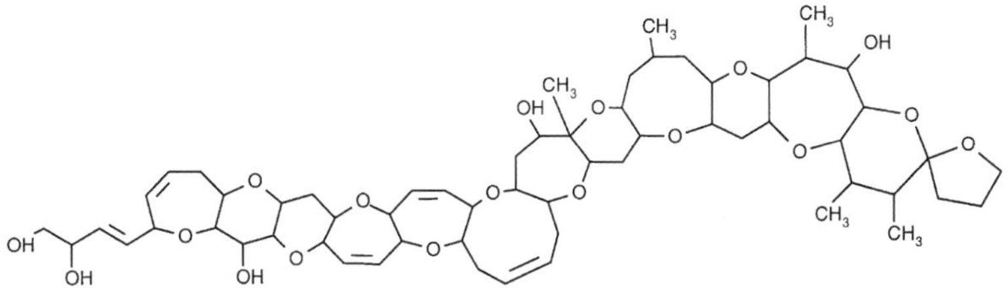

Fig. 5.2 Ciguatoxin structure.

5.4.2 Scombroid poisoning

Scombroid poisoning symptoms are:

- Metallic, sharp or peppery taste in the mouth
- Intense headache
- Dizziness
- Nausea and vomiting
- Facial swelling and flushing
- Epigastric pain
- Rapid and weak pulse
- Itching skin
- Burning throat and difficulty in swallowing

Usually recovery is within 12 hours. Fatalities are rare and are usually due to other predisposing factors. Initially it was believed that the symptoms were caused by histamine intoxication, the histamine being bacterially produced during storage. However, studies on human volunteers have failed to show a correlation between amounts of histamine in fish flesh and scombrotoxicosis. Therefore the causative agent is still unkown.

5.4.3 Paralytic shellfish poisoning

Paralytic shellfish poisoning symptoms are primarily neurological and include tingling, burning, numbness, drowsiness, incoherent speech and respiratory paralysis. Paralytic shellfish poisoning is due to 20 toxins which are all derived from saxitoxin produced by dinoflagellates (Fig. 5.3).

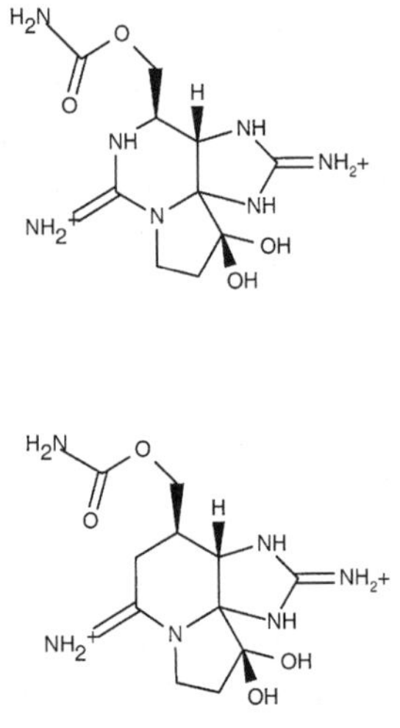

Fig. 5.3 The structure of saxitoxin (upper) and neosaxitoxin (lower).

5.4.4 Diarrhoeic shellfish poisoning

Diarrhoeic shellfish poisoning is normally a mild gastrointestinal disorder, that is, nausea, vomiting, diarrhoea and abdominal pain accompanied by chills, headache and fever. Onset of diarrhoeic shellfish poisoning may be as little as 30 minutes to 3 hours, depending on the dose of toxin ingested. The symptoms may last as long as 2 to 3 days. Recovery is complete with no after effects and the disease is generally not life-threatening. It is probably caused by high molecular weight polyethers, including okadaic acid, the dinophysis toxins, the pectenotoxins and yessotoxin produced by dinoflagellates.

5.4.5 Neurotoxic shellfish poisoning

Neurotoxic shellfish poisoning causes both gastrointestinal and neurological symptoms, including tingling and numbness of lips, tongue and throat, muscular aches, dizziness, reversal of the sensations of hot and cold, diarrhoea and vomiting. Onset of neurotoxic shellfish poisoning occurs within a few minutes to a few hours. The illness duration is fairly short, from a few hours to several days. Recovery is complete with few after effects; no fatalities have been reported. The poisoning is due to exposure to a group of polyethers called brevetoxins produced by dinoflagellates.

5.4.6 Amnesic shellfish poisoning

Amnesic shellfish poisoning is characterised by gastrointestinal disorders (vomiting, diarrhoea, abdominal pain) and neurological problems (confusion, memory loss, disorientation, seizure, coma). The gastroenteritis symptoms occur within 24 hours, whereas the neurological symptoms occur within 48 hours. The toxicosis is particularly serious in elderly patients, and includes symptoms reminiscent of Alzheimer's disease. All fatalities to date have involved elderly patients. The poisoning is caused by the presence of an unusual amino acid, domoic acid, as the contaminant of shellfish from diatoms (Fig. 5.4). Hence it is also known as domoic acid poisoning.

5.5 Foodborne pathogens: eucaryotes

5.5.1 Cyclospora cayetanensis

This coccidian parasite occurs in tropical waters worldwide and causes a watery and sometimes explosive diarrhoea in humans. The first known

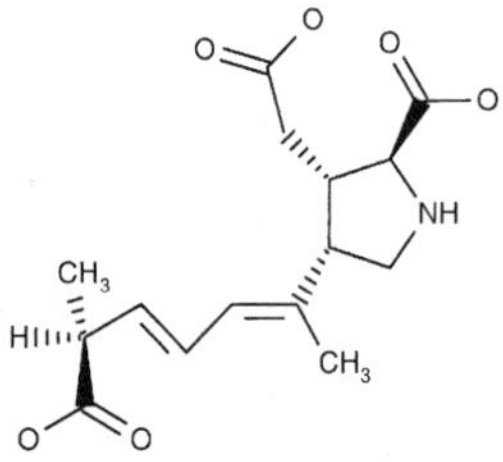

Fig. 5.4   The structure of domoic acid

human cases were reported in 1979. It was initially associated with waterborne transmission, but has also been linked to the consumption of raspberries, lettuce and fresh basil. *Cyclospora* infects the small intestine. The incubation period is 1 week after the ingestion of the contaminated food and the agent is shed in the faeces for more than 3 weeks. *Cyclospora* is spread by people who ingest contaminated water or food. The illness lasts from a few days to a month or longer. Relapses may occur one or more times.

Typical symptoms of cyclosporiasis are:

- Watery diarrhoea
- Frequent, sometimes explosive bowel movements
- Loss of appetite
- Substantial weight loss
- Bloating
- Increased gas and abdominal cramps
- Nausea
- Vomiting
- Muscle aches
- Low-grade fever
- Fatigue

5.5.2   Cryptosporidum parvum

The mode of transmission of ths coccidian protozoan is faecal to oral, including waterborne and foodborne means. The reservoirs include man and domestic animals, including cattle. Oocysts can survive in the environment for long periods of time, where they reman infective and are capable of resisting chemicals used to purify drinking water. They can, however, be removed from water supplies by filtration. Symptoms of crytosporosis in man include fever, diarrhoea, abdominal pain and anorexia. The disease usually subsides in less than 30 days, but may be prolonged in immunodeficient individuals and continue to death.

5.5.3 Anisakis simplex

Anisakiasis is an infection of the human intestinal tract caused by the ingestion of raw or undercooked fish containing larval stages of the nematodes *Anisakis simplex* or *Pseudoterranova decipiens*. Infections caused by the latter round worm are not a serious threat to human health, but those caused by *A. simplex* are more serious in that this agent penetrates the gastrointestinal tissue and causes disease that is difficult to diagnose. The primary hosts are warm-blooded marine mammals such as seals, walruses and porpoises. Their larvae pass via krill to fish such as cod, pollack, halibut, rockfish, flat fish, mackerel, salmon and herring.

5.5.4 Taenia saginata *and* T. solium

Tapeworm infections in man are caused by the beef tapeworm (*Taenia saginata*) and pork tapeworm (*T. solium*). Both tapeworms are obligate parasites of the human intestine. The organisms have a complex life cycle. The larval form is ingested in infected beef or pork meat and develops into the adult form (several metres in length) which attaches to the intestinal wall and produces hundreds of proglottids which are shed in the faeces. The proglottids produce eggs in the environment and in the intestine which are the main vector in cattle and pig infection. In the otherwise healthy adult, taeniasis is not severe and may not show any symptoms. Taeniasis is endemic in some countries such as Ethiopia, Kenya, Zaire, former Yugoslavia and central Asia. Breaking the life cycle of the organism is the main control measure, through thorough meat inspection and adequate cooking ($> 60°C$).

5.5.5 Toxoplasma gondii

T. gondii is the causative agent of toxoplasmosis which can be found in undercooked and raw meats such as pork, lamb, beef and poultry. The primary hosts are cats, and human infection takes place when contact is made with their faeces. This can also take place by the ingestion of raw or undercooked meat from intermediate hosts, such as rodents, swine, cattle, goats, chicken and birds. Toxoplasmosis in humans often produces mononucleosis-type symptoms, but transplacental infection can result in fetal death if it occurs early in pregancy. The organism causes hydrocephalus and blindness in children, the symptoms being less severe in adults. In immunocompromised individuals it can cause pneumonitis, myocarditis, meningoencephalitis, hepatitis or chorioretinitis, or combinations of these. Cerebral toxoplasmosis is often seen in AIDS patients. Proper cooking of meat will kill the organism. The incidence of the disease worldwide is unknown, but it is reported to be the most common parasitic infection in the UK.

5.5.6   Trichinella spiralis

T. spiralis causes trichinosis, also known as trichiniasis and trichinelliasis. It is mainly associated with the ingestion of contaminated pork. The organism is a round worm (nematode) which lives in the upper two-thirds of the small intestine. The female is viviparous, giving birth to living larvae which are deposited into the mucosa. About 1500 larvae are produced before the adult is expelled due to the host's immune system. The larvae are spread around the body via the bloodstream. Those which invade striated muscle (other than cardiac muscle) continue to develop. The larvae are digested by humans and subsequently invade the duodenal mucosa and become adult in 3 to 4 days, whereupon the life cycle continues.

The main symptoms of trichinosis are:

* First week, enteritis
* Second week, irregular fever (39-41°C), muscle pain, difficulty in breathing, talking or moving
* Third week, high fever, swollen eyelids, muscle pain
* Fourth week, fever and muscle pains subside

The larvae can be killed by a number of methods: heating to 65.5°C (150°F), freezing at −15°C (5°F) for 3 weeks or at −30°C (−22°F) for 1 day.

5.6   Mycotoxins

Mycotoxins are the toxic products of certain microscopic fungi which, in some circumstances, develop on or in foodstuffs of plant or animal origin. They are ubiquitous and widespread at all levels of the food chain. Hundreds of mycotoxins have been identified and are produced by some 200 varieties of fungi. Mycotoxins are secondary metabolites which have been responsible for major epidemics in man and animals. Mycotoxins are produced by the fungal genera *Aspergillus*, *Fusarium* and *Penicillium*. These fungi are ubiquitous and are part of the normal flora of plants. The aflatoxins (produced by *Aspergillus* spp.) range from single heterocyclic rings to six- or eight-membered rings (Fig. 5.5). *Penicillium* produces a range of 27 mycotoxins such as patulin (an unsaturated lactone) and penitrem A (nine adjacent rings composed of 4-8 atoms). Ergotism,

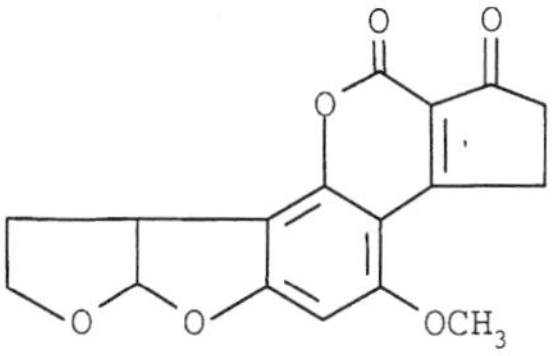

Aflatoxin B$_1$

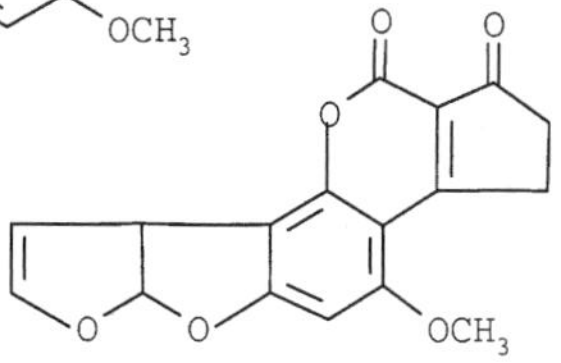

Aflatoxin B$_2$

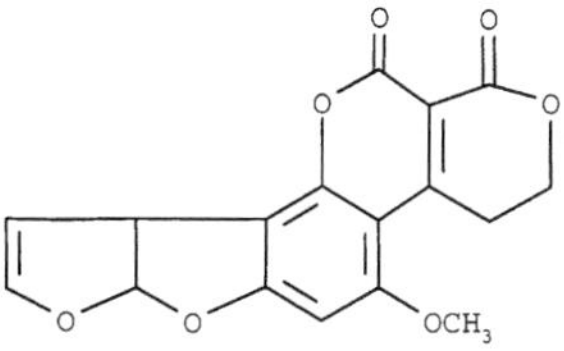

Aflatoxin G$_1$

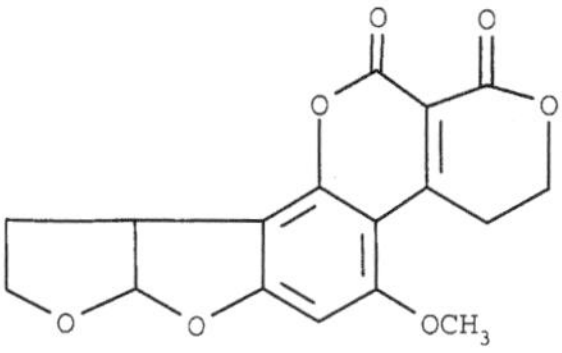

Aflatoxin G$_2$

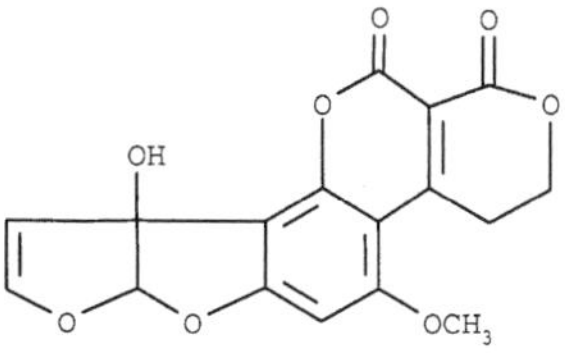

Aflatoxin M$_1$

Fig. 5.5 The structure of aflatoxins.

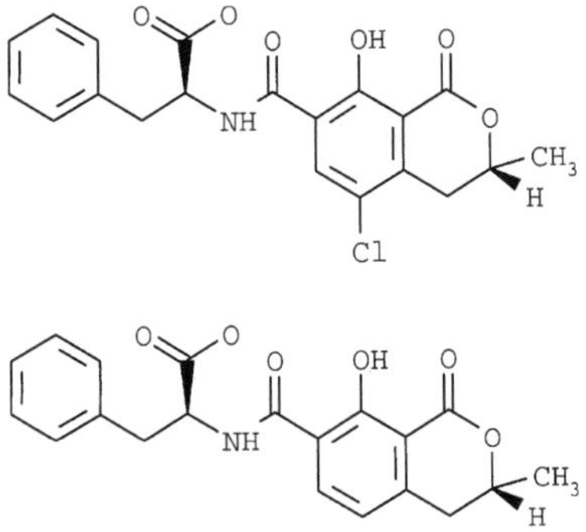

Ochrotoxin A (upper) and B (lower) structure

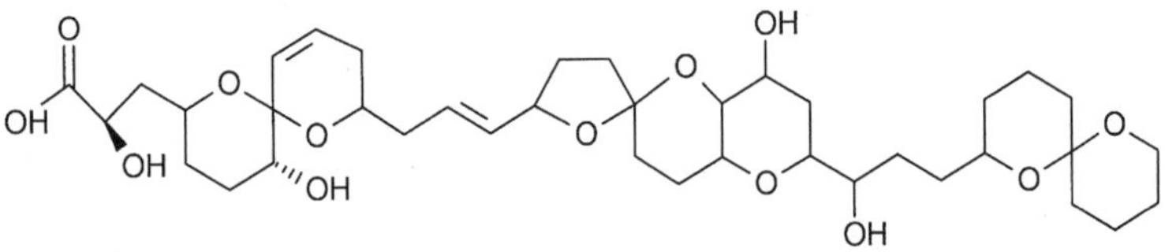

Okadaic acid structure

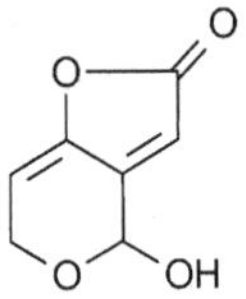

Patulin structure

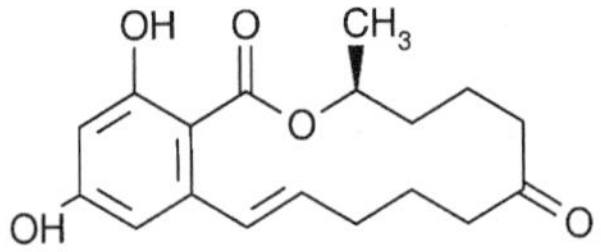

Zearalenone structure

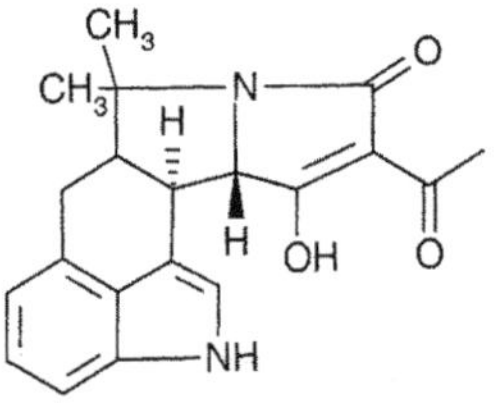

Cyclopaizonic acid structure

Fig. 5.6   The structure of mycotoxins, other than aflatoxins.

alimentary toxic aleukia, stachybotryotoxicosis and aflatoxicosis have killed thousands of humans and animals in the past century.

There are four types of toxicity:

• Acute, resulting in liver or kidney damage
• Chronic, resulting in liver cancer
• Mutagenic, causing DNA damage
• Teratogenic, causing cancer in the unborn child

5.6.1 Aflatoxins

The aflatoxins have been studied in more detail than other mycotoxins. The aflatoxins are a group of structurally related toxic compounds produced by certain strains of the fungi *Asp. flavus* and *Asp. parasiticus* under favourable conditions of temperature and humidity. These fungi grow on certain foods and feeds, resulting in the production of aflatoxins. The most pronounced contamination has been encountered in tree nuts, peanuts and other oilseeds, including corn and cottonseed. The major aflatoxins of concern are designated B_1, B_2, G_1 and G_2 by the blue (B) or green (G) fluorescence given when viewed under an ultraviolet lamp. These toxins are usually found together in various foods and feeds in various proportions. However, aflatoxin B_1 is usually predominant and is the most toxic. When a commodity is analysed by thin-layer chromatography, the aflatoxins separate into the individual components in the order given above; however, the first two fluoresce blue when viewed under ultraviolet light and the second two fluoresce green. Aflatoxin M, a major metabolic product of aflatoxin B_1 in animals, is usually excreted in the milk and urine of dairy cattle and other mammalian species that have consumed aflatoxin-contaminated food or feed.

Aflatoxins produce acute necrosis, cirrhosis and carcinoma of the liver in a number of animal species; no animal species is resistant to the acute toxic effects of aflatoxins, hence it is logical to assume that humans may be similarly affected. A wide variation in LD_{50} values has been obtained in animal species tested with single doses of aflatoxins. For most species, the LD_{50} value ranges from 0.5 to 10 mg/kg body weight. Animal species respond differently in their susceptibility to the chronic and acute toxicity of aflatoxins. The toxicity can be influenced by environmental factors, exposure level and duration of exposure, age, health and nutritional status of diet. Aflatoxin B_1 is a very potent carcinogen in many species, including nonhuman primates, birds, fish and rodents. In each species, the liver is the primary target organ of acute injury. Metabolism plays a major role in determining the toxicity of aflatoxin B_1. This aflatoxin requires metabolic activation to exert its carcinogenic effect and these effects can be modified by induction or inhibition of the mixed function oxidase system.

In well-developed countries, aflatoxin contamination rarely occurs in foods at levels that cause acute aflatoxicosis in humans. In view of this, studies on human toxicity from ingestion of aflatoxins have focused on their carcinogenic potential. The relative susceptibility of humans to aflatoxins is not known, even though epidemiological studies in Africa and Southeast Asia, where there is a high incidence of hepatoma, have revealed an association between cancer incidence and the aflatoxin content of the diet. These studies have not proved a cause–effect relationship, but the evidence suggests an association.

The discovery of aflatoxins in the 1960s led to extensive surveying of koji moulds (Section 4.4.7) for mycotoxin production. Although under laboratory conditions mycotoxins can be produced by *A. oryzae*, *A. sojae* and *A. tamari*, no aflatoxins have been demonstrated in commercial production strains (Table 4.1; Trucksess *et al.* 1987). The moulds used in cheese manufacture have also been tested for toxin production. *P. roqueforti* produces trace amounts of patulin, roquefortine C, whereas *P. camembertii* produces low levels of cyclopiazonic acid. These toxins are only produced under laboratory induced stress conditions and it is reported that levels in cheese are 'extremely low' (Rowan *et al.* 1998).

5.6.2 Ochratoxins

Ochratoxins are produced by *A. ochraceus*, *Penicillium verrucosum* and *P. viridicatum*. Ochratoxin A is the most potent of these toxins. The main dietary sources are cereals, but significant levels of contamination may be found in grape juice and red wine, coffee, cocoa, nuts, spices and dried fruits. Contamination may also carry over into pork and pig blood products and into beer. Ochratoxin is potentially nephrotoxic and carcinogenic, the potency varying markedly between species and sexes. It is also teratogenic and immunotoxic.

5.6.3 Fumonisins

Fumonisins are a group of *Fusarium* mycotoxins occurring worldwide in maize and maize-based products. Their casual role in several animal diseases has been established. Available epidemiological evidence has suggested a link between dietary fumonisin exposure and human oesophageal cancer in some locations with high disease rates. Fumonisins are mostly stable during food processing.

5.6.4 Zearalenone

Zearalenone is a fungal metabolite mainly produced by *Fusarium graminearium* and *F. culmorum*, which are known to colonise maize, barley, wheat, oats and sorghum. These compounds can cause hyperoestrogenism and severe reproductive and infertility problems in animals, especially in swine, but their impact in public health is hard to evaluate.

5.6.5 Trichothecenes

Trichothecenes are producd by many species of the genus *Fusarium*. They occur worldwide and infect many different plants, notable of which are the cereal grains, especially wheat, barley and maize. There are over 40 different trichothecenes but the best known are deoxynivalenol and nivalenol. In animals they cause vomiting and feed refusal, but also affect the immune system. In humans they cause vomiting, headache, fever and nausea.

Control of mycotoxins is very difficult as it is due to preharvest invasion by the fungi through seeds, soil or even air. Adequate drying and storage are useful provided there is good farm management practice beforehand. Ultraviolet screening procedures (for fluorescence of the aflatoxins) are of use in corn, cottonseed and figs, but not peanuts since they autofluoresce.

5.7 Emerging and uncommon foodborne pathogens

Some emerging foodborne diseases are well characterised, but are considered 'emerging' because the reporting of them has recently (in the last 10 to 15 years) become more common.

Emerging pathogens and toxins include:

- Bacteria: *E. coli* O157:H7, enteroaggregative *E. coli* (EAEC), *V. cholerae*, *V. vulnificus*, *Strep. parasanguinis*, *Mycobacterium paratuberculosis*, *L. monocytogenes*, *S. typhimurium* DT104, *S. enteritidis*, *C. jejuni*, *Arcobacter* spp., *Enterobacter sakazakii*.
- Viruses: hepatitis E, Norwalk virus and Norwalk-like virus.
- Protozoa: *Cyclospora cayetanensis*, *Toxoplasma gondi*, *Cryptosporidium parvum*.
- Helminths: *Anisakis simplex* and *Pseudoterranova decipiens*.
- Prions: bovine spongiform encephalitis, new variant CJD.
- Mycotoxins: fumonisins, zearalenone, trichothecenes, ochratoxins.

Further descriptions of these food pathogens and toxins can be found in Sections 3.8-9 and 5.2-6.

The emergence of certain food pathogens is due to a number of causes:

- Recent appearance in the microbial population.
- Dispersal to new vehicles of transmission.
- Rapidly increasing incidence or geographic range, e.g. *V. cholerae* in southern USA coastal waters in 1991.
- Recent identification due to increased knowledge or methods of identification, although previously widespread.

(adapted from Van de Venter 1999.)

The following factors can affect the epidemiology of emerging food pathogens:

- Microbial adaptation through natural selection; antibiotics usage can select for antibiotic resistant strains, e.g. *S. typhimurium* DT104.
- New foods and food preparation technologies, e.g. BSE and nvCJD.
- Changes in lifestyle such as an increased consumption of 'convenience food' and subsequent risk to *L. monocytogenes*.
- Increasing international trade and travel, facilitating the rapid spread of pathogens worldwide, e.g. *E. coli* O157:H7.
- Recognition of new food vehicles of transmission, e.g. *My. para-tuberculosis* (plausible).

5.7.1 Prions

Transmissible spongiform encephalopathies in animals and humans are caused by an unconventional virus or prion. These conditions include scrapie in sheep, bovine spongiform encephalopathy ('mad cow disease') in cattle and Creutzveld Jacob disease in humans. It is commonly accepted that BSE was first caused in Britain when cattle were fed carcass meal from scrapie-infected sheep. It is also accepted that humans contracted the nonclassic form of CJD called 'new variant CJD' (nvCJD) after consuming cattle meat, in particular nerve tissue.

On 20 March 1996 the Secretary of State for Health in the British Government announced the most likely cause of 10 cases of nvCJD in humans was the ingestion of meat from cattle suffering from BSE (Will *et al.* 1996). This resulted in the Specified Offal Ban, which stopped the recycling of potentially infectious material (such as the spinal cord) to cattle through food supplements and required more inspections of abattoirs. These restrictions together with the decrease in BSE cases means that there should now be almost no exposure of the UK public to the infectious agent.

Evidence that the infectious agent was a prion (abbreviation for 'proteinaceous infectious particle') came from the studies of Collinge *et al.* (1996), Bruce *et al.* (1997) and Hill *et al.* (1997). Prions are modified forms of a normal protein called PrP^c which is referred to as PrP^* or PrP^{Sc}. The proteins accumulate in the brain, causing holes or plaques and the subsequent clinical symptoms leading to death. The ultimate number of nvCJD cases in the UK is a matter of controversy. Some groups claim the number cannot be estimated (Ferguson *et al.* 1999) while Thomas & Newby (1999) estimate that the value will not exceed 'a few hundred, and is most likely to be a hundred or less' and Cousens *et al.* (1997) estimate a total of 80 000. Unfortunately, since the disease has an unknown incubation period and no diagnostic test for infection is available, only time will tell who is closest to the true number of nvCJD cases in the forthcoming years.

5.7.2 Enterobacter sakazakii: *dried infant formula*

Ent. sakazakii was designated a unique species in 1980 and was previously known as 'yellow-pigmented *Ent. cloacae*' (NararowecWhite & Farber 1997a). It has been implicated in a severe form of neonatal meningitis with a high mortality rate (40–80%). Dried infant formula has been implicated in two outbreaks and sporadic cases of *Ent. sakazakii* meningitis in Canada. Only one study has identified an environmental source (soil) for the organism (Wang *et al.* 1997). *Ent. sakazakii* has been isolated from 0–12% of dried infant formula samples (NararowecWhite & Farber 1997b). Minimum growth temperatures were 5.5 to 8.0°C. Generation times were 40 minutes at 23°C and 4.98 hours at 10°C. There was no growth at 4°C. A heat treatment of 68°C for 16 seconds gives a 5-log reduction in viable count (Section 2.4.2) and the organism is less heat resistant than *L. monocytogenes* (NararowecWhite *et al.* 1999).

5.7.3 Mycobacterium paratuberculosis *and milk, an emerging pathogen?*

Johne's disease is a chronic enteritis of cattle (and to a lesser extent of sheep and goats) and is caused by the mycobacterium *My. paratuberculosis* (commonly referred to as 'MparaTB'). The symptoms are diarrhoea, weight loss, debilitation and, since it is incurable, death. The prevalence of Johne's disease in the USA is 2.6% of the dairy herd, this is comparable to the 2% of cattle being clinically infected in England

(Çetinkaya *et al.* 1996). *My. paratuberculosis* is excreted at 10^8 cfu/g and has been isolated from milk of asymptomatic carriers at a level of 2–8 cfu/ 50 ml milk (Sweeney *et al.* 1992). It has been reported that the organism can survive pasteurisation (Chiodini & Hermon-Taylor 1993; Grant *et al.* 1996; Stabel *et al.* 1997). Hence it has been proposed that the organism is the causative agent of the human equivalent called 'Crohn's disease'. Crohn's disease is a gastrointestinal disease of humans in which the bowel becomes inflamed. The whole gastrointestinal tract may become infected and surgery required. This proposal is controversial because (i) the organism has not been detected from commercial pasteurised milk sources and (ii) it is not the consensus of opinion that Crohn's disease is caused by *My. paratuberculosis*. If it is proven, by research (still on-going at the publication of this book), that *My. paratuberculosis* is transferred from infected cattle to humans, resulting in Crohn's disease via pasteurised milk, then the standard pasteurisation time and temperature will have to be re-evaluated.

The problem with studying this organism is that it is very difficult to grow in the laboratory. Colony formation is only just visible after 4 weeks' incubation and requires confirmatory tests. Hence large-scale experiments are extremely time-consuming. The results of a large-scale surveillance of pasteurised milk in Northern Ireland should be available in 2000.

5.7.4   Arcobacter *genus*

The *Arcobacter* genus was formerly known as aerotolerant campylobacters and campylobacter-like organisms (CLO). Nowadays the organism has been formally recognised as a separate genus (Vandamme *et al.* 1991) and divided into four species: *A. butzleri, A. cryaerophilus, A. skirrowii* and *A. nitrofigilis*. *A. butzleri, A. cryaerophilus* and *A. skirrowii* are veterinary pathogens causing porcine abortions, whereas *A. nitrofigilis* has been reported only from the roots of *Spartina alterniflora*, a salt marsh plant. *A. butzleri* serotypes 1 and 5 are regarded as the primary human pathogens, however no epidemiological studies have yet shown the transmission of the organism through the food chain to humans. The situation is, however, reminiscent of *C. jejuni* and *L. monocytogenes* being initially recognised as veterinary pathogens.

Arcobacters appear resistant to antimicrobial agents typically used in the treatment of diarrhoeal illness caused by *Campylobacter* spp., for example erythromycin, other macrolide antibiotics, tetracycline and chloramphenicol. Isolation of arcobacters requires selective media such as mCCDA and CAT. Identification can subsequently be achieved using 16S rRNA probes (Mansfield & Forsythe 2000).

6

METHODS OF DETECTION

6.1 Prologue

Analysing food and environmental samples for the presence of food poisoning and food spoilage bacteria, fungi and toxins is standard practice for ensuring food safety and quality. The interpretation of results in food microbiology is far more difficult than is normally appreciated and the issue of sampling plans and statistical representation of samples is covered in Chapter 8. The reasons for caution in interpreting results are:

- Microorganisms are in a dynamic environment in which multiplication and death of different species occur at differing rates. This means that the result of a test is only valid for the time of sampling.
- Viable counts by plating out dilutions of food homogenate onto agar media can be misleading if no microorganisms are cultivated yet preformed toxins or viruses are present. For example, staphylococcal enterotoxin is very heat stable and will persist through the drying process in the manufacture of powdered milk.
- Homogeneity of food, however, is rare, especially with solid foods. Therefore the results for one sample may not necessarily be representative of the whole batch. However, it is not possible to subject a whole batch of the food to examination for microorganisms as there would be no product left to sell.
- Colony counts are only valid within certain ranges and have confidence limits (Table 6.1).

Because of the reasons above, microbiological counts obtained through random sampling can only form a small part of the overall assessment of the product.

There are a number of issues related to the recovery of microorganisms from food which must be addressed in any isolation procedure:

The Microbiology of Safe Food

Table 6.1   Confidence limits associated with numbers of colonies on plates (Cowell and Morisetti, 1989).

Colony count	95% confidence intervals for the count	
	Lower	Upper
3	< 1	9
5	2	12
10	5	18
12	6	21
15	8	25
30	19	41
50	36	64
100	80	120
200	172	228
320	285	355

(1)   If solid food, then a liquidised homogenate is necessary for dilution purposes.
(2)   The target organism is normally in the minority of the microbial population.
(3)   The target organism is present at low levels.
(4)   The target organism may be physically and metabolically injured.
(5)   The target organism may not be uniformly distributed in the food.
(6)   The food may not be of a homogenous composition

Plate counts are obtained for three purposes and groups of organisms:

(1)   The basic aerobic plate count (APC) indicates the general microbial load and hence the shelf life of the product. The APC is very useful in the food industry as the technique is easy to perform and can provide a threshold for acceptance or rejection decisions for samples taken regularly at the same point under the same conditions.
(2)   The presence of faecal organisms (i.e. coliforms) indicates whether the food has been inadequately heat processed or has been mis-handled and contaminated post-processing.
(3)   Specific pathogens may be associated with the raw ingredients of processed food.

A degree of assurance is only obtained when tests on uniform quantities of representative samples of the food by standard methods prove negative. The methods therefore must be reliable, robust and accredited. These aspects are considered in the following sections. Only representative examples of detection methods will be covered; full details of detection methods can be found in various sources and the reader should consult the most recent edition of these for up-to-date protocols.

Useful sources of approved protocols include:

- Association of Official Analytical Chemists, *Bacteriological Analytical Manual* (AOAC 1992)
- *Compendium of Methods for the Microbiological Examination of Foods* (Vanderzant & Splittstoesser 1992)
- *Practical Food Microbiology* (Roberts *et al.* 1995).

Methods of detection are often categorised into two groups: conventional and rapid. The terms are, however, misleading as some 'rapid' methods actually take 24 hours for a result to be obtained. Conventional refers to procedures that are in common use; they frequently involve homogenising the food sample, preparing a dilution series and inoculating specific agar plates for colony formation and subsequent enumeration. These methods may also be referred to as 'traditional methods'. Rapid methods are alternatives to the conventional method and are designed to obtain the end result in less laboratory time. This is highly desirable in the food industry, however the technique may be more costly and require more highly trained personnel.

6.2 Conventional methods

A number of steps are required in order to isolate a target organism from food:

(1) Homogenise solid ingredients/food using a Stomacher™ or Pulsifyer™ machine.
(2) Enrich the target organisms using enrichment media which encourages the growth of the target organisms and suppresses the growth of other microorganisms.
(3) The amount of food analysed is often in the range 1–25 g
(4) A pre-enrichment step may be required before (2) which allows all injured cells to repair their damaged membranes and metabolic pathways.
(5) Representative samples are required to test the batch of ingredients/food.
(6) Where practical, homogenise the food before sampling. Otherwise take representative samples from the different phases (liquid/solid).

Conventional methods are frequently plate counts obtained from homogenising the food sample, diluting and inoculating specific media to detect the target organism (Fig. 6.1). The first step is normally to prepare a 1:10 dilution of the food; typically 25 g food and 225 ml diluent. The sample is usually homogenised (Stomacher™ or Pulsifyer™) in order to release attached microorganisms from the food surface. The methods are very sensitive, relatively inexpensive (compare to rapid methods) but require incubation periods of at least 18 to 24 hours.

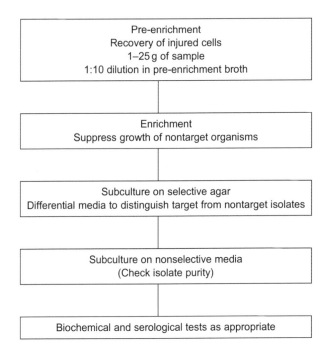

Fig. 6.1 General sequence of isolation of foodborne pathogens.

The target organism, however, is often in the minority of the food microbial flora and may be sublethally injured (Section 6.2.1) due to processing (cooking, etc.). Therefore the above procedure is frequently modified to allow a recovery stage for sublethally injured cells, or to enrich for the target organism. Therefore the recovery of *Salmonella* spp. from ready-to-eat foods is in stages: pre-enrichment, enrichment, selection and detection (see later, Fig. 6.14). As referred to above this approach is 'bacteriological' rather than 'microbiological' in that the presence of toxins, protozoa and viruses will not be revealed.

Specific examples of methods for the detection of key target organisms are given in Section 6.6.

6.2.1   Sublethally injured cells

Sublethal injury implies damage to structures within the cells which causes some loss or alteration of cellular functions and the leakage of intracellular material, making them susceptible to selective agents. Changes in cell wall permeability can be demonstrated by the leakage of compounds from the cytoplasm (increased absorbance at 260 nm of culture supernatants) and the influx of compounds such as ethidium bromide and propidium iodide.

'Metabolic' injury is often taken as the inability to form colonies on minimal salt media while retaining colony forming ability on complex nutrient media, whereas 'structural' injury can be taken as the ability to proliferate or survive in media containing selective agents that have no apparent inhibitory action upon nonstressed cells. Injury is reversible by repair, but only if the cells are exposed to favourable resuscitation conditions such as a nonselective, nutrient-rich medium under optimal growth conditions.

In practical analytical food microbiology the phenomenon of injury may present considerable problems, as many of the physical treatments, including heat, cold, drying, freezing, osmotic activity and chemicals (disinfectants, etc.), may generate injured cells causing variations in plate counts. The injured cells may remain undetected as selective media usually contains ingredients such as increasing salt concentrations, deoxycholate lauryl sulphate, bile salts, detergents and antibiotics. The injured cells are 'viable' but are not metabolically active enough to achieve cell division. Subsequently, microbiological examination for quality control can indicate low plate counts, when in fact the sample contains a high number of injured cells. An example of the difference between plate counts on selective and nonselective agar can be seen in Fig. 2.9, where food pathogens have been exposed to high pressure.

In food and beverage products, once the stress-causing injury is removed, these injured cells are often able to recover. The cells regain all of their normal capabilities, including pathogenic and enterotoxin properties. Therefore important food poisoning organisms may be undetected by analytical testing, but may cause a major food poisoning outbreak. For these reasons, substantial efforts need to be made to develop improved analytical procedures that will detect both injured and uninjured cells.

In salmonella detection (Section 6.5.2) the sample is incubated overnight in buffered peptone water (BPW) or lactose broth to allow injured salmonella to recover and multiply to detectable levels. However, it is uncertain whether BPW is the best recovery medium since other organisms can suppress the growth of low numbers of salmonellae and there is also a problem of 'how do you know injured salmonellae are present if you do not detect a colony on a plate?'.

For other organisms which might be sublethally injured it has been recommended that food samples should be resuscitated in a noninhibitory medium for an hour or two, allowing injured cells to resuscitate yet prevent the population size from increasing. This generalised approach is far from optimised and leaves plenty of opportunity for oversight in the detection of potentially pathogenic food poisoning organisms. Hence such techniques need to be validated urgently.

6.2.2   Viable but nonculturable bacteria (VNC)

It has been proposed that many bacterial pathogens are able enter a dormant state (Dodd *et al.* 1997). In this state the cells are not culturable yet remain viable (as demonstrated by substrate uptake) and virulent. Hence the term 'viable but nonculturable' or VNC was derived. This phenomenon has been shown in *Salmonella* spp., *C. jejuni, E. coli* and *V. cholerae*. For example, in the human intestine previously nonculturable vibrios were shown to regain their ability to multiply (Colwell *et al.* 1996). Therefore VNC bacterial pathogens pose a potential threat to health and are of considerable concern in food microbiology since a batch of food might be released due to the negative presence of pathogens, yet contain infectious cells.

The VNC state may be induced due to a number of extrinsic factors such as temperature changes, low nutrient level, osmotic pressure, water activity and pH. Of these the most important factor seems to be temperature changes. Hence current methods may not be recovering all the pathogens from foods and water. Therefore alternative end-detection methods need to be developed, such as those based on immunology (ELISA) and DNA sequences (PCR).

The VNC concept is not accepted by all microbiologists. Some argue that it is a matter of time before we design the most appropriate recovery media and others that the cells have self-destructed due to an oxidative burst causing DNA damage (Barer 1997; Bloomfield *et al.* 1998; Barer *et al.* 1998).

6.3   Rapid methods

Conventional procedures are by nature labour intensive and time-consuming. Therefore a plethora of alternative, rapid methods has been developed to shorten the time between taking a food sample and obtaining a result. These methods aim either to replace the conventional enrichment step with a concentration step (for example immunomagnetic separation) or to replace the end-detection method with one that requires a shorter time period (for example impedance microbiology and ATP bioluminescence).

Major improvements have been in three areas:

(1) Sample preparation
(2) Separation and concentration of target cell, toxins or viruses
(3) End detection

Sometimes a rapid technique will involve one or more of the above aspects, for example the hydrophobic grid membrane both concentrates the organisms and enumerates on specific detection agar media.

6.3.1 Sample preparation

Agar slides containing selective or nonselective agar can be pressed against the surface to be examined and directly incubated. This obviates the need for sampling and the errors inherent in releasing organisms from cotton wool swabs.

Another improvement in recent years in sample preparation is the automatic diluter. This enables the operator to take a food sample of approximately 25 g, and then an appropriate volume of diluent is added to give an accurate 1:10 dilution factor.

6.3.2 Separation and concentration of target

Separation and concentration of target organisms, toxin or viruses, can shorten the detection time and improve specificity of a test procedure. Common methods include:

• Immunomagnetic separation (IMS)
• Direct Epifluorescent Filter Technique (DEFT)
• Hydrophobic grid membrane

Immunomagnetic separation (IMS)
Immunomagnetic separation uses superparamagnetic particles coated with antibodies against the target organism. Hence the target organism is 'captured' in the presence of a mixed population due to the antigen-antibody specificity. This removes the need for an enrichment broth incubation period. A generalised procedure is given in Fig. 6.2. Commercially available IMS kits target key food and water pathogens: *Salmonella* spp., *E. coli* O157:H7, *L. monocytogenes* and *Cryptospor-*

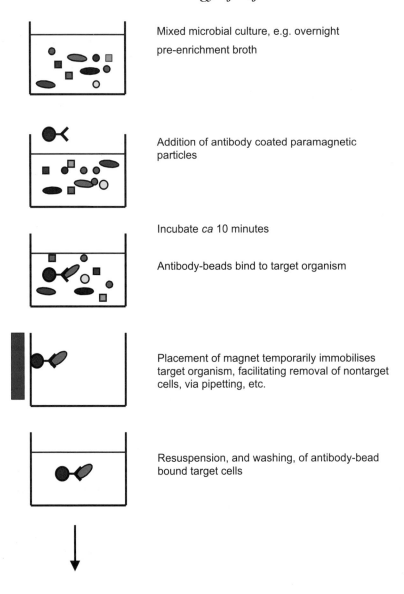

Mixed microbial culture, e.g. overnight pre-enrichment broth

Addition of antibody coated paramagnetic particles

Incubate *ca* 10 minutes

Antibody-beads bind to target organism

Placement of magnet temporarily immobilises target organism, facilitating removal of nontarget cells, via pipetting, etc.

Resuspension, and washing, of antibody-bead bound target cells

Pipetted onto selective media, ELISA, DNA techniques, etc.

Fig. 6.2   Immunomagnetic separation technique.

idium (Table 6.2). IMS can enrich for sublethally injured microorganisms which would otherwise be missed using the standard enrichment broth and plating procedures. These organisms might be killed in the enrichment broth due to changes in cell wall permeability (Section 6.2.1). Dead cells can be detected using a combined IMS and PCR procedure. For reviews of IMS in medical and applied microbiology see Olsvik *et al.* (1994) and Safarik *et al.* (1995).

Table 6.2 Applications of immunomagnetic separations (adapted from Safarik *et al.* 1995).

Organism	Application
E. coli O157	Food and water microbiology
Salmonella spp.	
Listeria monocytogenes	
St. aureus	
Cryptosporidium parvum	
Legionella spp.	
Yersinia pestis	Clinical microbiology
Chlamydia trachomatis	
HIV	
Erwinia chrysanthemi	Plant pathogen detection
Er. carotovora	
Sac. cerevisiae	Biotechnology
Mycobacterium spp.	

The IMS salmonella detection method is as efficient as the selenite broth selection stage, which is the most efficient of the BSI/ISO procedures (Mansfield & Forsythe 1996, 2000; Section 6.5.2). The selective enrichment step (overnight incubation) is replaced by the immunomagnetic separation (10 minutes). Hence the technique reduces the total time required for sampling and detection by one day. In addition IMS can have a greater recovery of stressed salmonellae than BSI/ISO protocols.

Direct Epifluorescent Technique (DEFT) and Hydrophobic Grid Membrane
Membrane filters can be used to shorten the overall detection time because:

(1) They can concentrate the target organism from a large volume to improve detection limits.
(2) They remove growth inhibitors.

(3)   Organisms may be transferred to a different growth medium without physical injury through centrifugation and resuspension.

The membranes can be made from nitrocellulose, cellulose acetate esters, nylon, polyvinyl chloride and polyester. Because they are only 10 μm in thickness they can be directly mounted on a microscope and the cells visualised.

The DEFT method concentrates cells on a membrane before staining with acridine orange (Fig. 6.3). Acridine orange fluoresces red when

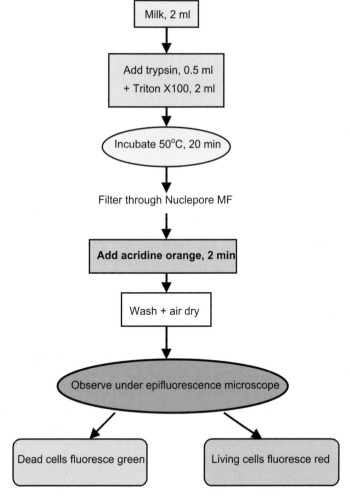

Fig. 6.3   Direct Epifluorescence Filter Technique (DEFT) for the detection of bacteria in milk.

interchelated with RNA, and green with DNA. Subsequently viable cells fluoresce orange-red whereas dead cells fluoresce green.

The DEFT count has gained acceptance as a rapid, sensitive method for enumerating viable bacteria in milk and milk products. The count is completed in 25 to 30 minutes and detects as few as 6×10^3 bacteria per ml in raw milk and other dairy products, which is 3 to 4 orders of magnitude better than direct microscopy. Because it is a microscopic technique, one is able to distinguish whether the microorganisms present are yeasts, moulds or bacteria.

The hydrophobic grid membrane filter (HGMF) is a filtration method which is applicable to a wide range of microorganisms (Entis & Lerner 1997). The pre-filtered food sample (to remove particulate matter $>5\,\mu m$) is filtered through a membrane filter which traps microorganisms on a membrane in a grid of 1600 compartments, due to hydrophobic effects. The membrane is then placed on an appropriate agar surface and the colony count determined after a suitable incubation period.

6.4   Rapid end-detection methods

Improvements in end-detection methods include:

- Better media design; chromogenic and fluorogenic substrates; motility enrichment; dipslides; Petrifilm system.
- Immunoassays; enzyme-linked immunosorbent assay (ELISA); latex agglutination.
- Impedance microbiology, also known as conductance microbiology.
- ATP bioluminescence.
- Gene probes linked to the polymerase chain reaction.

Improved media design
The advantage of incorporating fluorogenic and chromogenic substrates into growth media is that they generate brightly coloured or fluorescent compounds after bacterial metabolism. The main fluorogenic enzyme substrates are based on 4-methylumbelliferone such as 4-methylumbelliferyl-β-D-glucuronide (MUG; Fig. 6.4), whereas chromogenic substrates are commonly based on derivatives of phenol, for example 5-bromo-4-chloro-3-indolyl-β-D-glucuronide (BCIG).

Media such as the modified semi-solid Rappaport-Vassiliadis medium and diagnostic semisolid salmonella (DIASALM) agar have used bacterial motility as a means to enrich for the target organism. This principle has been applied to the improved detection of *Salmonella* serovars (as per

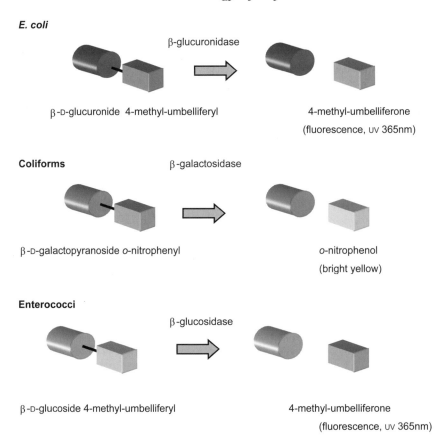

E. coli

β-glucuronidase

β-D-glucuronide 4-methyl-umbelliferyl

4-methyl-umbelliferone
(fluorescence, UV 365nm)

Coliforms   β-galactosidase

β-D-galactopyranoside o-nitrophenyl

o-nitrophenol
(bright yellow)

Enterococci

β-glucosidase

β-D-glucoside 4-methyl-umbelliferyl

4-methyl-umbelliferone
(fluorescence, UV 365nm)

Fig. 6.4   Fluorogenic substrates for specific detection of food pathogens.

above examples), *Campylobacter* spp. and the potentially emerging pathogen *Arcobacter* (de Boer *et al.* 1993, 1996; Wesley 1997). The semi-solid Rappaport medium isolates motile salmonellae as they migrate through the medium ahead of competing organisms. This medium, however, will not isolate nonmotile salmonella strains.

The Petrifilm system (manufactured by 3M) is an alternative to the conventional agar plate. The system uses a dehydrated mixture of nutri-ents and gelling agent on a film. The addition of 1 ml of sample rehydrates the gel, which facilitates the colony formation of the target organism. Colony counts are performed as per the standard agar plate method. The throughput of samples is estimated to be double that of conventional agar plates. Petrifilm systems are available for aerobic plate counts, coliforms and *E. coli*.

Enzyme-linked immunosorbent assay and antibody-based detection systems

Enzyme-linked immunosorbent assays (ELISAs) are widely used in food microbiology. ELISA is most commonly performed using McAb coated microtitre trays to capture the target antigen (Fig. 6.5). The captured antigen is then detected using a second antibody which may be conjugated to an enzyme. Addition of a substrate facilitates visualisation of the target antigen. ELISA methods offer specificity and potential automation.

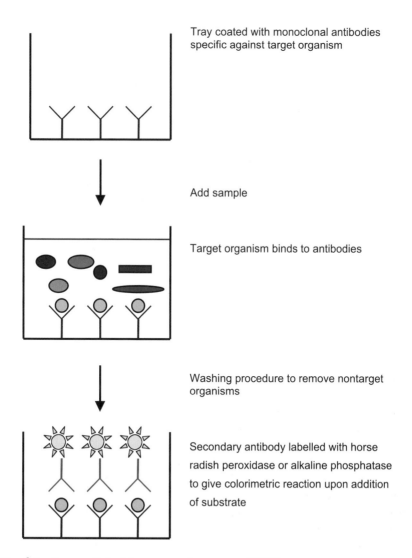

Tray coated with monoclonal antibodies specific against target organism

Add sample

Target organism binds to antibodies

Washing procedure to remove nontarget organisms

Secondary antibody labelled with horse radish peroxidase or alkaline phosphatase to give colorimetric reaction upon addition of substrate

Fig. 6.5 Enzyme-linked immunosorbent assay (ELISA).

A wide range of ELISAs is commercially available, especially for *Salmonella* spp. and *L. monocytogenes*. The technique generally requires the target organism to be 10^6 cfu/ml, although a few tests report a sensitivity limit of 10^4. Hence the conventional pre-enrichment, and even selective enrichment might be required prior to testing.

The VIDAS system (bioMerieux) has predispensed disposable reagent strips. The target organism is captured in a solid phase receptacle coated with primary antibodies and then transferred to the appropriate reagents (wash solution, conjugate and substrate) automatically. The end-detection method is fluorescence which is measured using an optical scanner. The VIDAS system can be used to detect most major food poisoning organisms.

Reversed passive latex agglutination
Reversed passive latex agglutination (RPLA) is used for the detection of microbial toxins such as the shiga toxins (from *Sh. dysenteriae* and EHEC), *E. coli* heat-labile (LT) and heat-stable (ST) toxins (Fig. 6.6). Latex particles are coated with rabbit antiserum which is reactive towards the target antigen. Therefore the particles will agglutinate in the presence of the antigen, forming a lattice structure. This settles to the bottom of a V-shaped microtitre well and has a diffuse appearance. If no antigen is present then a tight dot will appear.

6.4.1 Impedance (conductance) microbiology

Impedance microbiology is also known as conductance microbiology; impedance is the reciprocal of conductance and capacitance. It can rapidly detect the growth of microorganisms by two different methods (Fig. 6.7; Silley & Forsythe 1996):

(1) Directly due to the production of charged end products
(2) Indirectly from carbon dioxide liberation

In the direct method, the production of ionic end products (organic acids and ammonium ions) in the growth medium cause changes in the conductivity of the medium. These changes are measured at regular intervals (usually every 6 minutes) and the time taken for the impedance value to change is referred to as the 'time to detection'. The greater the number of organisms, the shorter the detection time. Hence a calibration curve is constructed and then the equipment can automatically determine the number of organisms in a sample.

The indirect technique is a more versatile method in which a potassium hydroxide bridge (solidified in agar) is formed across the electrodes. The test sample is separated from the potassium hydroxide bridge by a head

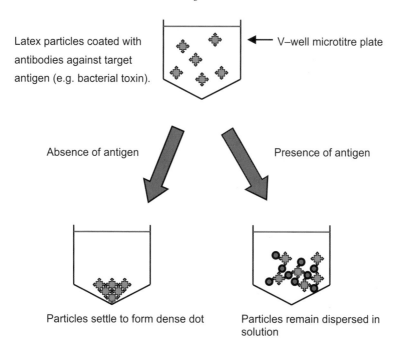

Appearance viewed from above microtitre plate:

Fig. 6.6   The principle of reversed passive latex agglutination (RPLA).

space. During microbial growth carbon dioxide accumulates in the head space and subsequently dissolves in the potassium hydroxide. The resultant potassium carbonate is less conductive and it is this decrease in conductance change which is monitored. Impedance changes of approximately 280 μS/μmol carbon dioxide are obtained at 30°C. The indirect technique is applicable to a wide range of organisms including *St. aureus, L. monocytogenes, Ent. faecalis, B. subtilis, E. coli, P. aeruginosa, A. hydrophila* and *Salmonella* serovars. Standard selective media or even an agar slant can be used for fungal cultures.

The time taken for a conductance change to be detectable ('time to detection') is dependent upon the inoculum size. Essentially the equip-

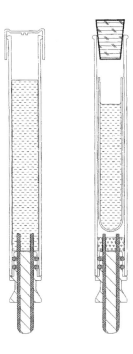

Fig. 6.7a Direct and indirect impedance tubes (diagram kindly supplied by Don Whitley Scientific Ltd, UK).

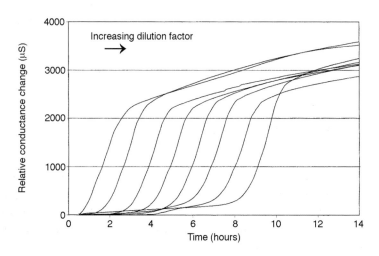

Fig. 6.7b Impedance curves obtained from *Escherichia coli* serial dilutions, direct method (data supplied by Don Whitley Scientific Ltd, UK).

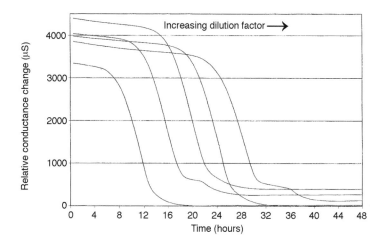

Fig. 6.7c Impedance curves obtained from serial dilutions of food spoilage yeast, indirect method (data supplied by Don Whitley Scientific Ltd, UK).

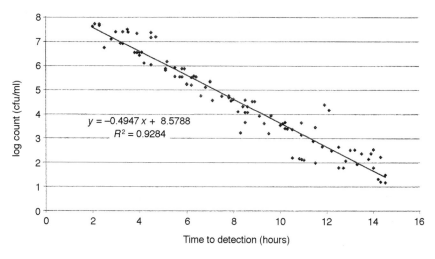

Fig. 6.7d Calibration graph for impedance determination of *Escherichia coli* in raw beef (data supplied by Don Whitley Scientific Ltd, UK).

ment has algorithms which determine when the rate of conductance change is greater than the preset threshold. Initially the reference calibration curve is constructed using known numbers of the target organism. Subsequently the microbial load of subsequent samples will be automatically determined. The limit of detection is a single viable cell since, by definition, the viable cell will multiply and eventually cause a detectable conductance change.

Microbes frequently colonise an inert surface by forming a biofilm (Section 4.6). Biofilms can be 10- to 100-fold more resistant to disinfectants than suspended cultures and therefore the efficacy of disinfectants for their removal is very important. Impedance microbiology can be used to monitor microbial colonisation and efficacy of biocides (Druggan *et al.* 1993).

6.4.2 Nucleic acid probes and the polymerase chain reaction

The use of DNA and RNA probes for selected target organisms is increasingly being used in the food industry (Scheu *et al.* 1998). The advantage is that food pathogens are detected without such an emphasis on selective media. However, the presence of DNA (or RNA) does not demonstrate the presence of a viable organism which is capable of multiplying to an infectious level. The key method is the use of the polymerase chain reaction (PCR) to amplify trace amounts of DNA and RNA to detectable levels (Fig. 6.8). Specificity is obtained by the design of appropriate DNA probes. The PCR technique uses a heat stable DNA polymerase, *Taq*, in a repetitive cycle of heating and cooling to amplify the target DNA.

The procedure is essentially:

(1) The sample is mixed with the PCR buffer, *Taq*, deoxyribonucleoside triphosphates and two primer DNA sequences (about 20 to 30 nucleotides long).

(2) The reaction mixture is heated to 94°C for 5 minutes to separate the double-stranded target DNA.

(3) The mixture is cooled to approximately 55°C for 30 seconds. During this time the primers anneal to the complementary sequence on the target DNA.

(4) The reaction temperature is raised to 72°C for 2 minutes and the *Taq* polymerase extends the primers, using the complementary strand as a template.

(5) The double-stranded DNA is separated by reheating at 94°C.

(6) The replicated target sites act as new templates for the next cycle of DNA copying.

(7) The cycle of heating and cooling is repeated 30 to 40 times. The PCR will have amplified the target DNA to a theoretical maximum of 10^9 copies, though usually the true amount is less due to enzyme denaturation. The amount of amplified DNA is approximately 100 µg.

(8) The DNA is stained with ethidium bromide and visualised by agarose gel electrophoresis with uv transillumination at 312 nm.

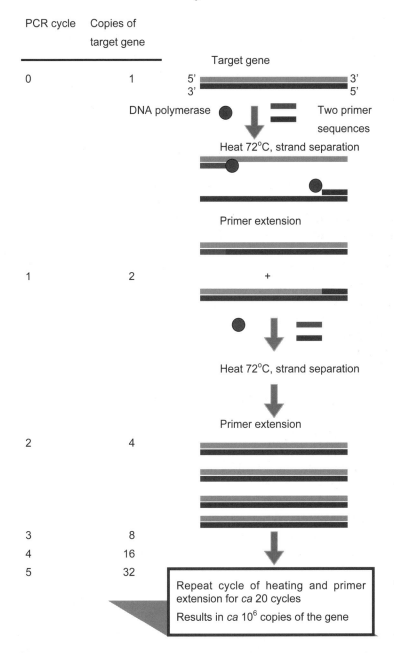

PCR cycle   Copies of
             target gene

0   1

1   2

2   4

3   8
4   16
5   32

Target gene

DNA polymerase   Two primer sequences

Heat 72°C, strand separation

Primer extension

+

Heat 72°C, strand separation

Primer extension

Repeat cycle of heating and primer extension for *ca* 20 cycles

Results in *ca* 10^6 copies of the gene

Fig. 6.8   The polymerase chain reaction (PCR).

Negative control samples omitting DNA must be used in order to check for contamination of the PCR reaction by extraneous DNA.

The ribosomal RNA (rRNA) molecule, especially the 16S rRNA, can be used for the generation of specific nucleic acid probes (Amann *et al.* 1995). The rRNA molecule contains regions which are highly conserved and other regions which are highly variable. If RNA is the target then a reverse transcriptase enzyme step is used in the above procedure to make a DNA copy.

Numerous detection kits have been developed for the detection of food pathogens. PCR is not directly performed on food samples since the reaction is inhibited by some food components and the target cell number may be too low for detection. Instead the target organism is usually detected after an enrichment broth step.

One variation on the PCR technique is 'DIANA', which stands for Detection of Immobilised Amplified Nucleic Acids. The main difference is that DIANA uses two sets of primers for PCR of which only the inner set of primers is labelled. One of the primers is biotinylated on the 5' end, the second one is labelled with a tail of a partial sequence of the *lac* operator (*lac*Op) gene. The target DNA is first amplified with the outer set of primers (30 to 40 cycles) to generate a large amount of the DNA. Then the inner set of labelled primers is amplified for 10 to 20 cycles. Streptavidin-coated magnetic beads are used to selectively isolate the amplified biotinylated primary DNA. After washing the magnetic particles the label is detected appropriately by the addition of a chromogenic substrate for the *lac* gene.

6.4.3   DNA chips and genomics

DNA chips are a combination of semiconductor technology and molecular biology (Fig. 6.9; Wallraff *et al.* 1997). In the future they will enable DNA sequences to be analysed quickly and cheaply. DNA chips consist of large arrays of oligonucleotides on a solid support (Schena *et al.* 1998; Graves 1999). They are prepared by one of three methods:

(1)   Growing oligonucleotides on the surface, base by base. This is called a Genechip™.
(2)   Linking presynthesised oligonucleotides or PCR products to a surface.
(3)   Attaching such materials within a small, three-dimensional spot of gel.

The applications of microarrays are (i) studies of genomic structure and (ii) studies of active-gene expression (Fig. 6.10). The array is exposed to

DNA sequence '**TGATGCCGCTTGGCTCGGCTCCAATTGA**' as shown by microarray analysis

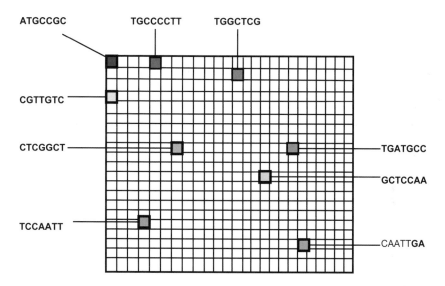

Fig. 6.9 The DNA chip (based on Graves 1999). Known sequences of 7 nucleotides are fixed to an array in a known position. The unknown DNA or RNA is then allowed to hybridise with the array and the complementary sequences visualised due to the fluorescent hybrid formation. The overall sequence (shown in bold) can be deduced from the overlapping short sequences.

labelled sample DNA and consensus sequences allowed to be hybridised, which takes between 1 and 10 hours (Ramsay 1998). The probe-target hybrid is detected either by direct fluorescence scanning or through enzyme-mediated detection (O'Donnell-Maloney *et al.* 1996). DNA chip technology also makes it possible to detect diverse individual sequences simultaneously in complex DNA samples. Therefore, it will be possible to detect and type different bacterial species in a single food sample. Development of this approach is continuing at a rapid pace and for the microbiologist, the DNA chip technology will be one of the major tools for the future. There are still many problems to solve, such as sample preparation, eliminating the effects of nonspecific binding and cross-hybridisation and increasing the sensitivity of the system (Graves 1999).

6.4.4 ATP bioluminescence techniques and hygiene monitoring

The molecule adenosine triphosphate (ATP) is found in all living cells (eucaryotic and procaryotic). Therefore the presence of ATP indicates that living cells are present. The limit of detection is around 1 pg ATP,

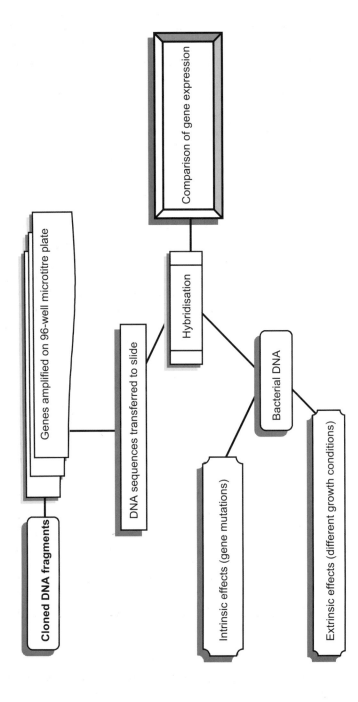

Fig. 6.10 Applications of bacterial genomics.

which is equivalent to approximately 1000 bacterial cells based on the assumption of 10^{-15} g ATP per cell. Since a sample is analysed in seconds to minutes it is considerably faster than conventional colony counts for the detection of bacteria, yeast and fungi. Additionally, food residues which act as ther loci of microbial growth will also be detected rapidly (Kyriakides 1992). Hence ATP bioluminescence is primarily used as a hygiene monitoring method and not for the detection of bacteria *per se*. In fact, in a food factory there will not necessarily be a correlation between plate counts and ATP values for identical samples, since the latter will additionally detect food residues (Fig. 6.11).

ATP is detected using the luciferase–luciferin reaction:

$$ATP + luciferin + Mg^{2+} \rightarrow oxyluciferin + ADP + light\ (562\,nm)$$

The firefly (*Photinus pyralis*) is the source of the luciferase and the reagents are formulated such that a constant yellow-green light (maximum 562 nm) is emitted.

ATP bioluminescence measurement requires a series of steps to sample an area (usually $10\,cm^2$). Many instruments currently have the extractants and luciferase–luciferin reagents encased with the swab in a 'single-shot' device. This saves the preparation of a series of reagents and the associated pipetting errors.

ATP bioluminescence can be used as a means of monitoring the cleaning regime, especially at a Critical Control Point of a Hazard Analysis

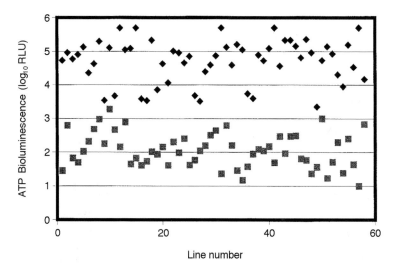

Fig. 6.11 Assessment of hygiene using ATP bioluminescence in beer lines before ◆ and after ■ cleaning.

Critical Control Point (HACCP) procedure (Section 7.3). See Table 6.3 for a list of examples of ATP bioluminescence applications. There are three food production processes which are not amenable to ATP bioluminescence, these are milk powder production, flour mixes and sugar, because the cleaning procedures do not remove all food residues.

Table 6.3 Application of ATP bioluminescence in the food industry.

(1) Hygiene monitoring
(2) Dairy industry
Raw milk assessment
Pasteurised milk, shelf life prediction
Detection of antibiotics in milk
Detection of bacterial proteases in milk
(3) Assessing microbial load
Poultry carcasses
Beef carcasses
Minced meat
Fish
Beer

It has been noted that the luciferase–luciferin reaction can be affected by residues containing sanitisers (free chlorine), detergents, metal ions, acid and alkali pH, strong colours, many salts and alcohol (Calvert *et al.* 2000). Hence the commercially available ATP bioluminescence kits may contain detergent neutralisers such as lecithin, Tween 80 and cyclodextrin. Enhancement and inhibition of the luciferase–luciferin reaction can lead to errors of decision and hence an ATP standard should be used to test the activity of the luciferase. A recent improvement has been the use of caged ATP as an internal ATP standard, whereby a known quantity of ATP is released into solution upon exposing the swab to high intensity blue light (Calvert *et al.* 2000).

6.4.5 Protein detection

An alternative to ATP detection for hygiene monitoring is the detection of protein residues, using the Biuret reaction (Fig. 6.12). There are many simple kits available now which are able to detect approximately 50 µg protein on a work surface within 10 minutes (Table 6.4). The surface is sampled either by swabbing or by a dipstick, and reagents added. The development of a green colour indicates a clean, hygienic surface, grey is 'caution' and purple is 'dirty'. The technique is more rapid than conventional microbiology and less expensive than ATP bioluminescence since

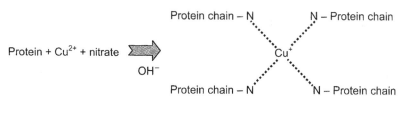

Fig. 6.12 The Biuret reaction.

Table 6.4 Biuret reaction with protein.

Protein (μg)	Colour	Absorbance (OD$_{562}$)
0–25	Lime green	0.07–0.25
55–150	Bluish grey	0.50–1.15
200–420	Light purple	1.25–2.20
600–1300	Dark purple	2.70–4.00

no capital equipment is required. It is, however, less sensitive than ATP bioluminescence.

6.4.6 Flow cytometry

Flow cytometry is based on light scattering by cells and fluorescent labels which discriminate the microorganisms from background material such as food debris (Fig. 6.13). Fluorescence-labelled antibodies have been produced for the major food poisoning organisms such as *Salmonella* serovars, *L. monocytogenes*, *C. jejuni* and *B. cereus*. The level of detection of bacteria is limited to approximately 10^4 cfu/ml due to interference and autofluorescence by food particles. Fluorescent labels include fluorescein isothiocyanate (FITC), rhodamine isothiocyanate and phycobiliproteins such as phycoerythrin and phycocyanin. These emit light at 530 nm, 615 nm, 590 nm and 630 nm, respectively. Viable counts are obtained using carboxyfluorescein diacetate which intracellular enzymes will hydrolyse, releasing a fluorochrome. Fluorescent-labelled nucleic acid probes, designed from 16S rRNA sequences, enable a mixed population to be identified at genus, species or even strain level (Section 6.4.3). How-ever, as the organism might be non-culturable (Section 6.2.2) it is uncer-tain whether the organism was viable in the test sample and subsequently questions whether its detection is of any significance. The method has been used for the detection of viruses in sea water (Marie *et al.* 1999).

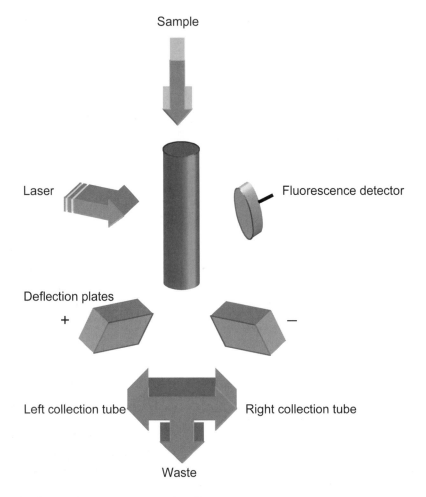

Fig. 6.13   Flow cytometry with cell sorting.

6.5   Specific detection procedures

Standardised protocols for the isolation of most foodborne pathogens have been defined by various regulatory and accreditation bodies. There is, however, no single ideal method for each pathogen and countries vary with the preferred technique. Subsequently, an overview of techniques will be given. A consequence of the variation in methodology is the uncertainty of whether food poisoning statistics can be compared between countries (Section 3.5).

Because batches of media can vary in their composition, as a means of monitoring personnel proficiency good laboratory management requires that positive and negative control organisms are used to confirm the selectivity of the media (Table 6.5). The control organisms originate from national and international culture collections such as the National Collection of Type Cultures (NCTC) and American Type Culture Collection (ATCC). This is indicated by the culture collection index numbering, for example *St. aureus* ATCC® 25923. These are well characterised strains available to all quality control laboratories and act as international standards for referencing.

6.5.1   Aerobic plate count

The aerobic plate count (APC) is commonly used to determine the general microbial load of the food and not specific organisms. It is a complex growth medium containing vitamins and hydrolysed proteins which enables the growth of nonfastidious organisms. The agar plates may be inoculated by a variety of techniques (Miles-Misra, spread plate, pour plate) which vary in the volume of sample (20 µl to 0.5 ml) applied. The plates are generally incubated at 30°C for 48 hours before enumeration of colonies. The accuracy of colony numbers is given in Table 6.1.

6.5.2   Salmonella *spp.*

Standardised procedures for the isolation of *Salmonella* spp. by different regulatory bodies are summarised in Table 6.6. The detection criterion for ready-to-eat foods is the isolation of 1 salmonella cell from 25 g of food. Subsequently the protocols require a number of steps which are designed to recover salmonella cells from low initial numbers. Additionally the cells may have been injured during processing and hence the initial step is resuscitation (see Section 6.3.2 on IMS technique). The flow charts for BSI/ISO and FDA/AOAC are given (Fig. 6.14). The general outline for both procedures is:

(1)   *Pre-enrichment*, to enable injured cells to resuscitate. Resuscitation requires a nutritious, nonselective medium such as buffered peptone water and lactose broth. These may be modified if there are large numbers of Gram-positive bacteria by the addition of 0.002% brilliant green or 0.01% malachite green. Since milk products are so highly nutritious the resuscitation broth can be distilled water plus 0.002% brilliant green. Normally 25 g of food is homogenised and added to 225 ml pre-enrichment broth and incubated overnight. If the food is highly bacteriostatic this can be overcome by the addition of sodium thiosulphate (in the case of onion) or increased dilution factor, for example 25 g in 2.25 litres (Table 6.7).

Table 6.5 Control organisms for media quality control.

Target organism	Media	Positive control	Negative control
Campylobacter jejuni	Campylobacter agar	*C. jejuni* ATCC® 29428	*E. coli* ATCC® 25922
Salmonella spp.	Buffered peptone water	*S. typhimurium* ATCC® 14028	Uninoculated broth
	Selenite-cystine broth base	*S. typhimurium* ATCC® 14028	*E. coli* ATCC® 25922
	Tetrathionate broth	*S. typhimurium* ATCC® 14028	*E. coli* ATCC® 25922
	Rappaport-Vassiliadis broth	*S. typhimurium* ATCC® 14028	*E. coli* ATCC® 25922
	Modified semi-solid RV	*S. typhimurium* ATCC® 14028 giving straw colonies at site of inoculation surrounded by halo of growth	*E. coli* ATCC® 25922
		S. enteritidis ATCC® 13076 giving straw colonies at site of inoculation surrounded by halo of growth	
		C. freundii ATCC® 8090 giving restricted or no growth	
	Brilliant green agar	*S. typhimurium* ATCC® 14028	*E. coli* ATCC® 25922, *P. vulgaris* ATCC® 13315
	XLD agar	*S. typhimurium* ATCC® 14028	*E. coli* ATCC® 25922
Salmonella and *Shigella*	SS agar	*S. enteritidis* NCTC® 13076	*Ent. faecalis* ATCC® 29212
		Sh. sonnei ATCC® 25931	
Coliforms	MacConkey broth	*E. coli* ATCC® 25922	*St. aureus* ATCC® 25923
	Lactose broth	*E. coli* ATCC® 25922	Uninoculated broth
		Ent. aerogenes ATCC® 13048	
	Minerals modified medium	*E. coli* (acid + gas) ATCC® 25933	*Ent. aerogenes* (acid only) ATCC® 13048
	Violet red bile lactose agar	*E. coli* ATCC® 25922	*St. aureus* ATCC® 25923

(Contd)

Table 6.5 *(Contd)*

Target organism	Media	Positive control	Negative control
	MacConkey agar	E. coli ATCC® 25922 Ent. faecalis ATCC® 29212 Sh. sonnei ATCC® 25931 St. aureus ATCC® 25923	Uninoculated broth
E. coli	MUG reagent	E. coli ATCC® 25922	P. mirabilis ATCC® 110975
L. monocytogenes	Listeria enrichment broth Fraser broth Oxford agar PALCAM	L. monocytogenes ATCC® 19117 L. monocytogenes ATCC® 19117 L. monocytogenes ATCC® 19117 L. monocytogenes ATCC® 19117	St. aureus ATCC® 25923 Ent. faecalis ATCC® 29212 St. aureus ATCC® 25923 E. coli ATCC® 25922, St. aureus ATCC® 25923, Step. faecalis ATCC® 29212
St. aureus	Baird-Parker agar	St. aureus giving grey-black shiny convex, narrow white entire margin surrounded by zone of clearing	
Cl. perfringens	Perfringens agars	Cl. perfringens ATCC® 13124	Cl. sordellii ATCC® 9714, Cl. bifermentans ATCC® 638
B. cereus	PEMBA	B. cereus ATCC® 10876	B. coagulans ATCC® 7050

ATCC® = American Type Culture Collection.
NCTC® = National Collection of Type Cultures
Information kindly supplied by OXOID Ltd., Basingstoke, UK.

Table 6.6 *Salmonella* isolation procedures approved by different regulatory bodies.

Body	Culture media		
	Pre-enrichment	Enrichment	Plating
ISO	Buffered peptone water	Rappaport-Vassiliadis (RV) broth Selenite cystine broth	Brilliant green agar (Edel & Kampelmacher) Any other solid selective medium[a]
APHA	Lactose broth	Selenite cystine broth Tetrathionate broth (USP)	SS agar Bismuth sulphite agar Hektoen agar
AOAC/FDA	Lactose broth Tryptone soya broth Nutrient broth	Selenite cystine broth Tetrathionate broth (USP)	Brilliant green agar Hektoen agar XLD agar Bismuth sulphite agar
IDF	Buffered peptone water Distilled water plus brilliant green 0.002%[b]	Muller-Kauffman tetrathionate broth Selenite cystine broth	Brilliant green agar (Edel & Kampelmacher) Bismuth sulphite agar
BSI	Buffered peptone water	Rappaport-Vassiliadis (RV) broth Selenite cystine broth	Brilliant green agar (Edel & Kampelmacher) Any other solid selective medium[a]

(Contd)

Table 6.6 *(Contd)*

Body	Culture media		
	Pre-enrichment	Enrichment	Plating
AFNOR	Buffered peptone water	Rappaport-Vassiliadis (RV) broth Selenite cystine broth	Brilliant green agar (Edel & Kampelmacher) XLD agar Hektoen agar Deoxycholate-citrate-lactose agar
Nordic Committee on Food Analysis Number 71, 4th Edition	Buffered peptone water	Rappaport-Vassiliadis (RV) broth	XLD agar Brilliant green agar (Edel & Kampelmacher)
Australian Standard Method 2.5 A.S. 1766.2.5	Buffered peptone water	Mannitol-selenite-cystine broth Rappaport-Vassiliadis (RV) broth	XLD agar Bismuth sulphite agar
Health Canada Analytical Methods	Nutrient broth	Selenite cystine broth Tetrathionate broth (USP)	Bismuth sulphite agar Brilliant green sulfa agar (BGS)

[a]The choice of the second medium is discretionary unless a specific medium is named in an International Standard relating to the product to be examined.
[b]The choice of pre-enrichment medium is dependent on the product under examination. The standard methods of the appropriate body should be consulted for details.
Information kindly supplied by OXOID Ltd, Basingstoke, UK.

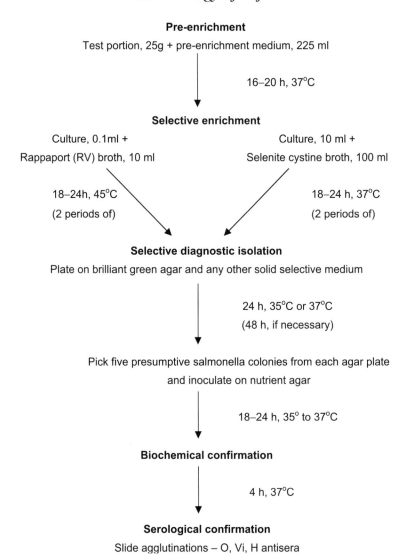

Pre-enrichment

Test portion, 25g + pre-enrichment medium, 225 ml

16–20 h, 37°C

Selective enrichment

Culture, 0.1ml +
Rappaport (RV) broth, 10 ml

Culture, 10 ml +
Selenite cystine broth, 100 ml

18–24h, 45°C
(2 periods of)

18–24 h, 37°C
(2 periods of)

Selective diagnostic isolation

Plate on brilliant green agar and any other solid selective medium

24 h, 35°C or 37°C
(48 h, if necessary)

Pick five presumptive salmonella colonies from each agar plate
and inoculate on nutrient agar

18–24 h, 35° to 37°C

Biochemical confirmation

4 h, 37°C

Serological confirmation

Slide agglutinations – O, Vi, H antisera

Fig. 6.14a BSI/ISO salmonella isolation procedure.

(2) *Selective enrichment*, to suppress the growth of nonsalmonella cells
and enable salmonella cells to multiply. This is achieved by the
addition of inhibitors such as bile, tetrathionate, sodium biselenite
(with care, as this compound is very toxic) and either brilliant green
or malachite green dyes. Selectivity is enhanced by incubation at
41–43°C. Selective broths are selenite cystine broth, tetrathionate
broth and Rappaport broth. More than one selective broth is used

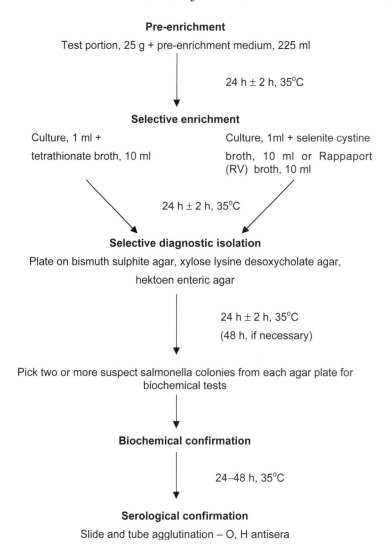

Pre-enrichment

Test portion, 25 g + pre-enrichment medium, 225 ml

24 h ± 2 h, 35°C

Selective enrichment

Culture, 1 ml +
tetrathionate broth, 10 ml

Culture, 1ml + selenite cystine broth, 10 ml or Rappaport (RV) broth, 10 ml

24 h ± 2 h, 35°C

Selective diagnostic isolation

Plate on bismuth sulphite agar, xylose lysine desoxycholate agar, hektoen enteric agar

24 h ± 2 h, 35°C
(48 h, if necessary)

Pick two or more suspect salmonella colonies from each agar plate for biochemical tests

Biochemical confirmation

24–48 h, 35°C

Serological confirmation

Slide and tube agglutination – O, H antisera

Fig. 6.14b   FDA/AOAC BAM salmonella isolation procedure.

because the broths have different selectivities towards the 2000+ *Salmonella* serovars.

(3)   *Selective diagnostic isolation*, to isolate salmonella cells on an agar medium to enable single colonies to be isolated and identified. The media contain selective agents similar to the selective broths such as bile salts and brilliant green. Salmonella colonies are differentiated from nonsalmonella by detection of lactose fermentation and H_2S production. Selective agars include brilliant green agar, MLCB agar

Table 6.7 Selection of pre-enrichment media.

Medium	Commodity
Buffered peptone water (BPW)	General purpose
BPW + casein	Chocolate, etc.
Lactose broth	Egg and egg products; frog legs; food dyes pH > 6
Lactose broth + Tergitol 7 or Triton X-100	Coconut; meat; animal substances – dried or processed
Lactose broth + 0.5% gelatinase	Gelatin
Non-fat dry milk + brilliant green	Chocolate; candy and candy coatings
Tryptone soya broth	Spices; herbs; dried yeast
Tryptone soya broth + 0.5% potassium sulphate	Onion and garlic powder, etc.
Water + brilliant green	Dried milk

and XLD agar. As for the selective enrichment stage, more than one agar medium is used since the media differ in their selectivities.

(4) *Biochemical confirmation*, to confirm the identity of presumptive salmonella colonies.

(5) *Serological confirmation*, to confirm the identity of presumptive salmonella colonies and to identify the serotype of the salmonella isolate (useful in epidemiology).

6.5.3 Campylobacter

Campylobacter cells can be stressed during processing and hence a pre-enrichment stage to enable injured cells to be resuscitated is commonly used prior to selection for the organism. Frequently, selective agents are added as supplements and lowered incubation temperatures are used. In order to aid the growth of the organism ferrous sulphate, sodium meta-bisulphite and sodium pyruvate (FBP) are added to growth media to quench toxic radicals and increase the organism's aerotolerance. The organism is microaerophilic and is unable to grow in normal air levels of oxygen. The preferred atmosphere is 6% oxygen and 10% carbon dioxide. This is achieved in gas jars by using gas sachets which generate the required gases. There has been a considerable variety of growth media developed for campylobacter. The major methods and protocols are given in Table 6.8a–c and Fig. 6.15a–e.

Table 6.8a Culture media specified by some national bodies for detection of campylobacter in foods.

Country	Organisation responsible	Culture media: enrichment	Culture media: plating	Other media specified in procedure
Australia	Standards Australia Committee FT/4 Food Microbiology	Preston broth	(1) Preston agar (2) Skirrow's agar	Nutrient agar Nutrient broth
North America/ Canada	Food and Drug Administration (FDA) Bacteriological Analytical Manual (BAM) 1992	Hunt and Radle broth (BAM M29) Add antibiotic formulae 1, 2 or 3 (See Table 6.8b)	(1) Isolation agar A (2) Isolation agar B	Blood agar Heart infusion agar Peptone diluent Brucella semi-solid medium Triple-sugar iron agar MacConkey agar Cary-Blair medium
France	AFNOR: General Guidance for Detection of Thermotolerant Campylobacter. Norme Francaise ISO/DIS 10272	(1) Preston broth (2) Park and Sanders broth	(1) Karmali agar (2) Skirrow agar (3) Campylobacter blood-free agar (4) Preston agar	Brucella broth Columbia blood agar Mueller-Hinton blood agar Triple-sugar iron agar
UK	MAFF/DoH Steering Group on the Microbiological Safety of Food	(1) Park and Sanders broth (2) Exeter broth	Campylobacter blood-free agar Exeter agar	
UK/international	BS 5763: ISO/DIS 10272 Methods for Microbiological Examination of Food and Animal Feeding Stuffs. Detection of Thermotolerant Campylobacters	(1) Preston broth (2) Park and Sanders broth	(1) Karmali agar (mandatory) (2) Skirrow agar or (3) Campylobacter blood-free agar or (4) Preston agar or (5) Butzler agar	Brucella broth Columbia blood agar Mueller-Hinton blood agar Triple-sugar iron agar

Information kindly supplied by OXOID Ltd, Basingstoke, UK.

Table 6.8b Selective agents used in FDA Bacteriological Analytical Manual methods for testing foods, environmental samples and dairy products (AOAC 1992).

For foods		
Antibiotic formula 1 (modified Park formula)		
		mg/litre
First addition:	Cefoperazone	15
	Trimethoprim	12.5
	Vancomycin	10.0
	Cycloheximide	100.0
Second addition:	Cefoperazone	15
For water and environmental swabs		
Antibiotic formula 2 (modified Humphrey formula)		
		mg/litre
Cefoperazone		15
Trimethoprim		12.5
Vancomycin		10.0
Cycloheximide		100.0
For dairy products		
Antibiotic formula 3 (modified Preston formula)		
		mg/litre
Rifampicin		10
Cefoperazone		15
Trimethoprim lactate		12.5
Cycloheximide		100

Information kindly supplied by Oxoid Ltd, UK.

6.5.4   *Enterobacteriaceae, coliforms and* E. coli

E. coli and coliforms are often initially detected together in liquid media and then differentiated by secondary tests of indole production, lactose metabolism, gas production and growth at 44°C. *E. coli* produces acid and gas at 44°C within 48 hours. MacConkey broth is a commonly used medium for the presumptive detection of coliforms from water and milk. It selects for lactose fermenting, bile tolerant organisms. Acid formation from lactose metabolism is shown by a yellow coloration of the broth (due to a pH indicator dye, neutral red or bromocresol purple) and gas formation is indicated by gas trapped in an upturned Durham tube. Lauryl tryptose broth (also known as lauryl sulphate broth) can be used for the detection of coliforms from food. Initially the inoculated medium is incubated at 35°C and afterwards presumptive positive tubes are used to

Table 6.8c   The selective agents incorporated in some campylobacter enrichment broth media (concentrations in mg/litre unless otherwise stated).

Medium	Cefoperazone	Colistin	Cyclohexatimide	Polymyxin B	Rifampicin	Trimethoprim	Vancomycin
Bolton	20		50				20
Doyle & Roman			50	20000 iu		5	15
Exeter	15	4				10	10
Hunt & Radle:							
(a) Antibiotic formula 1	30		100			12.5	10
(b) Antibiotic formula 2	15		100			12.5	10
(c) Antibiotic formula 3	15		100		10	12.5	10
Park & Sanders:							
(1) Antibiotic solution A						10	10
(2) Antibiotic solution B	32		100				
Preston			100	5000 iu	10	10	10

Information kindly supplied by Oxoid Ltd, UK.

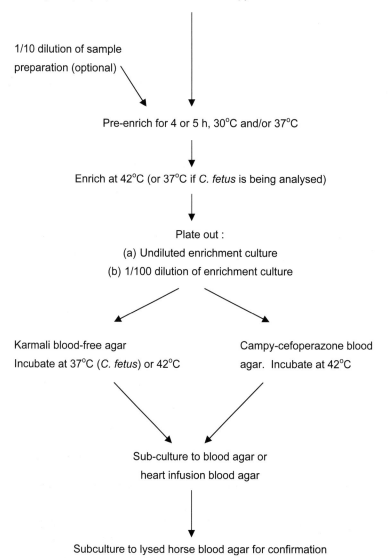

Fig. 6.15a   FDA/BAM method for detecting *Campylobacter* spp. in foods.

inoculate duplicate tubes, one for incubation at 35°C and the other at 44°C. Both broths can be supplemented with 4-methylumbelliferyl-β-D-glucuronide (Section 6.4) to enhance *E. coli* detection. EE broth, also known as buffered glucose-brilliant green bile broth, is an enrichment medium for Enterobacteriaceae from food. The broth is inoculated with

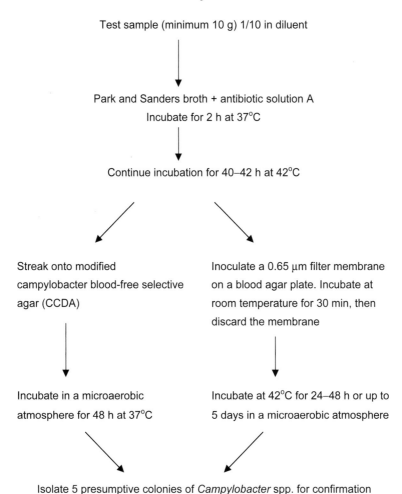

Test sample (minimum 10 g) 1/10 in diluent

↓

Park and Sanders broth + antibiotic solution A
Incubate for 2 h at 37°C

↓

Continue incubation for 40–42 h at 42°C

Streak onto modified
campylobacter blood-free selective
agar (CCDA)

↓

Incubate in a microaerobic
atmosphere for 48 h at 37°C

Inoculate a 0.65 µm filter membrane
on a blood agar plate. Incubate at
room temperature for 30 min, then
discard the membrane

↓

Incubate at 42°C for 24–48 h or up to
5 days in a microaerobic atmosphere

Isolate 5 presumptive colonies of *Campylobacter* spp. for confirmation

Fig. 6.15b Detection of *Campylobacter* spp., Park and Sanders method.

samples which have been incubated at 25°C in aerated tryptone soya broth (1:10 dilution) for resuscitation of any injured cells. There are numerous solid media employed for the detection of *E. coli*, coliforms and Enterobacteriaceae (Table 6.5). For example, there are two types of violet red bile agars: (1) violet red bile lactose agar for coliforms in food and dairy products and (2) violet red glucose agar for detection of Enterobacteriaceae. These select for bile-tolerant organisms, a predicted trait for intestinal bacteria. Other media include MacConkey agar, china blue lactose agar, desoxycholate agar and eosin methylene blue agar. These have different differentiation efficiencies and regulatory approval. A recent trend has been the inclusion of chromogenic and fluorogenic

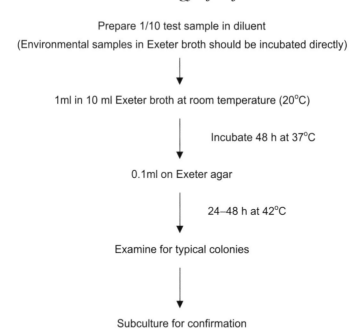

Prepare 1/10 test sample in diluent
(Environmental samples in Exeter broth should be incubated directly)

1ml in 10 ml Exeter broth at room temperature (20°C)

Incubate 48 h at 37°C

0.1ml on Exeter agar

24–48 h at 42°C

Examine for typical colonies

Subculture for confirmation

Fig. 6.15c   Detection of *Campylobacter* spp., Exeter method.

substrates, in particular, to detect β-glucuronidase which is produced by approximately 97% of *E. coli* strains (Section 6.4).

6.5.5   Pathogenic E. coli, including E. coli O157:H7

Since *E. coli* is a commensal organism in the human large intestine there is a problem isolating and differentiating pathogenic strains from the more numerous nonpathogenic varieties (Sections 3.9.1 and 5.2.3). The key differentiation traits have been based on the observation that, unlike most nonpathogenic *E. coli* strains, *E. coli* O157:H7 does not ferment sorbitol, does not possess β-glucuronidase and does not grow above 42°C. Subsequently MacConkey agar was modified to include sorbitol in place of lactose as the fermentable carbohydrate (SMAC). This medium has been further modified by the inclusion of various other selective agents such as tellurite and cefixime. Pre-enrichment in a modified buffered peptone water broth or modified tryptone soya broth is used to resuscitate injured cells before plating onto solid media (see Fig. 6.16a–c). Because the cell surface antigen (O157:H7) is indicative of pathogenicity (although not 100%), the immunomagnetic separation technique (Section 6.3.2) greatly increases the recovery of *E. coli* O157:H7 (Chapman & Siddons 1996). The

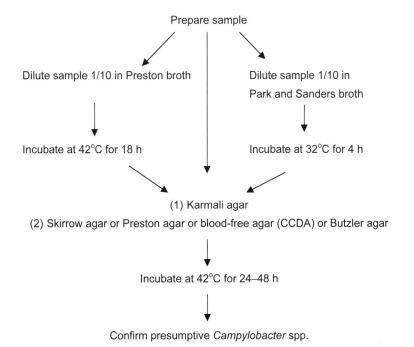

Fig. 6.15d General guidance for detection of thermotolerant campylobacter. Norme Francaise ISO/DIS 10272.

IMS technique is used worldwide and is recognised as the most sensitive method for *E. coli* O157:H7 and has been approved in Germany and Japan.

The toxins of *E. coli* O157:H7 can be detected using cultured Vero cells and reversed passive latex agglutination (RPLA) which are sensitive to 1–2 mg/ml culture filtrate (Fig. 6.6). Polymyxin B is added to the culture to facilitate the release of the verocytotoxins/shiga toxins. ELISA methods specific for pathogenic strains of *E. coli* have been developed. DIANA (Section 6.4.2) has been applied to a number of immunomagnetic separation assays including enterotoxigenic *E. coli*. The assay can detect five ETEC cells in 5 ml without interference from a 100-fold excess of SLI negative strains.

6.5.6 Shigella *spp.*

The differentiation between *E. coli* and *Shigella* is troublesome since the two organisms are so similar. In fact *Shigella* and enteroinvasive strains of *E. coli* are closely related phenotypically and they have a close antigenic relationship. Serologically *Sh. dysenteriae* 3 and *E. coli* O124, *Sh. boydii* 8

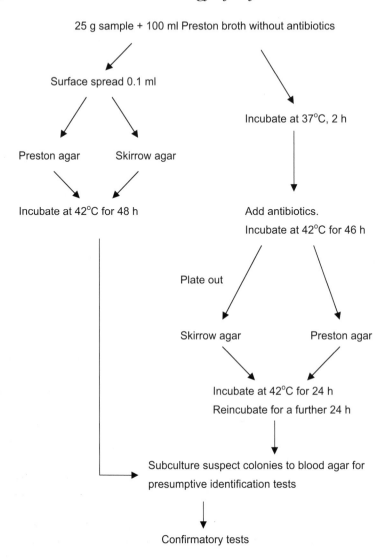

Fig. 6.15e   Australian standard method for detection of campylobacter in foods.

and *E. coli* 0143 and *Sh. dysenteriae* 12 and *E. coli* 0152 appear identical. Distinguishing traits are given in Table 6.9. Other useful differential traits are:

- *Shigella* cultures are always non-motile.
- Cultures that ferment mucate, utilise citrate or produce alkali on acetate agar are likely to be *E. coli*.

- Cultures that decarboxylate ornithine are most likely to be *Sh. sonnei*.
- Cultures that ferment sucrose are likely to be *E. coli*.

A three-stage procedure of resuscitation, selective enrichment and selective plating is normally required for the isolation of *Shigella* spp. from food samples. However, there is little consensus of opinion regarding the optimal protocols. An abbreviated version of the FDA-BAM method is given in Fig. 6.17 for illustrative purposes. The FDA-BAM method uses anaerobic incubation as under these conditions shigella cells can compete against coliforms. Direct plating onto selective media is unlikely to be successful due to the close relationship of shigella and *E. coli*. Hence the detection of shigella is generally through use of the distinguishing biochemical differences (Table 6.9). For example, on MacConkey agar shigella colonies initially appear as non-lactose fermenters.

Shiga toxins are virtually identical to the verocytotoxin produced by EHEC (Sections 3.8.4, 3.9.3 and 3.9.4) and therefore detection methods are applicable to both groups of foodborne pathogens.

6.5.7 Listeria

Unlike the isolation procedures for salmonella and *E. coli*, pre-enrichment is not commonly used for the isolation of *Listeria* spp. This is because other organisms present will outgrow listeria cells. Various enrichment media have been developed (Table 6.10a and b) with different regulatory approval. A common enrichment broth is Fraser broth (modified from UVM broth), which employs aesculin hydrolysis coupled with ferrous iron as an indicator of presumptive *Listeria* spp. Enrichments are streaked onto agars such as Oxford and PALCAM; Oxford agar is often incubated at 30°C, whereas PALCAM agar is incubated at 37°C under microaerophilic conditions (Fig. 6.18a and b). A large number of selective agents are used in listeria media such as acriflavin, cycloheximide, colistin and polymixin B, as *L. monocytogenes* can be quickly outgrown by competing flora (Table 6.11). Typical *L. monocytogenes* colonies are surrounded by a black zone due to black iron phenolic compounds. On PALCAM agar the centre of the colony may have a sunken centre after 48 hours' incubation.

Presumptive *L. monocytogenes* colonies are confirmed using biochemical and serological testing. Most nonlisteria isolates can be eliminated using the motility test, catalase test and Gram staining. *Listeria* spp. are short, Gram-positive rods, catalase-positive, and are nonmotile if incubated above 30°C. The motility of cultures grown at room temperature is characterised by a tumbling action. *L. monocytogenes* is β-haemolytic on horse blood agar. The CAMP test (named after Christie, Atkins, Munch and Peterson) is used for species differentiation. The

The Microbiology of Safe Food

listeria isolates are streaked on sheep blood agar and *St. aureus* NCTC 1803 and *Rhodococcus equi* NCTC 1621 are streaked in parallel close to the listeria streaks (Fig. 6.19). The phenomenon of enhanced zones of haemolysis is observed (Table 6.12).

6.5.8 St. aureus

Since small numbers of *St. aureus* in food are of little significance, enrichment is not used for the organism's isolation. Additionally large numbers of cells are required to produce sufficient amounts of heat-stable toxin. Therefore tests for viable cells are applicable for samples before heat treatment, and tests for the enterotoxin and heat-stable thermonuclease for heat-treated samples. The Baird-Parker agar (Tables 6.13 and 6.14) is the most widely accepted selective agar for *St. aureus*. This

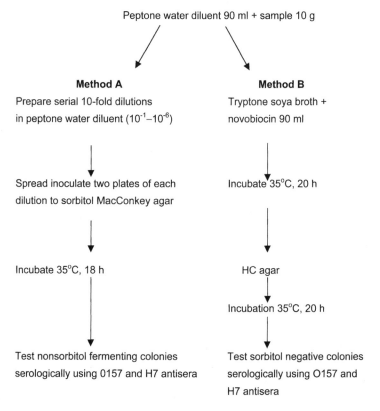

Fig. 6.16a FDA/BAM methods for isolation of enterohaemorrhagic *E. coli* (EHEC).

Method C

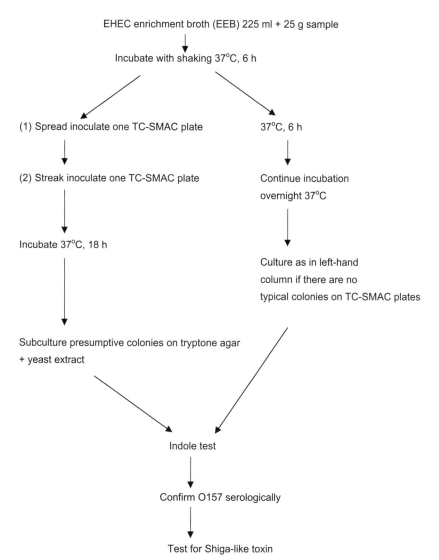

EHEC enrichment broth (EEB) 225 ml + 25 g sample

Incubate with shaking 37°C, 6 h

(1) Spread inoculate one TC-SMAC plate

37°C, 6 h

(2) Streak inoculate one TC-SMAC plate

Continue incubation
overnight 37°C

Incubate 37°C, 18 h

Culture as in left-hand
column if there are no
typical colonies on TC-SMAC plates

Subculture presumptive colonies on tryptone agar
+ yeast extract

Indole test

Confirm O157 serologically

Test for Shiga-like toxin

Fig. 6.16a *(Contd)*

medium includes sodium pyruvate to aid the resuscitation of injured cells. The selectivity is due to the presence of tellurite, lithium chloride and glycine. *St. aureus* forms black colonies due to tellurite reduction and clearance of egg yolk due to lipase activity (Table 6.14). Glycine acts as a growth stimulant and is an essential component of the staphylococcal cell wall. An alternative medium, mannitol salt agar, has better recovery

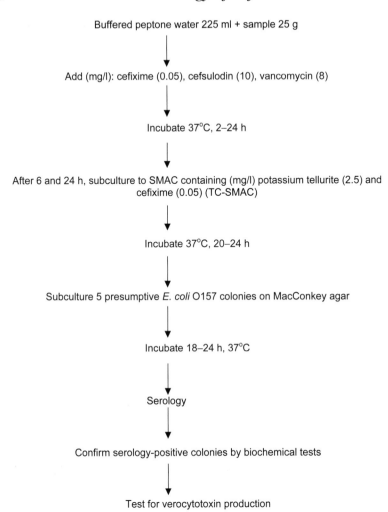

Buffered peptone water 225 ml + sample 25 g

Add (mg/l): cefixime (0.05), cefsulodin (10), vancomycin (8)

Incubate 37°C, 2–24 h

After 6 and 24 h, subculture to SMAC containing (mg/l) potassium tellurite (2.5) and cefixime (0.05) (TC-SMAC)

Incubate 37°C, 20–24 h

Subculture 5 presumptive *E. coli* O157 colonies on MacConkey agar

Incubate 18–24 h, 37°C

Serology

Confirm serology-positive colonies by biochemical tests

Test for verocytotoxin production

Fig. 6.16b   PHLS (UK) Method 1 for isolation of *E. coli* O157:H7 from foods.

efficiency of *St. aureus* from cheese. The selective agent is salt (7.5%) and mannitol fermentation is indicated by the pH indicator phenol red (reddish-purple zones surrounding *St. aureus* colonies).

The coagulase test (clotting of diluted mammalian blood plasma) is a reliable test for pathogenic *St. aureus*. The DNAse reaction (on DNAse agar) tests for deoxyribonuclease enzymes and is indicative of pathogenicity. DNA and toluidine blue or methyl green are included in the agar medium (Table 6.14). The dyes form coloured complexes with the DNA and hence there are zones of decolourisation surrounding DNA degrading

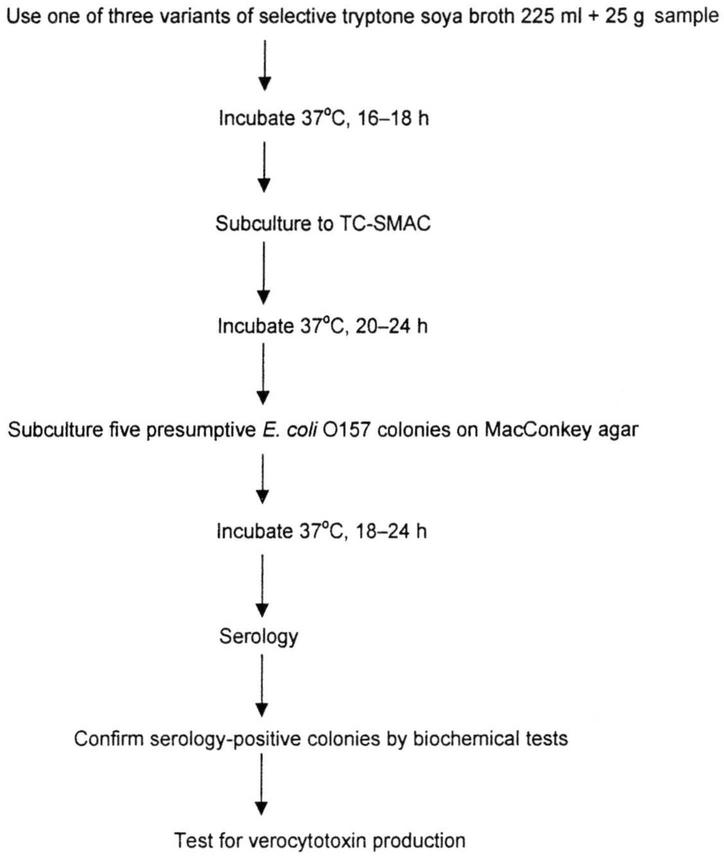

Use one of three variants of selective tryptone soya broth 225 ml + 25 g sample

↓

Incubate 37°C, 16–18 h

↓

Subculture to TC-SMAC

↓

Incubate 37°C, 20–24 h

↓

Subculture five presumptive *E. coli* O157 colonies on MacConkey agar

↓

Incubate 37°C, 18–24 h

↓

Serology

↓

Confirm serology-positive colonies by biochemical tests

↓

Test for verocytotoxin production

Fig. 6.16c PHLS (UK) Method 2 for isolation of *E. coli* O157:H7 from foods.

Table 6.9 Major differentiation characteristics between *E. coli* and *Shigella* spp.

Property	*E. coli*	*Shigella* spp.
Motility	+	—
Lactose fermentation	+	—
Indole fermentation	+	—
Gas from glucose	+	—

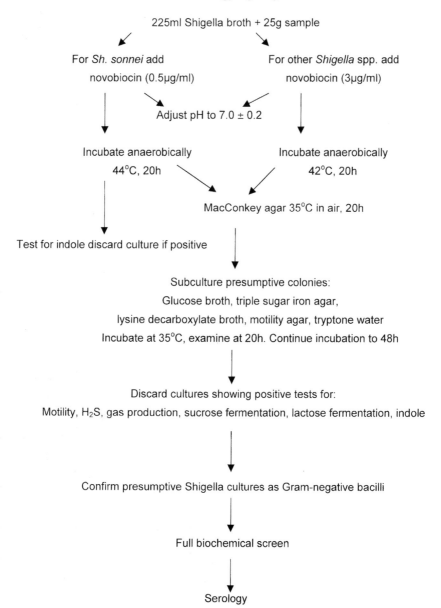

Fig. 6.17   FDA-BAM method for enrichment culture of *Shigella* spp. in foods.

Table 6.10a Culture media specified by some national standards bodies for detection of *Listeria monocytogenes*.

Country	Organisation responsible for the procedure	Culture media: enrichment	Culture media: plating	Other media specified in procedure
Australia	Standards Australia Committee on Food Microbiology	Listeria enrichment broth	Oxford agar	Blood agar base No 2 Tryptone soya agar Tryptone soya broth
Canada	Health and Welfare Canada MFHPB-30 (Sept 93)	(1) Fraser broth (2) UVM 1 broth (Add 1 g aesculin to 1 litre before sterilisation)	(1) Oxford agar, MOX agar (2) PALCAM agar (3) LPM agar	Tryptone soya agar. MRVP medium. Triple sugar iron agar. Tryptone soya broth
France	AFNOR DGAL/SDHA/N93/8105 June 1993	(1) AFNOR half strength Fraser broth *1 vial per litre of medium*. (Add 0.25 g of ferric ammonium citrate to 1 litre of medium) (2) Fraser broth	(1) Oxford agar (2) PALCAM agar	Blood agar base No 2
Germany	BGA Official Collection of Methods 35 LMBG Part L 00 00-22 December 1991	Listeria enrichment broth	(1) Oxford agar (2) PALCAM agar	Columbia agar. Tryptone soya yeast extract broth. Tryptone soya yeast extract agar.
Italy	Institute Superiore di Sanità with the Italian Health Ministry	Fraser broth	Oxford agar	Buffered peptone water. Tryptone soya yeast extract broth. Tryptone soya yeast extract agar
New Zealand	New Zealand Meat Board	(1) UVM 1 broth (2) Fraser broth	Modified Oxford agar (MOX)	
Spain	Instituto de Salud Carlos III Centro Nacional de Alimentación	Pre-enrichment: listeria enrichment broth base. Enrichment: listeria enrichment broth. Listeria UVM 1 broth. Listeria UVM 2 broth	(1) Oxford agar (2) Modified McBride agar	Brain heart infusion. Blood agar base. Columbia agar. Tryptone soya agar. Urea agar. SIM medium. MRVP medium. Triple sugar iron agar

Table 6.10b Culture media recommended by some regulatory bodies for detection of *Listeria monocytogenes*.

Regulatory body	Enrichment media	Plating media
FDA	Listeria enrichment broth (LEB)	Modified McBride agar (MMA). Oxford agar. LPM agar
USDA	UVM 1 broth Fraser broth	Modified Oxford agar (MOX)
BSI/ISO/IDF/AOAC for milk and milk products	Listeria enrichment broth (LEB)	Oxford agar
BSI/ISO Provisional proposal for foods	Half strength Fraser broth Fraser broth	Oxford agar
AFNOR to be confirmed. Derived from Ministry of Agriculture National Standard Method	Half strength Fraser broth Fraser broth	Oxford agar PALCAM agar

Information kindly supplied by OXOID Ltd, Basingstoke, UK.

(presumptive pathogenic) *St. aureus* colonies. Staphylococcal enterotoxins (Sections 3.8.2 and 5.2.7) can be detected using reverse phase latex agglutination (RPLA; Section 6.4). The limit of sensitivity is about 0.5 ng of enterotoxin per gram food. A number of enzyme immunoassays are available for staphylococcal enterotoxin detection. ELISA kits are also available which have a detection limit of > 0.5 μg toxin per 100 g food and require 7 hours to obtain the result.

6.5.9   Cl. perfringens

Cl. perfringens is a strict anaerobe which produces spores that can survive heating processes. Therefore enrichment is required to detect low numbers of clostridia cells which may be outnumbered by other organisms (Fig. 6.20). Numerous media include sulphite and iron which result in a characteristic blackening of *Cl. perfringens* colonies (Table 6.15). However, this blackening reaction is not limited to *Cl. perfringens* and hence the term 'sulphite reducers' is often used instead of *Cl. perfringens*. The lecithinase (phospholipase C; Section 3.8.3) activity of *Cl. perfringens* is also a common test in diagnostic media, resulting in opaque zones surrounding the colonies. Selectivity is by the inclusion of cycloserine or

Table 6.11 *Listeria* selective enrichment broths (concentration of selective agents in mg/litre unless stated g/litre).

Selective agent	FDA listeria enrichment broth	Buffered FDA enrichment broth	Modified FDA listeria enrichment broth	UVM 1	UVM 2	Fraser	Half strength Fraser	Organon-teknika Fraser
Acriflavine HCl	15	15	10	12	25	25	12.5	12.5
Nalidixic acid	40	40	40	20	20	20	10	20
Cycloheximide	50	50	50	—	—	—	—	—
Lithium chloride	—	—	—	—	—	3 g	3 g	3 g

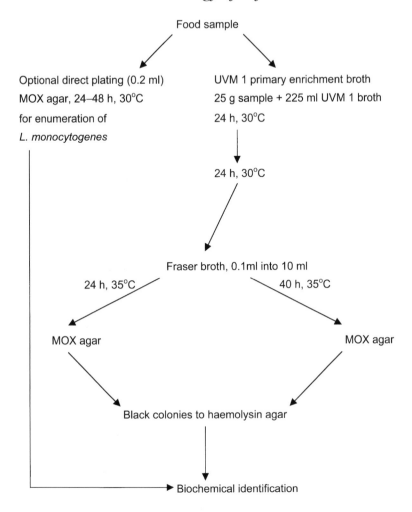

Fig. 6.18a USDA procedure for isolation of *L. monocytogenes*.

neomycin. All media incubation is under anaerobic conditions which are generated using either an anaerobe jar or an anaerobic cabinet.

Tests to distinguish *Cl. perfringens* from other anaerobic sulphite reducers are microscopy of the Gram stain, metronidazole sensitivity, the Nagler reaction and the reversed CAMP test. The Nagler test uses *Cl. perfringens* type A antitoxin to neutralise lecithinase activity. The reversed CAMP test involves streaking sheep blood agar with *St. aga-lactiae* and the test isolate at right angles, without touching. After anaerobic incubation at 37°C for 24 hours, a positive result is indicated by arrow-shaped areas of synergistic enhanced haemolysis at the junction of the two streaks (Fig. 6.21). Production of 'stormy clot' in litmus milk and

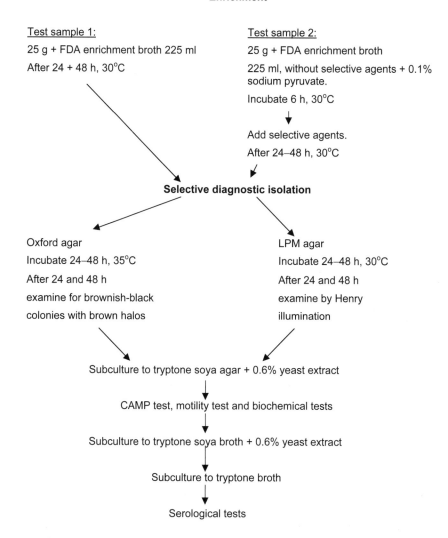

Enrichment

Test sample 1:
25 g + FDA enrichment broth 225 ml
After 24 + 48 h, 30°C

Test sample 2:
25 g + FDA enrichment broth
225 ml, without selective agents + 0.1%
sodium pyruvate.
Incubate 6 h, 30°C

Add selective agents.
After 24–48 h, 30°C

Selective diagnostic isolation

Oxford agar
Incubate 24–48 h, 35°C
After 24 and 48 h
examine for brownish-black
colonies with brown halos

LPM agar
Incubate 24–48 h, 30°C
After 24 and 48 h
examine by Henry
illumination

Subculture to tryptone soya agar + 0.6% yeast extract

CAMP test, motility test and biochemical tests

Subculture to tryptone soya broth + 0.6% yeast extract

Subculture to tryptone broth

Serological tests

Fig. 6.18b FDA/BAM *Listeria* isolation procedure.

the detection of acid phosphatase are useful confirmatory tests. A number of biological methods are available, including the rabbit ligated ileal loop test which, although very effective and widely used, does require live animal testing. To date few commercially produced kits are available for the detection of the extracellular toxins produced by *Cl. perfringens*. Reversed passive latex agglutination (RPLA; Section 6.4) is available for *Cl. perfringens* enterotoxin.

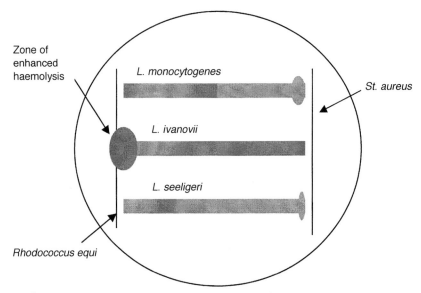

Fig. 6.19   The CAMP test for haemolysis testing of *L. monocytogenes*.

Table 6.12   CAMP reactions for *Listeria* spp.

Listeria spp.	*St. aureus*	*Rh. equi*
L. monocytogenes	+	−
L. seeligeri	+	−
L. ivanovii	−	+

6.5.10   B. cereus, B. subtilis *and* B. licheniformis

Enrichment is not normally used for the detection of *B. cereus* since low numbers of the organism are not regarded as being of significance (Fig. 6.22). Direct plating onto selective media containing the antibiotic polymyxin B is often used (Table 6.16). The two key distinguishing features incorporated into media design are the demonstration of phospholipase C (Section 3.8.3) activity and the inability to produce acid from the sugar mannitol. If it is necessary only to count spores the vegetative cells must be killed by heat treatment (1:10 dilution, 15 minutes, 70°C) or alcohol treatment (1:1 dilution in 95% ethyl alcohol, 30 minutes at room temperature).

B. subtilis and *B. licheniformis* can be isolated easily using routine nonselective media. They have a similar appearance on PEMBA medium, but are distinguishable from *B. cereus*. ELISA and RPLA tests are commercially available for *Bacillus* diarrhoeal enterotoxin. They have a

Table 6.13 Approved *St. aureus* detection media.

Body	Enrichment	Plating	Other media
American Public Health Association	Tryptone soya broth + 10% or 20% sodium chloride and sodium pyruvate	Baird-Parker or Baird-Parker plasma-fibrinogen agar	Toluidine blue-DNA agar Brain-heart infusion
Australian standard AS1766.2.4 1994	Giolitti-Cantoni broth	Baird-Parker	
Canada: Health Protection Branch MFHPB – 21 1985		Baird-Parker	Brain-heart infusion Tryptone soya agar Blood agar, nutrient agar Toluidine-blue-DNA agar Phenol red-Carbohydrate broth
France: AFNOR V08–057–1		Baird-Parker	Brain-heart infusion
Italy: Istituto Superiore di Sanita Rapporti ISTISAN 96/35		Baird-Parker	Brain-heart infusion
ISO/European/British Standard 4285 and 5763	Giolitti-Cantoni broth	Baird-Parker	Brain-heart infusion
Nordic Committee on Food Analysis, Number 66, 2nd edition 1992		Baird-Parker	Blood agar

Table 6.14   Selective agents and presumptive diagnostic features for *S. aureus.*

Medium	Selective agents	Diagnostic features
Mannitol salt	Sodium chloride	Mannitol fermentation
Lipovitellin-salt-mannitol agar (LSM)	Sodium chloride	Mannitol fermentation Egg yolk reaction
Baird-Parker	Potassium tellurite Glycine Lithium chloride	Black colonies Egg yolk reaction
Rabbit plasma- fibrinogen (RPF)	Potassium tellurite Glycine Lithium chloride	Coagulase reaction White or grey or black colonies
Vogel and Johnson	Potassium tellurite Glycine	Black colonies Mannitol fermentation
Improved Vogel and Johnson (PCVJ)	Potassium tellurite Glycine Lithium chloride	Black colonies Mannitol fermentation DNAse production
Staph 110	Sodium chloride Sodium azide	Mannitol fermentation Gelatin liquefaction Pigment production Egg yolk reaction
Staph 4S	Sodium chloride Potassium tellurite Incubation at 42°C	Egg yolk reaction Grey/dark-grey colonies
KRANEP	Potassium thiocyanate Lithium chloride Sodium azide Cycloheximide	Mannitol fermentation Egg yolk reaction Pigment production
Egg yolk-azide	Sodium azide	Egg yolk reaction
Columbia CNA (Staph/ Strep selective medium)	Nalidixic acid Colistin sulphate	Pigment Haemolysis
Polymyxin-coagulase- mannitol	Polymyxin B	Coagulase reaction Mannitol fermentation
DNAse medium		Deoxyribonuclease production
Phosphatase medium	Polymyxin B	Phosphatase production
Tellurite-polymyxin-egg yolk (TPEY)	Potassium tellurite Polymyxin B	Egg yolk reaction Black colonies

Information kindly supplied by OXOID Ltd, Basingstoke, UK.

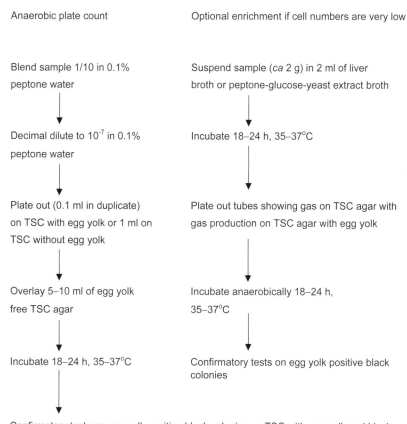

Anaerobic plate count

Optional enrichment if cell numbers are very low

Blend sample 1/10 in 0.1% peptone water

Suspend sample (*ca* 2 g) in 2 ml of liver broth or peptone-glucose-yeast extract broth

Decimal dilute to 10^{-7} in 0.1% peptone water

Incubate 18–24 h, 35–37°C

Plate out (0.1 ml in duplicate) on TSC with egg yolk or 1 ml on TSC without egg yolk

Plate out tubes showing gas on TSC agar with gas production on TSC agar with egg yolk

Overlay 5–10 ml of egg yolk free TSC agar

Incubate anaerobically 18–24 h, 35–37°C

Incubate 18–24 h, 35–37°C

Confirmatory tests on egg yolk positive black colonies

Confirmatory tests on egg yolk positive black colonies on TSC with egg yolk and black colonies on TSC without egg yolk

Fig. 6.20 Procedure for the isolation and quantification of *Cl. perfringens*.

sensitivity limit of 1 ng toxin/ml of material and take approximately 4 hours to obtain a result. However, no test has been developed for the emetic toxin due to problems of purification.

6.5.11 Mycotoxins

ELISA and latex agglutination assays are commercially available for the detection of aflatoxins. The toxins are visualised under uv light since the four naturally produced aflatoxins, B_1, B_2, G_1 and G_2, are named after their blue and green fluorescence colours produced. ELISA assays can be used for the quantitative detection of moulds in foods. Compared to ELISA tests, which require 5 to 10 hours to complete, the latex agglutination method is much faster, taking only 10 to 20 minutes. Trichothecenes, fumonisins

Table 6.15 Approved media for the detection of *Cl. perfringens*.

Regulatory body	Detection media	Other media
AOAC/FDA Bacteriological Analytical Manual (BAM) (1992)	Tryptose-sulphite-cycloserine agar (TSC)	Thioglycollate medium Modified cooked meat medium or chopped liver broth Iron-milk medium Lactose-gelatin medium Buffered motility-nitrate medium Sporulation broth Spray's fermentation medium AE sporulation medium Duncan-Strong sporulation medium
Agriculture Canada (1988) Methods Manual	Tryptose-sulphite-cycloserine agar (TSC)	Modified cooked meat medium Fluid thioglycollate medium Nitrate-motility medium Lactose-gelatin medium
German Institute of Normalisation (DIN 10165) adopted by the Official Collection of Investigative Procedures L06-00-20 (1984)	Tryptose-sulphite-cycloserine agar (TSC)	Thioglycollate medium Nitrate-motility medium Lactose-gelatin medium
British Standards Institution (BSI) BS5763: Part 9: 1986 (1992) and ISO 7937 BS4285: Part 3: Section 3–13 (1990)	Tryptose-sulphite-cycloserine agar (SC agar)	Fluid thioglycollate medium Nitrate-motility medium Lactose-gelatin medium

(Contd)

Table 6.15 *(Contd)*

Regulatory body	Detection media	Other media
Standards Australia Committee on Food Microbiology AS 1766.2.8. (1991)	(a) Tryptose-sulphite-cycloserine (TSC) agar with egg yolk (b) TSC agar without egg yolk	(a) Neomycin-cooked meat medium (b) Lactose-egg yolk agar (LEY) (c) Lactose-gelatin medium (LG) (d) Nitrate-motility medium (NM)
Normalisation Française (AFNOR) V08-056 (1994)	Tryptose-sulphite-cycloserine agar (TSC)	Fluid thioglycollate medium Lactose-sulphite medium (LS)
Spanish Ministeria de Sanidad y Consumo Instituto de Salidad, Carlos III. Technical Manual	(a) Tryptose-sulphite-cycloserine agar (TSC) (b) TSC agar with egg yolk (c) Oleandomycin-polymyxin-sulphadiazine Perfringens agar (OPSP) (d) Tryptose-sulphite-neomycin agar (TSN)	(a) Lactose fermentation agar (b) Tryptone-yeast extract agar (c) Lactose-milk-egg yolk agar (d) Reinforced clostridial medium (RCM) with added neomycin sulphate (100 µg/ml) (e) Cooked meat-neomycin medium (f) Neomycin-blood agar

Information kindly supplied by OXOID Ltd, Basingstoke, UK.

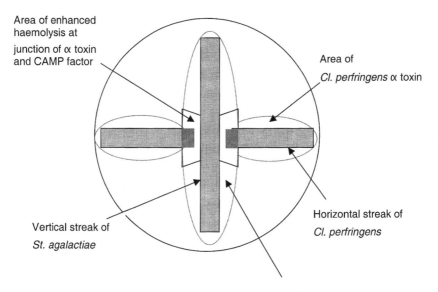

Area of enhanced haemolysis at junction of α toxin and CAMP factor

Area of *Cl. perfringens* α toxin

Vertical streak of *St. agalactiae*

Horizontal streak of *Cl. perfringens*

Area of CAMP factor β haemolysis

Fig. 6.21 Reversed CAMP test for *Cl. perfringens* haemolysis.

and moniliformin (from *Fusarium* spp.) can be detected using high performance liquid chromatography.

6.5.12 Viruses

Large volumes (25–100 g) of food need to be analysed for enteric viruses since the level of food contamination is assumed to be low. The use of tissue cultures to detect enteric viruses requires the viruses to be separated from the bulk of the food. The viruses may be enumerated using plaque assays to dilution end point assays. Cell culture technique, however, is not applicable for hepatitis A virus and Norwalk-like virus due to the lack of host cell line. The sequencing of Norwalk-like virus genomes has facilitated the development of the reverse transcriptase polymerase chain reaction (RT-PCR; Fig. 6.8) for detection and characterisation of the viruses. RT-PCR is applicable to faecal, vomit and shellfish samples. The Norwalk virus can be characterised by sequencing an amplified product of the test, and this has enabled outbreaks in different locations to be linked to a single source. Electron microscopic analysis of samples is only of use with samples during the first 2 days of symptoms, whereas RT-PCR is usable within 4 days of symptoms appearing.

Table 6.16 Approved media for *B. cereus* isolation.

Regulatory body	Detection media	Other media
Spanish Ministeria de Sanidad y Consumo Instituto de Salidad Carlos III, Technical Manual	Mannitol-egg yolk agar (MYA) Mannitol-egg yolk polymyxin agar (MYP)	Anaerobic agar without glucose and indicator Glucose broth without phosphate Glucose agar Nutrient agar
German Institute of Normalisation (DIN 10198) adopted by the Official Collection of Investigative Procedures 00.00–25 (1992)	Polymyxin-pyruvate-egg yolk-mannitol bromothymol blue agar (PEMBA) Mannitol-egg yolk polymyxin agar (MYP)	Glucose medium Voges-Proskauer medium Nitrate medium
Standards Australia AS 1766.2.6–1991 Food Microbiology Method 2.6	Polymyxin-pyruvate-egg yolk-mannitol-bromothymol blue agar (PEMBA) Tryptone-soy-polymyxin broth (MPN technique)	
AOAC/FD BAM (1992)	Mannitol-egg yolk-polymyxin agar (MYP) Tryptone-soy-polymyxin	Phenol red-glucose broth Nitrate broth Modified VP medium Tyrosine agar Lysozyme broth
Agriculture Canada (1988) Methods Manual	Mannitol-egg yolk-polymyxin agar (MYP)	Tryptone soya agar Sheep blood agar Modified VP medium Nitrate broth
BS 5763: Part II (1988) ISO 7932 (1987) BS 4285: Part 3: Section 3.12 1989 (1994)	Mannitol-egg yolk-polymyxin agar (MYP)	Glucose agar Voges-Proskauer (VP) medium Nitrate medium
Normalisation Française (AFNOR) XPV 08-058 (1995)	Mannitol-egg yolk-polymyxin agar (MYP)	Blood agar Motility agar Voges-Proskauer medium Nitrate broth

Vegetative cells

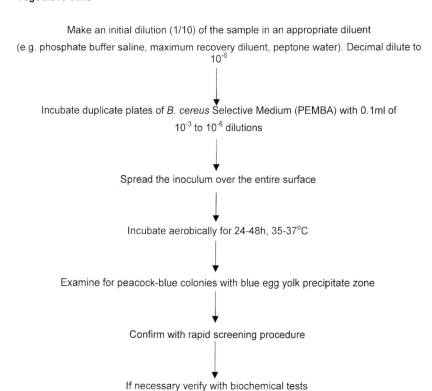

Make an initial dilution (1/10) of the sample in an appropriate diluent
(e.g. phosphate buffer saline, maximum recovery diluent, peptone water). Decimal dilute to
10^{-6}

Incubate duplicate plates of *B. cereus* Selective Medium (PEMBA) with 0.1ml of
10^{-3} to 10^{-6} dilutions

Spread the inoculum over the entire surface

Incubate aerobically for 24-48h, 35-37°C

Examine for peacock-blue colonies with blue egg yolk precipitate zone

Confirm with rapid screening procedure

If necessary verify with biochemical tests

Spores

If necessary to count spores, then first treat with heat or alcohol to destroy the vegetative cells.

Heat treatment: Heat the initial 1/10 suspension for 15 min, 70°C. Proceed as for detection of vegetative cells

Alcohol treatment: Dilute the initial suspension 1:1 in 95% ethyl alcohol. Leave for 30 min, room temperature. Proceed as for detection of vegetative cells, adjusting the dilutions to account for the 1:1 dilution of the sample suspension in alcohol.

Fig. 6.22   Typical procedure for detection of *B. cereus*.

6.6   Accreditation schemes

In order to have confidence that the method of choice could have detected the target organism the methods used should be accredited. Therefore the method must be validated against standard tests using collaborative studies. Validation of laboratory procedures will form part of a company's quality control system. There are various international bodies which validate detection methods. The Association of Analytical Chemists (AOAC) international guidelines (Anon 1999c) are the most widely accepted; other relevant bodies include UKAS (UK), EMMAS (Europe), AFNOR (France), DIN (Germany), the European MICROVAL and EFSIS.

7

FOOD SAFETY
MANAGEMENT TOOLS

Although industry and national regulators strive for production and processing systems which ensure that all food is 'safe and wholesome', complete freedom from risks is an unattainable goal. Safety and wholesomeness are related to a level of risk that society regards as reasonable in the context, and in comparison with other risks in everyday life.

The microbiological safety of foods is principally assured by:

- Control at the source
- Product design and process control
- The application of good hygienic practices during production, processing (including labelling), handling, distribution, storage, sale, preparation and use
- The above, in conjunction with the application of the Hazard Analysis Critical Control Point (HACCP) system. This preventative system offers more control than end-product testing, because the effectiveness of microbiological examination in assessing the safety of food is limited

Consideration of safety needs to be applied to the complete food chain, from food production on the farm, or equivalent, through to the consumer. To achieve this an integration of food safety tools is required (see back to Fig. 1.1):

- Good Manufacturing Practice (GMP)
- Good Hygienic Practice (GHP)
- Hazard Analysis Critical Control Point (HACCP)
- Microbiological Risk Assessment (MRA)
- Quality management; ISO series
- Total Quality Management (TQM)

These tools can be implemented worldwide, which can ease communication with food distributors and regulatory authorities, especially at port of entry.

7.1 Microbiological safety of food in world trade

It is important to understand that the ratification of the World Trade Organisation Agreement is a major factor in developing new hygiene measures for the international trade in food. There has been a noted requirement for quantitative data on the microbial risks associated with different classes of foods, and traditional Good Manufacturing Practice based food hygiene requirements (i.e. end-product testing) are being challenged. Subsequently risk assessment (Chapter 9) as a decision-making criterion for risk management will put more emphasis on predictive microbiology for the generation of exposure data and establishing Critical Limits for HACCP schemes (Section 2.8).

The Final Act of the Uruguay Round of multilateral trade negotiations established the World Trade Organisation (WTO) to succeed the General Agreement on Tariffs and Trade (GATT, Section 10.3). The Final Act led to the 'Agreement on the Application of Sanitary and Phytosanitary Measures' (SPS Agreement, Section 10.3) and the 'Agreement to Technical Barriers to Trade' (TBT Agreement). These are intended to facilitate the free movement of foods across borders, by ensuring that means established by countries to protect human health are scientifically justified and are not used as nontariff barriers to trade in foodstuffs. The Agreement states that SPS measures based on appropriate standards, codes and guidelines developed by the Codex Alimentarius Commission (Section 10.2) are deemed to be necessary to protect human health and consistent with the relevant GATT provisions.

The SPS Agreement is of particular relevance to food safety. It provides a framework for the formulation and harmonisation of sanitary and phytosanitary measures. These measures must be based on science and implemented in an equivalent and transparent manner. They cannot be used as an unjustifiable barrier to trade by discriminating among foreign sources of supply or providing an unfair advantage to domestic producers. To facilitate safe food production for domestic and international markets, the SPS Agreement encourages governments to harmonise their national measures or base them on international standards, guidelines and recommendations developed by international standard setting bodies.

The purpose of the TBT Agreement is to prevent the use of national or regional technical requirements, or standards in general, as unjustified technical barriers to trade. The agreement covers all types of standards including quality requirements for foods (except requirements related to sanitary and phytosanitary measures), and it includes numerous measures designed to protect the consumer against deception and economic fraud. The TBT Agreement also places emphasis on international standards. WTO members are obliged to use international standards or parts of them, except where the international standard would be ineffective or inappropriate in the national situation.

The WTO Agreement also states that risk assessment should be used to provide the scientific basis for national food regulations on food safety and SPS measures, by taking into account risk-assessment techniques developed by international organisations.

Because of SPS and WHO the Codex standards, guidelines and other recommendations have become the baseline for safe food production and consumer protection. Hence the Codex Alimentarius Commission has become the reference for international food safety requirements.

7.2   The management of hazards in food which is in international trade

The management of microbiological hazards for foods in international trade can be divided into five steps (ICMSF 1997):

(1) *Conduct a risk assessment.* The risk assessment and consequential risk management decisions provide a basis for determining the need to establish microbiological safety objectives.

(2) *Establish food safety objectives.* A microbiological food safety objective is a statement of the maximum level of a microbiological hazard considered acceptable for consumer protection. These should be developed by governmental bodies with a view to obtaining consensus with respect to a food in international trade.

(3) *Achievable food safety objectives.* The food safety objectives should be achievable throughout the food chain. They can be applied through the general principles of food hygiene and any product specific codes and HACCP systems. The HACCP requirements must be developed by the food industry.

(4) *Establish microbiological criteria, when appropriate.* This must be performed by an expert group of food microbiologists.

(5) *Establish acceptance procedures for the food at port of entry.* A list of approved suppliers as determined by inspection of facilities and operations, certification, microbiological testing and/or other testing such as pH and water activity measurements.

Therefore an understanding of HACCP (Section 7.3), Microbiological Risk Assessment (Section 9.1), Food Safety Objectives (Section 9.4) and Microbiological Criteria (Chapter 8) is required.

7.3 Hazard Analysis Critical Control Point (HACCP)

Most food poisoning can be prevented by the application of the basic principles of food hygiene throughout the food chain. This is achievable through:

(1) Education and training of food handlers and consumers in the application of safe food production practices.
(2) Inspection of premises to ensure consistent hygienic practices are adhered to.
(3) Microbiological testing for the presence or absence of foodborne pathogens and toxins.

Traditionally, the safety of food being produced was through end-product testing for the presence of food pathogens or their toxins. This retrospective approach, however, does not guarantee safe food for several reasons, as will be covered in Section 8.6.3. Hygienic practice can be consistently achieved through the adoption of HACCP. This approach to safe food production has been accepted worldwide. Therefore factory inspection has changed in emphasis towards inspection of HACCP implementation.

The HACCP system for managing food safety was derived from two major developments:

• The HACCP system in the 1960s, as pioneered by the Pillsbury Company, the United States Army and NASA as a collaborative development for the production of safe foods for the USA space programme. NASA required 'zero defects' in food production to guarantee the astronauts' food (Bauman 1974).
• Total Quality Management systems which emphasise a total system approach to manufacturing that could improve quality while lowering costs.

HACCP is a scientifically based protocol. It is systematic, identifies specific hazards and measures for their control to ensure the safety of food. It is interactive, involving the food plant personnel. HACCP is a tool to assess hazards and establish control systems. Its focus is on the prevention of problems occurring, rather than a reliance on end-product testing. HACCP schemes can accommodate change, such as advances in equipment design, processing procedures or technological developments (Codex Alimentarius Commission 1997a).

HACCP can be applied throughout the food chain from primary production to final consumption and its implementation should be guided by scientific evidence of risks to human health. As well as enhancing food safety, implementation of HACCP can provide other significant benefits. In addition, the application of HACCP systems can aid inspection by regulatory authorities and promote international trade by increasing confidence in food safety.

The successful application of HACCP requires the full commitment and involvement of management and the work force. It also requires a multidisciplinary approach. The application of HACCP is compatible with the implementation of quality management systems, such as the ISO 9000 series, and is the system of choice in the management of food safety within such systems.

Guidance for the establishment of HACCP based systems is detailed in *Hazard Analysis and Critical Control Point System and Guidelines for its Application* (Codex Alimentarius Commission 1997a).

Table 7.1 outlines the implementation of HACCP in a food company. The seven steps in bold are known as the 'Seven Principles of HACCP' and are covered in Section 7.4.

Table 7.1 Establishing and implementing HACCP.

Decision by management to use the HACCP system
Training and formation of the HACCP team
Development of the HACCP plan document, including the following parts (CRC 1997a):
Assemble the HACCP team
Describe the food product and its distribution
Identify the intended use and consumers
Develop and verify the flow diagram for the production process
On-site confirmation of the flow diagram
(1) Conduct a hazard analysis
(2) Determine the critical control points (CCPs) (see Fig. 7.1)
(3) Establish critical limits
(4) Establish monitoring procedures
(5) Establish corrective actions
(6) Establish verification procedures
(7) Establish documentation and record-keeping procedures

7.4 Outline of HACCP

In order to produce a safe food product with negligible levels of food-borne food pathogens and toxins, three controlled stages must be established:

(1) Prevent microorganisms from contaminating food through hygienic production measures. This must include an examination of ingredients, premises, equipment, cleaning and disinfection protocols and personnel.
(2) Prevent microorganisms from growing or forming toxins in food. This can be achieved through chilling, freezing or other processes such as reduction of water activity or pH. These processes, however, do not destroy microorganisms.
(3) Eliminate any foodborne microorganisms, for example by using a time and temperature processing procedure or by the addition of suitable preservatives.

These principles are given by the Codex Alimentarius Commission (1993) and the National Advisory Committee on Microbiological Criteria for Foods (NACMCF 1992, 1997). Hence it is an internationally recognised procedure. Differences arise, however, with the interpretation and implementation of the seven principles. The approach here will adhere to the Codex, WHO and NACMCF (1997) format. The Codex principles are given in bold lettering. It should be noted that the Codex document reverses principles 6 and 7.

7.4.1 Food hazards

A hazard is defined as:

A biological, chemical, or physical agent in a food, or condition of a food, with the potential to cause an adverse health effect.

Biological hazards are living organisms, including microbiological organisms, bacteria, viruses, fungi and parasites.

Chemical hazards are in two categories: naturally occurring poisons and chemicals or deleterious substances. The first group covers natural constituents of foods that are not the result of environmental, agricultural, industrial or other contamination. Examples are aflatoxins and shellfish poisons. The second group covers poisonous chemicals or deleterious substances which are intentionally or unintentionally added to foods at some point in the food chain. This group of chemicals can include pesticides and fungicides and well as lubricants and cleaners.

A physical hazard is any physical material not normally found in food which causes illness or injury. Physical hazards include glass, wood, stones and metal which may cause illness and injury.

Examples of hazards are given in Table 7.2.

Table 7.2 Hazards associated with food (adapted from Snyder 1995).

Biological	Chemical	Physical
Macrobiological	Vertinary residues: antibiotics,	Glass
Microbiological	growth stimulants	Metal
Pathogenic bacteria	Plasticisers and packaging	Stones
sporeforming	migration	Wood
organisms	Chemical residues: pesticides,	Plastic
non-sporeforming	cleaning fluids	Parts of pests
Parasites and protoza	Allergens	Insulation
Viruses	Toxic metals: lead, cadmium	Bone
Mycotoxins	Food chemicals: preservatives,	Fruit pits
	processing aids	
	Polychlorinated biphenyls (PCBs)	
	Printing inks	
	Prohibited substances	

7.4.2 Pre-HACCP principles

Before the HACCP seven principles can be applied there is the need to:

(1) *Assemble the HACCP team.* The food operation should ensure that the appropriate product specific knowledge and expertise is available for the development of an effective HACCP plan. Optimally, this may be accomplished by assembling a multidisciplinary team. Where such expertise is not available on site, expert advice should be obtained from other sources.

(2) *Describe the product.* A full description of the product should be drawn up, including relevant safety information such as composition, physical/chemical structure (including a_w, pH, etc.), microcidal/static treatments (heat treatment, freezing, brining, smoking, etc.), packaging, durability and storage conditions and method of distribution.

(3) *Identify the intended use.* The intended use should be based on the

expected uses of the product by the end user or consumer. In specific cases, vulnerable groups of the population, e.g. institutional residents, may have to be considered.

(4) *Construct a flow diagram*. The flow diagram should be constructed by the HACCP team. It should cover all steps in the operation. When applying HACCP to a given operation, consideration should be given to steps preceding and following the specified operation.

(5) *On-site confirmation of the flow diagram*. The HACCP team should confirm the processing operation against the flow diagram during all stages and hours of operation, and amend the flow diagram where appropriate.

7.4.3　Principle 1

Conduct a hazard analysis. Prepare a list of steps in the process where significant hazards occur and describe the preventative measures.

The HACCP team should list all of the hazards that may reasonably be expected to occur at each step from primary production, processing, manufacture and distribution until the point of consumption. The evaluation of hazards should include:

• The likely occurrence of hazards and severity of their adverse health effects.
• The qualitative and/or quantitative evaluation of the presence of hazards.
• Survival or multiplication of the microorganisms of concern.
• Production or persistence in foods of toxins, chemicals or physical agents.
• Conditions leading to the above.

The hazard analysis should identify which hazards can be eliminated or reduced to acceptable levels as required for the production of safe food.

Table 7.3 is an example from the Food and Drug Administration (USA) of a worksheet for use in documenting the identified hazards.

7.4.4　Principle 2

Identify the Critical Control Points (CCPs) in the process.

The HACCP team must identify the Critical Control Points (CCP) in the production process which are essential for the elimination or acceptable

Table 7.3 Example of a hazard analysis worksheet.

Firm Name: _____   Product Description: _____

Firm Address: _____   Method of Storage and Distribution: _____

_____   Intended Use and Consumer: _____

_____

(1)	(2)	(3)	(4)	(5)	(6)
Ingredient/processing step	Identify potential hazards introduced, controlled or enhanced at this step(1)	Are any potential food-safety hazards significant? (Yes/No)	Justify your decisions for column 3.	What preventative measures can be applied to prevent the significant hazards?	Is this step a critical control point? (Yes/No)
	Biological				
	Chemical				
	Physical				
	Biological				
	Chemical				
	Physical				
	Biological				
	Chemical				
	Physical				
	Biological				
	Chemical				
	Physical				

reduction of the hazards which were identified in principle 1. These CCPs are identified through the use of decision trees such as that given in Fig. 7.1 (NACMCF 1992). A series of questions is answered, which lead to the decision as to whether the control point is a CCP. Other decision trees can be used if appropriate.

A CCP must be a quantifiable procedure in order for measurable limits and monitoring to be achievable in principles 3 and 4. If a hazard is identified for which there is no control measure in the flow diagram, then the product or process should be modified to include a control measure. Some groups have differentiated CCPs into CCP1 as primary CCPs which eliminate hazards and CCP2 which only reduce hazards. This approach has the advantages of identifying which hazards are of crucial importance. For example, milk pasteurisation would be CCP1, whereas assessment of raw milk on receipt would be CCP2.

7.4.5 Principle 3

Establish Critical Limits for preventative measures associated with each identified CCP.

Critical Limits must be specified and validated if possible for each CCP. The Critical Limit will describe the difference between safe and unsafe products at the CCP. The Critical Limit must be a quantifiable parameter: temperature, time, pH, moisture or a_w, salt concentration or titratable acidity, available chlorine.

7.4.6 Principle 4

Establish CCP monitoring requirements. Establish procedures from the results of monitoring to adjust the process and maintain control.

Monitoring is the scheduled measurement or observation of a CCP relative to its critical limits. The monitoring procedures must be able to detect loss of control at the CCP. Monitoring should provide the information in time (ideally on-line) for correcting the control measure. Ideally by following the trend in measured values then correction can take place before deviation from the critical limits.

The monitoring data must be evaluated by a designated person with knowledge and authority to carry out corrective actions when indicated. If monitoring is not continuous, then the amount or frequency of monitoring must be sufficient to guarantee the CCP is in control. Most monitoring procedures for CCPs will need to be done rapidly because they

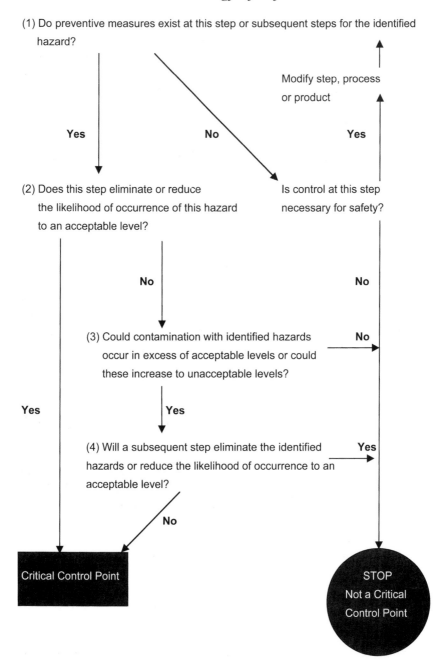

Fig. 7.1   Critical Control Point decision tree.

relate to on-line processes and there will not be time for lengthy analytical testing. Physical and chemical measurements are often preferred to microbiological testing because they may be done rapidly and can often indicate the microbiological control of the product. Microbiological testing based on single samples or sampling plans will be of limited value in monitoring those processing steps which are CCPs, this is mainly because the conventional microbiological methods are too time consuming for effective feedback. All records and documents associated with monitoring CCPs must be signed by the person(s) doing the monitoring and by a responsible reviewing official(s) of the company.

7.4.7 *Principle 5*

Establish corrective actions to be taken when monitoring indicates a deviation from an established Critical Limit.

Specific corrective actions must be developed for each CCP in order to deal with deviations from the Critical Limits. The remedial action must ensure that the CCP is under control and that the affected product is appropriately recycled or destroyed as appropriate.

7.4.8 *Principle 6*

Establish procedures for verification that the HACCP system is working correctly.

Verification procedures must be established. These will ensure that the HACCP plan is effective for the current processing procedure. NACMCF (1992) gives four processes in the verification of HACCP:

(1) Verify that Critical Limits at CCPs are satisfactory.
(2) Ensure that the HACCP plan is functioning effectively.
(3) Document periodic revalidation, independent of audits or other verification procedures.
(4) It is the government's regulatory responsibility to ensure that the HACCP system has been correctly implemented.

The frequency of verification should be sufficient to confirm that the HACCP system is working effectively. Verification and auditing methods, procedures and tests, including random sampling and analysis, can be used to determine whether the HACCP system is working correctly.

Examples of verification activities include:

- Review of the HACCP system and its records
- Review of deviations and product dispositions
- Confirmation that CCPs are kept under control.

Verification should be conducted:

- Routinely, or on an unannounced basis, to ensure that CCPs are under control.
- When there are emerging concerns about the safety of the product.
- When foods have been implicated as a vehicle of foodborne disease.
- To confirm that changes have been implemented correctly after an HACCP plan has been modified.
- To assess whether a HACCP plan should be modified due to a change in the process, equipment, ingredients, etc.

7.4.9   Principle 7

Establish effective record-keeping procedures that document the HACCP system.

HACCP procedures should be documented. Records must be kept to demonstrate safe product manufacture and that appropriate action has been taken for any deviations from the Critical Limits.
Examples of documentation are:

- Hazard Analysis
- CCP determination
- Critical Limit determination

Examples of records are:

- CCP monitoring activities
- Deviations and associated corrective actions
- Modifications to the HACCP system

Table 7.4 is a useful worksheet for documenting the HACCP system.

7.5   Microbiological criteria and HACCP

Since HACCP is meant to be the internationally accepted means of ensuring food safety, then it may be argued that end-product testing is no longer required. However, food distributors (supermarket chains, etc.)

Table 7.4 Example of an HACCP form.

Firm Name: _____   Product Description: _____

Firm Address: _____   Method of Storage and Distribution: _____

_____   Intended Use and Consumer: _____

(1)	(2)	(3)	(4)	(5)	(6)	(7)	(8)	(9)	(10)
Critical Control Point (CCP)	Significant Hazard(s)	Critical Limits for each Preventive Measure	Monitoring				Corrective Action(s)	Records	Verification
			What	How	Frequency	Who			

Signature of Company Official: _____   Date: _____

Page 1 of ___

have continued to insist on microbiological criteria (Chapter 8). The main role of end-product testing should be during the production of a new food and for verification of the HACCP scheme. In the context of HACCP it must be emphasised that only those microbiological criteria which refer to foodborne hazards can be considered since HACCP systems are specifically designed to control hazards which are significant for food safety. Microbiological criteria which occur in legislation and in the published literature may provide a reference point to assist food manufacturers in evaluating their data during hazard analysis (Section 8.9).

It should be noted that single sample analysis is limited as an HACCP verification procedure. Basically, one can only be 100% certain the food does not contain a hazard by analysing 100% of the food. The statistical confidence of sample analysis and sampling plans is covered in Section 8.6.3. The microbiological analysis of food using sampling plans results in the acceptance or rejection of a 'lot'. This includes the statistical probability that a lot may be falsely accepted or rejected, known as 'producer's risk' and 'consumer's risk'. Hence statistical analysis demonstrates that microbiological testing is not a stand-alone verification tool for hazard analysis systems.

7.6   Microbiological hazards and their control

7.6.1   Sources of microbiological hazards

The common sources of foodborne pathogens are:

(1) *The raw ingredients.* The microbial flora of raw ingredients entering a food factory can be controlled by using reputable suppliers, certificates of quality, temperature monitoring on receipt, etc. Ingredients may be rejected on receipt if they fail to comply with agreed standards.

(2) *Personnel.* It has been estimated that approximately 1 in 50 employees is shedding 10^9 pathogens per gram faeces without showing any clinical symptoms (Snyder 1995; Table 7.5). Subsequently poor personal hygiene such as failing to wash hands after going to the toilet can leave 10^7 pathogens under the finger nails. The movement of personnel in a factory needs to be controlled. Frequently there will be a low risk (or low care) area partitioned from the high risk (or high care) area. Essentially the low risk area is where the ingredients are stored, weighed, mixed and cooked. After cooking, the food enters the high risk area. Increased diligence is required since there will not be any further heat treatment to destroy

Table 7.5 Human carriage of foodborne pathogens (Snyder, 1995).

Faeces	*Salmonella* spp., *E. coli*, *Shigella* spp., Norwalk-like virus, hepatitis A, *Giardia lamblia*	1 in 50 employees are highly infective and shedding 10^9 pathogens/g faeces
Vomit	Norwalk-like virus	10 viral particles are infectious
Skin, nose, boils and skin infections	*St. aureus*	60% of the population are carriers, there are 10^8 organisms per drop of pus
Throat and skin	*Streptococcus* group A	10^5 *Strep. pyogenes* in a cough

any bacteria, etc., from personnel or environmental contamination and also to prevent the growth of surviving organisms.

(3) *The environment (air, water and equipment).* The microbial quality of water needs to be monitored frequently as this can have severe repercussions if it is contaminated with potential foodborne pathogens. The accumulation of food residues can result in biofilm formation (Section 4.6), which requires physical removal as sanitisers will be neutralised by the organic material.

The severity of the illness caused by the organisms can be determined from standard texts, especially the ICMSF books, and is simplified in Table 5.2. The likely occurrence of the foodborne pathogen can also be determined from ICMSF and related literature (Table 7.6). A list of Web pages is given in the Appendix section, where further information on foodborne pathogens can be obtained; a standard reference source is the FDA Bad Bug book. There are numerous Web sites giving information on current food poisoning outbreaks worldwide which may be accessed without any charges. National food poisoning statistics can be obtained from regulatory authorities.

Details of the food's pH, a_w, heat treatment process, etc., can be used to predict the microorganisms of concern in the foodstuff (Section 2.5 and Table 2.7). Although the infectious dose (Table 7.7; see Section 9.1.4) has been determined for a number of foodborne pathogens, this should be used only for indicative purposes since the susceptibility of the consumer will vary according to their immunocompetence, age and general health.

The microbial load of a product after processing can be predicted using storage tests and microbiological challenge testing, supplemented with predictive modelling (Sections 2.8 and 4.3). The shelf life of the product

Table 7.6   Sources of foodborne pathogens.

Food	Pathogen	Incidence (%)
Meat, poultry and eggs	*C. jejuni*	Raw chicken and turkey (45-64)
	Salmonella spp.	Raw poultry (40-100), pork (3-20), eggs (0.1) and shellfish (16)
	St. aureus[a]	Raw chicken (73), pork (13-33) and beef (16)
	Cl. perfringens[b]	Raw pork and chicken (39-45)
	Cl. botulinum	
	E. coli O157:H7	Raw beef, pork and poultry
	B. cereus[b]	Raw ground beef (43-63)
	L. monocytogenes	Red meat (75), ground beef (95)
	Y. enterocolitica	Raw pork (48-49)
	Hepatitis A virus	
	Trichinella spiralis	
	Tapeworms	
Fruit and vegetables	*C. jejuni*	Mushrooms (2)
	Salmonella spp.	Artichoke (12), cabbage (17), fennel (72), spinach (5)
	St. aureus[a]	Lettuce (14), parsley (8), radish (37)
	L. monocytogenes	Potatoes (27), radishes (37), bean sprouts (85), cabbage (2), cucumber (80)
	Shigella spp.	
	E. coli O157:H7	Celery (18) and coriander (20)
	Y. enterocolitica	Vegetables (46)
	A. hydrophila	Broccoli (31)
	Hepatitis A virus	
	Norwalk-like virus	
	Giardia lamblia	
	Cryptosporidium spp.	
	Cl. botulinum	
	B. cereus[b]	
	Mycotoxins	
Milk and dairy products	*Salmonella* spp.	
	Y. enterocolitica	Milk (48-49)
	L. monocytogenes	Soft cheese and pate (4-5)
	E. coli	
	C. jejuni	
	Shigella spp.	
	Hepatitis A virus	
	Norwalk-like virus	
	St. aureus[a]	
	Cl. perfringens[b]	

(Contd)

Table 7.6 *(Contd)*

Food	Pathogen	Incidence (%)
	B. cereus[b]	
	Mycotoxins	
Shellfish and	*Salmonella* spp.	
fin fish	*Vibrio* spp.	Raw seafood (33–46)
	Shigella spp.	
	Y. enterocolitica	
	E. coli	
	Cl. botulinum[b]	
	Hepatitis A virus	
	Norwalk-like virus	
	Giardia lamblia	
	Cryptosporidium spp.	
	Metabolic byproducts	
	Algal toxins	
Cereals, grains,	*Salmonella* spp.	
legumes and	*L. monocytogenes*	
nuts	*Shigella* spp.	
	E. coli	
	St. aureus[a]	
	Cl. botulinum[b]	
	B. cereus	
	Mycotoxins[a]	
Spices	*Salmonella* spp.	
	St. aureus[a]	
	Cl. perfringens[b]	
	Cl. botulinum[b]	
	B. cereus[b]	
Water	*Giardia lamblia*	Water (30)

[a] Toxin not destroyed by pasteurisation.
[b] Sporeforming organism. Not killed by pasteurisation.
Sources: various including Synder (Hospitality Institute of Technology and Management, Web address in Appendix), ICMSF (1998).

can be determined according to chemical, physical and microbiological parameters (Section 4.2).

7.6.2 Temperature control of microbiological hazards

The cooking step is an obvious CCP for which Critical Limits of temperature and time can be set, monitored and corrected. The time and temperature of the cooking process should be designed to give at least a 6

Table 7.7 Infectious dose of enteric pathogens.

Organism	Estimated minimum infectious dose
Non-sporeforming bacteria	
C. jejuni	1000
Salmonella spp.	10^4-10^{10}
Sh. flexneri	10^2->10^9
Sh. dysenteriae	10-10^4
E. coli	10^6->10^7
E. coli O157:H7	10-100
St. aureus	10^5->10^6/g[a]
V. cholerae	1000
V. parahaemolyticus	10^6-10^9
Y. enterocolitica	10^7
Sporeforming bacteria	
B. cereus	10^4-10^8
Cl. botulinum	10^{3a}
Cl. perfringens	10^6-10^7
Viruses	
Hepatitis A	< 10 particles
Norwalk-like virus	< 10 particles

[a] Viable count able to produce sufficient toxin to elicit a physiological response.

log kill of vegetative cells, that is 10^7 cells/g reduced to 10 cells/g (Section 2.4.2). It will not kill spores and hence the time required to cool the food to a safe temperature needs to be monitored to prevent spore outgrowth. A list of time and temperature equivalencies is given in Table 7.8.

The objective of a 6 log kill is based upon two tenets:

(1)   Food pathogens are in the minority of the microbial flora on meat. Subsequently any raw meat kept at temperatures that would allow food pathogens to multiply to 1×10^6/g would be rejected as spoilt. The cooking time and temperature regime is sufficient to kill 1×10^6/g of an enteric pathogen in the coldest area of the product. Since the number of food pathogens should be less than 1×10^6/g, the chance of vegetative cells of a pathogen surviving is reduced to a negligible level.

(2)   Since blood coagulates at 73.9°C, visual observation can indicate whether the correct temperature has been reached. Therefore clear juices indicate that the blood proteins have been denatured and likewise bacterial vegetative cells. Hence the phrases 'cook until well done' and 'cook until the juices run clear' are often quoted. However, bacterial spores do survive this time and temperature regime.

Table 7.8 Equivalent cooking time and temperature regimes (Department of Health, UK).

Temperature ($°C$)[a]	Time
60	45 minutes
65	10 minutes
70	2 minutes
75	30 seconds
80	6 seconds

[a] To convert to $°F$ use the equation $°F = (9/5)°C + 32$. As a guidance: $60°C = 140°F$.

An example of the effect of cooking temperature on the survival of *E. coli* O157:H7 is given in Table 7.9. It can be seen that accurate temperature monitoring is necessary since the relationship of cell death to temperature is logarithmic. In other words, a small decrease in cooking temperature can result in considerable numbers of cells surviving the process.

Table 7.9 Effect of cooking temperature on the survival of *E. coli* O157:H7.

Temperature ($°C$)	Time for viable count to be reduced by 6 log cycles[a] (min)
58.2	28.2
62.8	2.8
67.5	0.28

[a] For example 10^7 cfu/g to 10 cfu/g.
Data taken from Table 2.3; $D_{62.8} = 0.4$, $z = 4.65$.

The cooling period must be short enough to prevent spore outgrowth and germination from mesophilic *Bacillus* and *Clostridium* spp. Notably, the cooking process would create an anaerobic environment in the food which is ideal for *Clostridium* spp. The temperature growth range of *Cl. perfringens* is 10 to 52°C (Table 2.7), hence cooling regimes should be designed to minimise the time the food is between these temperatures. A lower limit of 20°C is normally adopted, since it only multiplies slowly below this value.

The control of the holding temperature of foods after processing and before consumption is crucial for safe food production. General recommended holding temperatures are:

Cold served foods < 8°C
Hot served foods > 63°C

The microbiological 'danger zone' is often given as temperatures between 8 and 63°C (Section 2.5.4). Ready to serve food should be at these temperatures for the minimum time since it is proposed that any surviving mesophilic pathogens or post-processing contaminants are able to multiply rapidly at these temperatures to infectious levels. There is, however, no microbiological reason for the high (63°C) hot serve temperature since no foodborne pathogen grows between 53 and 63°C. The best explanation is that hot servers do not maintain the temperature of food accurately and the extra 10°C is a safety barrier. These holding temperatures do vary between countries and, where necessary, local regulatory authorities should be consulted.

At cold serve temperatures (< 8°C) *Clostridium* spp. are not a problem since they do not grow below 10°C. *St. aureus* produces toxins above 10°C, but is able to multiply down to 6.1°C. Subsequently temperature abuse can lead to *St. aureus* growth and toxin production. Since *St. aureus* toxin is not inactivated by reheating to 72°C, *St. aureus* toxin production must be prevented. Chilled foods should be stored below 3.3°C to assure safety from spore germination and subsequent toxin production by *Cl. botulinum* E. In contrast, *Cl. botulinum* types A and B only multiply and produce toxin above 10°C. *L. monocytogenes* and *Y. enterocolitica* have a minimum growth temperature of −0.4 and −1.3°C, respectively. Therefore the cold holding time must be limited for foods which are not going to be reheated.

A P value (length of cooking period at 70°C; Section 2.4.1) of 30 to 60 minutes will give a shelf life of at least 3 months, depending upon the risk factors involved.

7.6.3   Non-temperature control of microbiological hazards

In addition to temperature, the water activity (a_w), pH and the presence of preservatives are important factors in the control of microbiological hazards. The limits of microbial growth are covered in Section 2.5. The pH and water activities of various foods are given in Table 2.7 and can be used to predict the relevant microbiological hazards.

7.7   HACCP plans

As explained in Section 7.3, HACCP plans require a team of company personnel who understand the details of the food's production. The following outlines are given as examples to be modified accordingly. HACCP has not been applied extensively in the retail trade, however there is a source of HACCP-TQM technical guidelines available on the Web (see Appendix for URL address).

7.7.1 Production of pasteurised milk

Figure 7.2 outlines the production of pasteurised milk. Receipt of raw milk is a CCP since it must be from certified tuberculosis-free herds and must be checked for total microbial load. The crucial steps are the pasteurisation process (72°C, 15 seconds), which is designed to eliminate milk-borne pathogens, and the cooling period (less than 6°C in 1.5 hours) which prevents the growth of surviving organisms. Therefore pasteurisation is a

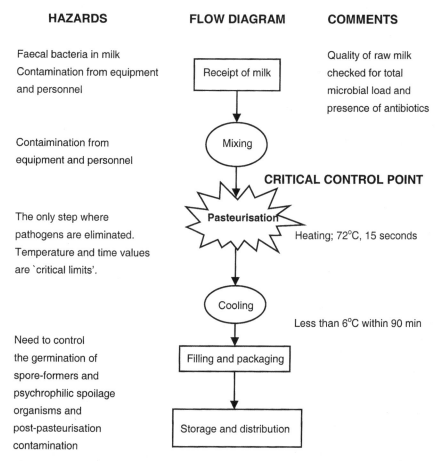

HAZARDS

Faecal bacteria in milk
Contamination from equipment
and personnel

Contaimination from
equipment and personnel

The only step where
pathogens are eliminated.
Temperature and time values
are `critical limits'.

Need to control
the germination of
spore-formers and
psychrophilic spoilage
organisms and
post-pasteurisation
contamination

FLOW DIAGRAM

Receipt of milk

Mixing

Pasteurisation

Cooling

Filling and packaging

Storage and distribution

COMMENTS

Quality of raw milk
checked for total
microbial load and
presence of antibiotics

CRITICAL CONTROL POINT

Heating; 72°C, 15 seconds

Less than 6°C within 90 min

Fig. 7.2 HACCP flow diagram for the pasteurisation of milk.

CCP and the times and temperatures are the Critical Limits. Since there are no further heat treatments, contamination of the pasteurised milk (especially from raw milk) must be avoided and therefore aseptic packaging is also a CCP.

7.7.2 Swine slaughter in the abattoir

Animal slaughter must be controlled to minimise contamination of the meat with intestinal bacteria, which may include foodborne pathogens (Fig. 7.3). Common pathogens associated with pigs are *C. jejuni*, *C. coli*,

Process step	Hygiene aspect	Preventive actions	CCP
Lairge ↓ Stunning	Contamination between animals	Cleaning & disinfection	
Killing ↓	Contamination from tools	Cleaning & disinfection	
Scalding ↓	Reduction of bacterial levels	Time–temperature	
Dehairing ↓	Contamination from machines	Cleaning & disinfection	
Singeing/flaming ↓	Reduction of bacterial levels	Time–temperature	
Polishing ↓	Contamination from machines	Cleaning & disinfection	
Evisceration ↓	Contamination from gut material Contamination from tools	Enclosure of rectum Working instructions Disinfection of tools	YES
Splitting ↓	Contamination via splitter/saw	Line speed Water temperature	
Meat inspection ↓	Contamination from inspection	Disinfection of tools	YES
Deboning of head	Contamination from head	Working instructions Disinfection of tools	YES

Fig. 7.3 Control of pathogens during swine slaughter (adapted from Borch *et al.* 1996).

Salmonella spp and *Y. enterocolitica*. Pathogenic microorganisms commonly found in the processing environment are *L. monocytogenes*, *St. aureus* and *A. hydrophila*. Since the meat will subsequently be heat treated to eliminate pathogens, it is not necessary to produce a sterile animal carcass at the abattoir stage of food production. Therefore the HACCP approach is to minimise pathogen contamination and not necessarily to eliminate it. The key steps are enclosure of the rectum to prevent faecal contamination of the meat and disinfection of tools which can be contaminated with pathogenic organisms and vehicles of cross-contamination.

7.7.3 Chilled food manufacture

Chilled foods are typically multi-component foods and therefore the HACCP flow diagram is complicated since each ingredient is represented. Figure 7.4 shows the flow diagram for a chicken salad (Anon 1993b). There are nine identified CCPs. These are primarily after the mixing step and are more fully explained in Table 7.10. Since there is no heat treatment careful temperature monitoring during storage is required to prevent significant multiplication of any foodborne pathogens present.

The process flow diagram for a bakery is reproduced in Fig. 7.5. The monitoring of cleaning and fumigation, pest control and glass-perspex control are separate from the production process.

7.7.4 Generic models

HACCP implementation has mainly been product specific. However, a more amenable approach for food manufacturers with large numbers of products is generic HACCP (Anon 1993a and b). A number of generic HACCP models are available which have been produced by Canada, New Zealand and the USA (see Web pages in Appendix). The generic models from the USDA for freshly squeezed orange juice and dried meats (beef jerky) are given in Tables 7.11 and 7.12.

7.8 Sanitation Standard Operating Procedures (SSOPs)

In America the regulatory use of HACCP also added 'Sanitation Standard Operating Procedures', or SSOPs, as part of the necessary prerequisite programmes for the HACCP system (see Table 7.12). These are written procedures determining how a food processor will meet sanitation conditions and practices in a food plant. This has emphasised the need for adequate cleaning, etc., in a food production plant where cross-

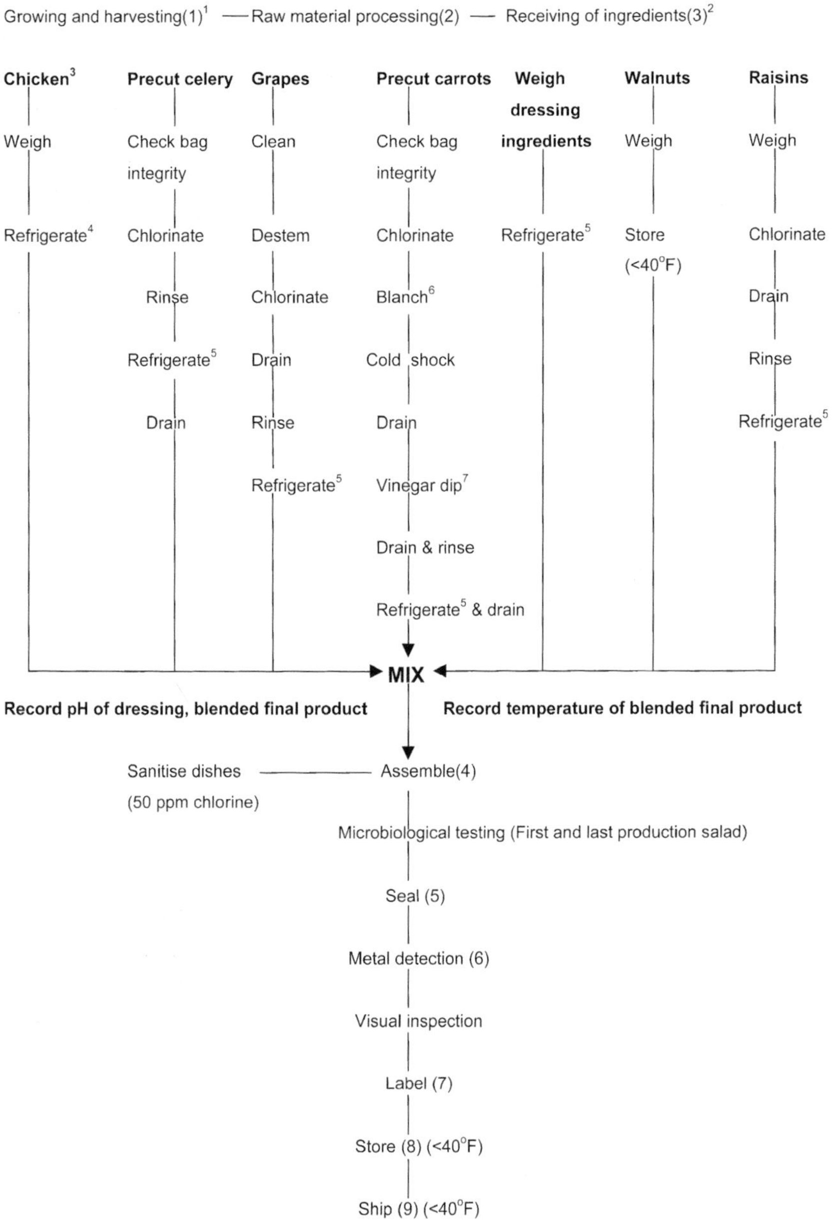

Growing and harvesting(1)[1] —— Raw material processing(2) —— Receiving of ingredients(3)[2]

Chicken[3]	Precut celery	Grapes	Precut carrots	Weigh dressing ingredients	Walnuts	Raisins
Weigh	Check bag integrity	Clean	Check bag integrity		Weigh	Weigh
Refrigerate[4]	Chlorinate	Destem	Chlorinate	Refrigerate[5]	Store (<40°F)	Chlorinate
	Rinse	Chlorinate	Blanch[6]			Drain
	Refrigerate[5]	Drain	Cold shock			Rinse
	Drain	Rinse	Drain			Refrigerate[5]
		Refrigerate[5]	Vinegar dip[7]			
			Drain & rinse			
			Refrigerate[5] & drain			

▶ MIX ◀

Record pH of dressing, blended final product   **Record temperature of blended final product**

Sanitise dishes —————— Assemble(4)
(50 ppm chlorine)

Microbiological testing (First and last production salad)

Seal (5)

Metal detection (6)

Visual inspection

Label (7)

Store (8) (<40°F)

Ship (9) (<40°F)

Fig. 7.4 Chunky chicken salad HACCP flow diagram (Anon 1993b, adapted and reprinted with permission from *Journal of Food Protection*, **56**, 1077-84). Footnotes: [1] Numbers in parentheses are item numbers in Table 7.10. [2] Refers to all raw ingredients of the next stage (chicken, carrots, etc.). [3] < 28°F. [4] 0-10°F, < 24 hours. [5] < 40°F, < 24 hours. [6] 15 seconds. [7] < 40°C, 30 minutes.

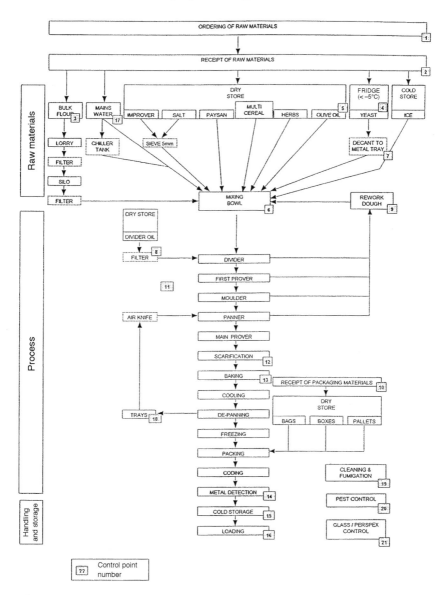

Fig. 7.5 HACCP flow diagram for a bakery.

contamination and post-processing contamination are potential hazards. Other countries have not adopted a separate SSOPs approach, but kept cleaning within the company's HACCP scheme. *The key to good sanitary practice is that sanitisers should be applied to areas that are already visually clean.* Otherwise the organic material will neutralise the active agent in the sanitiser, rendering it ineffective.

Table 7.10 HACCP worksheet for Critical Control Points: chilled chicken salad (Anon 1993b, adapted and reprinted with permission from *Journal of Food Protection*, **56**, 1077–84).

Item	Hazard	Control	Limit	Monitoring freq./ documentation	Action (for exceeding limit)	Personnel responsible
(1) Growing and harvesting	Antibiotic, chemical	Supplier compliance to raw material specifications	Regulatory approved chemicals and antibiotics; specified tolerances	Certificates of conformance for each lot. Random annual monitoring by QC	Reject lot	Shipping and receiving operator
(2) Raw material processing	Chemical, physical and microbiological	Certified supplier HACCP programme	Free of pathogens and foreign material	Monitor supplier HACCP programme	Reject as supplier	QC must monitor HACCP programme
(3) Raw ingredient storage temperatures	Microbiological	Compliance to raw material specifications	Precooked chicken 0–10°F. Fruits/ vegetables ≤40°F (4.4°C). All other ≤80°F (24.4°C)	Check recorders in coolers daily	Report to shift supervisor and QC. Investigate time/ temperature abuse and evaluate risk	Line operator
(4) Filling/ assembly	Microbiological	Temperature control specifications	Components/ finished product temperature ≤40°F (4.4°C)	Check temperatures once/shift	Report to shift supervisor and QC. Adjust temperature according to specification and evaluate risk	Line operator
(5) Sealer	Microbiological	Proper sealer settings to ensure all trays hermetically sealed and have consistent seal bead	Upper limit tolerance of sealer	Sealer settings checked every 15 min. Visual inspection every 2 h	Examine all packages since last check. Peel off membrane and reseal	Seal inspector

(Contd)

Table 7.10 *(Contd)*

Item	Hazard	Control	Limit	Monitoring freq./documentation	Action (for exceeding limit)	Personnel responsible
(6) Labeller	Incorrect code, dates, traceability	Legible and correct dates on all sleeves and shippers	Use proper labels	Each batch and at product changeover	Destroy incorrect/illegible code dated sleeves and shippers	Packaging operator
(7) Metal detector	Physical (metal)	On-line metal detector	No metal detected	Calibrate before each batch or every 2 h	Locate source of contamination. Destroy product	Packaging operator
(8) Plant storage of finished product	Microbiological	Product storage temperature ≤ 40°F (4.4°C)	≤ 40°F (4.4°C)	Check continuous recorders in coolers daily	Report to shift supervisor and QC. Place product on hold. Investigate time/temperature abuse and evaluate risk.	Shipping and receiving operator
(9) Shipping temperature	Microbiological	Ship ≤ 40°F (4.4°C)	≤ 40°F throughout shipment	Each load temperature monitored by continuous recorder	Report to shift supervisor and QC. Place product on hold. Investigate time/temperature abuse and evaluate risk	Shipping operator/stock handler

Table 7.11 Generic HACCP plan for fresh squeezed orange juice.

(1) Ingredient/ processing step	(2) Identify potential hazards introduced, controlled or enhanced at this step	(3) Are any potential food safety hazards significant? (Yes/No)	(4) Justify your decision for column 3	(5) What preventative measure(s) can be applied to prevent the significant hazards?	(6) Is this step a critical control point? (Yes/No)
		HAZARD ANALYSIS WORKSHEET			
Receiving	BIOLOGICAL Pathogens CHEMICAL Pesticides PHYSICAL	Y	Environmental contamination Spraying	Control source. Letter of warrantee. No raw fertilisation. No drops. No harvesting before pre-harvest interval.	N
Grading	BIOLOGICAL Pathogens CHEMICAL Pesticides PHYSICAL	Y	Assumed to enter on product	Wash and sanitise remove culls, decayed product, spills, etc.	N
Wash/sanitising	BIOLOGICAL Pathogens CHEMICAL Pesticides PHYSICAL	Y	Assumed to be on product at receipt	Using effective sanitisation	Y

(Contd)

Table 7.11 *(Contd)*

HAZARD ANALYSIS WORKSHEET

(1) Ingredient/ processing step	(2) Identify potential hazards introduced, controlled or enhanced at this step	(3) Are any potential food safety hazards significant? (Yes/No)	(4) Justify your decision for column 3	(5) What preventative measure(s) can be applied to prevent the significant hazards?	(6) Is this step a critical control point? (Yes/No)
Extraction	BIOLOGICAL Pathogens CHEMICAL Pesticides PHYSICAL	N	Controlled by sanitising	SSOP and GMP	N
Filling	BIOLOGICAL Pathogens CHEMICAL Pesticides PHYSICAL	N	Controlled by sanitising	SSOP and GMP	N
Chilling/holding	BIOLOGICAL Pathogens CHEMICAL Pesticides PHYSICAL	N	Need to maintain at refrigeration temperatures (41°F)	SSOP and GMP	N

Source: FAMFES, web address in Appendix.

Table 7.12 Generic HACCP plan for dried meats (beef jerky).

Hazard description	Critical limits	Monitoring procedures	Deviation procedures	Verification procedure	HACCP records
Step: receiving; fresh and frozen meat					
Bacterial growth	Meat shall exhibit no unusual colours or odours	For each lot, meat Receiver shall: perform sensory evaluation of meat prior to unloading He will inspect each combo of fresh meat or select 2 boxes of frozen meat per pallet for evaluation	Meat Receiver: Fresh or frozen do not unload truck, notify Production Manager	HACCP Coordinator will run a Lab analysis on a sample of meat: once a week 3 samples (3 cases composite per sample) HACCP coordinator to verify once a week for proper monitoring at receiving (organoleptic inspection & temperature) (Specification to be established)	Beef Jerky Meat Receiving Log Beef Jerky Processing Record
Bacterial growth	Meat temp. shall be less than or equal to 4°C for fresh or solid frozen (frozen) Microbiological criteria to be established by each plant	For each lot, meat receiver shall: record temp. of meat at centre and on surface of each combo of fresh meat or freezing condition of 2 boxes of frozen meat per pallet	If meat temp. is 4 to 7°C; Meat Receiver shall notify QC, use 2nd thermometer and retest If meat temperature is >7°C, then hold lot and notify QC who will investigate possible causes, and decide on proper disposition	HACCP Coordinator will run a Lab analysis on a sample of meat: once a week 3 samples, (3 cases composite per sample) HACCP coordinator to verify once a week for proper monitoring at receiving (organoleptic inspection & temperature) (Specification to be established)	Beef Jerky Meat Receiving Log

(Contd)

Table 7.12 (Contd)

Hazard description	Critical limits	Monitoring procedures	Deviation procedures	Verification procedure	HACCP records
Foreign particles in meat	Product must be from listed supplier with contractual specification (supplier shall do an acceptable boneless meat resinspection programme)	Receiver shall: Ensure product is received from a listed supplier	If product is arriving from registered establishment that is not a QC listed supplier the lot is identified, inspected and is used for other products if found unacceptable	Inspect the cold storage to ensure all non-listed suppliers have been identified	Receiving Log
Step: receiving: dry ingredients					
Spices have an excessive bacterial load or contain foreign material	Specifications to be developed by the Establishment. Each lot certified by supplier	Receiver shall monitor each lot received for supplier's certification (lab. analysis)	Receiver holds the lot until appropriate receiving certification & notify QC	QC samples one lot every 10 lots & run analyses vs critical limits	Receiving records QC lot results
Step: receiving: receiving packaging material					
Packaging material not of food quality	All packaging material must be 'approved' by AAFC to be received	Receiver ensures that product is approved before allowing receiving	Return lot and notify QC	Once every 3 months QC inspect stocks of packaging material to ensure all material is approved QC review receiving bills once a month	Receiving records

(Contd)

Table 7.12 *(Contd)*

Hazard description	Critical limits	Monitoring procedures	Deviation procedures	Verification procedure	HACCP records
Step: thawing of meat					
Bacterial pathogens growth	Thawing of meat shall be done in a room not exceeding 10°C	Designated employee records thawing room temperature every 6 hours	Designated employee reports out of spec. room temp. to the Maintenance Foreperson and holds batch for QC evaluation	QC audits temp. records weekly QC takes temp. & correlates results with records on a weekly basis Audits employees' work in monitoring time/temp weekly QC takes (monthly) samples to evaluate microbial acceptability of thawed product Criteria to be developed by plant	Plant temperature record Beef Jerky Processing Record
Bacterial pathogens growth	Meat surface temp. max. 7°C	Designated employee records meat temp. every 6 hours and before meat is removed from the room	If meat surface temperature > 7°C, hold the batch, submit samples for analysis. Process product, but hold until results are available. If APC > 5 × 10⁵/g, test finished product, otherwise release product. If *St. aureus* > 5 × 10⁵/g, test for pathogens, if AAPC/HPB guidelines are not met, condemn	QC takes temp. & correlate results weekly Audit employees' work in monitoring time/temp weekly QC takes (monthly) samples to evaluate microbial acceptability of thawed product Criteria to be developed by plant	Meat Temp. Record-Thaw Room Beef Jerky Processing Record

(Contd)

Table 7.12 *(Contd)*

Hazard description	Critical limits	Monitoring procedures	Deviation procedures	Verification procedure	HACCP records
Step: weighing of nitrite					
Excessive amount may be toxic. Insufficient amount may allow for germination of *C. botulinum* spores	Minimum concentration is 100 ppm. Maximum concentration is 200 ppm.	Designated employee keeps records on number of bags prepared and their use, write amount of nitrite in each bag	Designated employee notifies QC for follow-up and assessment	QC verifies operator's records, inventory of nitrite, weight of nitrite bags	Beef Jerky Ingredient Record Nitrite Conc. Out of Compliance Sheet Beef Jerky Complete Processing Record
Step: formulation mixing with spices and massaging					
Lack or excess Na/K Nitrite	100–200 ppm	Foreperson ensures that Operator checks off chart for ingredients Correct ingredient and product identification tags	Product since last satisfactory monitoring is held by foreperson for rework	Foreperson to verify operator's records 'X' times/week QC verify operator's records & laboratory test records 'X' times/month QC verifies once/week a sample for nitrite content	Spice room records Operator's records Foreperson verification records Chemical lab records QC verification records

(Contd)

Table 7.12 (*Contd*)

Hazard description	Critical limits	Monitoring procedures	Deviation procedures	Verification procedure	HACCP records
Step: screening of spice					
Foreign material > 2 mm	No contamination is acceptable	Foreperson ensures that the operator on duty screens spice for every batch and notifies QC of any finding If foreign material is found, remove it from product, label sample bag with manufacturer's codes and send to QC	Foreperson re-instruct the employee on normal procedures and completes record Quality Team shall submit material to supplier. All remaining spice from the lot in question shall be returned to the manufacturer for rescreening or credit	QC audit employee practices weekly	Jerky Spice Check Sheet Beef Jerky Processing Record
Step: smoking of spice					
Pathogen growth due to incorrect come up time. Bacterial survival due to improper time/temp.	Exact Cooking Cycles to be defined here (come up time) *House #A* 'X' hrs at 71°C *House #B* 'Y' hour at 70°C, 'Z' hrs at 75°C	Smokehouse Operator on Duty: Checks cooking cycles against critical limits and initials smokehouse chart recorders	Smokehouse Operator on Duty is to add more time if time is insufficient. If temperature is insufficient, hold batch. Take a sample of the product. Notify QC	Maintenance Foreman: Is to check accuracy of smokehouse thermometer using a mercury based reference thermometer Also check chart recorder accuracy (time and temperature) monthly QC reviews and initials smokehouse charts weekly	Chart Recorders Beef Jerky Moisture Level Sheet Beef Jerky Processing Record

(*Contd*)

Table 7.12 *(Contd)*

Hazard description	Critical limits	Monitoring procedures	Deviation procedures	Verification procedure	HACCP records
Step: a_w check					
Product not shelf stable due to an excessive a_w	Final product have a_w less than or equal to 0.85	Foreperson checks at a frequency 'X' that the designated employee tests all lots prior to release	Hold batch, notify quality team	QC audits at a frequency 'X', test procedures, test tool calibration, log book QC audits operator's practices for every smokehouse load of beef jerky for 5 minutes	a_w records
			Hold all suspect product until QC assessment adjust/reject lot		
			QC shall submit samples for analysis. Process product, but hold until results are available. If results APC > 5×10^5/g, test finished product. If *St. aureus* > 1×10^4/g, test for toxin, if positive, condemn product, if finished product APC > 5×10^3/g, test for pathogens, if AAPC/HPB guidelines are not met, condemn product		

(Contd)

Table 7.12 *(Contd)*

Hazard description	Critical limits	Monitoring procedures	Deviation procedures	Verification procedure	HACCP records
Step: metal detector					
Presence of undetected metal particles	No contamination is acceptable for ferrous and non-ferrous particles > 2 mm	Operator on duty sets up the metal detector using two dummy bags of jerky each containing a metal particle (one contains ferrous, the other, non-ferrous) 2 mm in diameter. The dummy bags shall be passed through the detector every hour to ensure that detector is functioning	Operator on Duty Sets rejected package aside. Pass rejected package through detector. If package fails test, set aside and notify quality team, open package and examine each piece of meat QC will examine the identified metal to locate the most likely source of metal, check the equipment and take further action as needed	QC audits the operation by passing the dummy bags (same as for monitoring) through the detector once per shift	Metal Detector Check Sheet Rejected Jerky Metallic Particle Record Beef Jerky Processing Record

Date: _____   Approved by: _____

Source: Canadian Food Inspection Agency, Web address in Appendix.

7.9 Good Manufacturing Practice (GMP) and Good Hygiene Practice (GHP)

GMP covers the fundamental principles, procedures and means needed to design an environment suitable for the production of food of acceptable quality. GHP describes the basic hygienic measures which establishments should meet and which are the prerequisite(s) to other approaches, in particular HACCP. GMP/GHP requirements have been developed by governments, the Codex Alimentarius Committee on Food Hygiene (FAO/WHO) and the food industry, often in collaboration with other groups and food inspection and control authorities.

General GHP requirements usually cover the following:

(1) The hygienic design and construction of food manufacturing premises.
(2) The hygienic design, construction and proper use of machinery.
(3) Cleaning and disinfection procedures (including pest control).
(4) General hygienic and safety practices in food processing including:
   • The microbial quality of raw foods
   • The hygienic operation of each process step
   • The hygiene of personnel and their training in the hygiene and safety of food.

GMP codes and the hygiene requirements they contain are the relevant boundary conditions for the hygienic manufacture of foods. They should always be applied and documented. *No food processing methods should be used to substitute for GMPs in food production and handling.*

7.10 Quality Systems

The quality of a product may be defined as its measurement against a standard regarded as excellent at a particular price which is satisfactory both to the producer and to the consumer. The aim of quality assurance is to ensure that a product conforms as closely as possible and consistently to that standard at all times. Quality can be measured in terms of the senses (for example taste panels), chemical composition, physical properties and the microbiological flora, both quantitative and qualitative.

An 'excellent' quality at a specific price can only be achieved by answering in the affirmative the questions asked by Quality Assurance and Quality Control: 'Yes, we are doing the right thing and we are doing things right'. To answer thus means that a successful quality assurance scheme (and likewise a food hygiene scheme with all its ramifications) must be operating with the full and sincere support of top management and all those concerned with the implementation of the scheme. There must, of course, be full control over all aspects of production so that a consistency of product quality is maintained. This necessitates strict control over the initial quality of raw materials, over the process itself and over packaging and storage conditions. In microbiological terms the build-up of bacteria during a process run must be monitored at critical processing steps.

7.11　Total Quality Management (TQM)

TQM represents the 'cultural' approach of an organisation; it is centred on quality and based on the participation of all members of the organisation and the concept of continuous improvement. It aims at long-term success through customer satisfaction, benefits to the members of the organisation and benefits to society in general.

Total Quality Management (TQM) is similar in emphasis to quality assurance and has been defined as 'a continual activity, led by management, in which everybody recognises personal responsibility for safety and quality' (Shapton & Shapton 1991). This requires the company as a whole achieving uniformity and quality of a product and thus safety is maintained. Hence TQM is broader in scope than HACCP, including quality and customer satisfaction in its objectives (Anon 1992).

Quality systems cover organisational structure, responsibilities, procedures, processes and the resources needed to implement comprehensive quality management. They apply to, and interact with, all phases of a product cycle. They are intended to cover all quality elements.

A combination of HACCP, quality systems, TQM and business excellence provides a total systems approach to food production, which embraces quality, productivity and food safety. TQM and quality systems provide the philosophy, culture and discipline necessary to commit every member of an organisation to the achievement of all managerial objectives related to quality. Within this framework, the inclusion of HACCP as the key specific safety assurance plan provides the necessary confidence that products will conform to safety needs and that no unsafe or unsuitable product will leave the production site. Collectively, these tools provide a comprehensive and proactive approach to further reduce the risk of food safety problems.

7.12　ISO 9000 Series of standards

In 1987 the International Organization for Standardization in Geneva Switzerland published the *ISO 9000* standards. They are equivalent to the European standards EN29000 series and the British Standards *BS*

5750:1987. The *ISO 9000* series is composed of five standards:

- *ISO 9000* Quality management and quality assurance standards – guidelines for selection and use.
- *ISO 9001* Quality systems – model for quality assurance in design/ development, production, installation and servicing.
- *ISO 9002* Quality systems – model for quality assurance in production and installation.
- *ISO 9003* Quality systems – model for quality assurance in final inspection and test.
- *ISO 9004* Quality management and quality system elements – guidelines.

These standards can be used as a starting point for designing TQM programmes and should be used for managing the HACCP system.

8

MICROBIOLOGICAL CRITERIA

8.1 International Commission on Microbiological Specifications for Foods

The ICMSF book *Microorganisms in Foods. Vol. 2. Sampling for Microbiological Analysis: Principles and Specific Applications* was first published in 1974. It recognised the need for scientifically based sampling plans for foods in international trade. The sampling plans were originally designed for application at port of entry, that is when there is no prior knowledge on the history of the food. This pioneering work set forth the principles of sampling plans for the microbiological evaluation of foods and is also known as attributes and variables sampling, depending on the extent of microbiological knowledge of the food. Subsequent books were published by the ICMSF to assist in the interpretation of microbiological data, such as *Microbiological Ecology of Foods* and *Factors Affecting Growth and Death of Microorganisms*.

The second edition of the book in 1986 took note of the successful application of the acceptance sampling plans on a worldwide basis, not only at an international level but at national and local levels by both industry and regulatory agencies. Additionally, Harrigan & Park (1991) wrote an excellent book on the practical mathematics of sampling plans.

Microbiological criteria should be established according to these principles, and be based on scientific analysis and advice, and where sufficient data are available, on a risk analysis appropriate to the foodstuff and its use (Codex Alimentarius Commission 1997b). These criteria may be relevant to the examination of foods, including raw materials and ingredients of unknown or uncertain origin, or when no other means of verifying the efficacy of HACCP based systems and good hygienic practices are available. Microbiological criteria may also be used to determine whether processes are consistent with the General Principles of Food Hygiene. Microbiological criteria are not normally suitable for monitoring Critical Limits as defined in the HACCP system.

The purpose of establishing microbiological criteria is to protect the public's health by providing food which is safe, sound and wholesome and to meet the requirements of fair trade practices. The presence of criteria, however, does not protect the consumer's health since it is possible for a food lot to be accepted which contains defective units. Microbiological criteria may be applied at any point along the food chain and can be used to examine food at the port of entry and at the retail level.

8.2 Codex Alimentarius principles for the establishment and application of microbiological criteria for foods

As described in Chapter 10 Section 2, the Codex Alimentarius Commission (CAC) has become the reference for international food safety requirements. The CAC (1997b) definition of a microbiological criterion is thus:

A microbiological criterion for food defines the acceptability of a product or a food lot, based on the absence or presence, or number of microorganisms including parasites, and/or quantity of their toxins/ metabolites, per unit(s) of mass, volume, area or lot.

By this definition, a microbiological criterion consists of:

(1) A statement of the microorganisms of concern and/or their toxins/ metabolites and the reason for that concern (see Section 3.9 and Chapter 5).
(2) The analytical methods for their detection and/or quantification (see Chapter 6).
(3) A plan defining the number of field samples to be taken and the size of the analytical unit (see Sections 8.3 and 8.4).
(4) Microbiological limits considered appropriate to the food at the specified point(s) of the food chain (see Section 8.7).
(5) The number of analytical units that should conform to these limits (see Section 8.6).

A microbiological criterion should also state:

• The food to which the criterion applies
• The point(s) in the food chain where the criterion applies
• Any actions to be taken when the criterion is not met

A microbiological criterion should be established and applied only where there is a definite need and where its application is practical. Such need is demonstrated, for example, by epidemiological evidence that the food under consideration may represent a public health risk and that a criterion is meaningful for consumer protection, or as the result of a risk assessment. The criterion should be technically attainable by applying Good Manufacturing Practices (Codes of Practice). Criteria should be reviewed periodically for relevance with respect to emerging pathogens (Section 5.7), changing technologies and new understandings of science.

A sampling plan includes the sampling procedure and the decision criteria to be applied to a lot, based on examination of a prescribed number of sample units and subsequent analytical units of a stated size by defined methods. A well-designed sampling plan defines the probability of detecting microorganisms in a lot, but it should be borne in mind that no sampling plan can ensure the absence of a particular organism. Sampling plans should be administratively and economically feasible (CAC 1997b and c). In particular, the choice of sampling plans should take into account:

(1) Risks to public health associated with the hazard.
(2) The susceptibility of the target group of consumers.
(3) The heterogeneity of distribution of microorganisms where variables sampling plans are employed.
(4) The Acceptable Quality Level (AQL) and the desired statistical probability of accepting a nonconforming lot.

The AQL is the percentage of nonconforming sample units in the entire lot for which the sampling plan will indicate lot acceptance for a prescribed probability (usually 95%; Section 8.6). For many applications, two- or three-class attribute plans may prove useful (Section 8.5).

The statistical performance characteristics or operating characteristics curve should be provided in the sampling plan (Section 8.6.2). Performance characteristics provide specific information to estimate the probability of accepting a nonconforming lot. The time between taking the field samples and analysis should be as short as reasonably possible, and during transport to the laboratory the conditions (e.g. temperature) should not allow increase or decrease of the numbers of the target organism, so that the results reflect, within the limitations given by the sampling plan, the microbiological conditions of the lot.

8.3 Sampling plans

Just as it is impractical to test a sample for every possible food pathogen, so it is also impractical to test 100% of an ingredient or end-product.

Therefore there is a need to use sampling plans to test a batch of material appropriately and give a statistical basis for acceptance or rejection of a food lot.

Microbiological sampling plans are frequently used in food production, import control and in contractual agreements with suppliers and customers. Sampling plans are used to check the microbiological status of a commodity, its compliance to safety requirements and adherence to Good Hygiene Practices (GHP; Section 7.9) during or after manufacture. The results from single sample examinations may give valuable baseline data which can be used for trend analysis, particularly where samples form part of a specific survey. However, statistical principles should be observed when sampling particular food commodities many of which (usually end products) are heterogeneous, even when they have a similar formulation. In situations where a food inspector might be concerned about a particular food, a sample taken for microbiological analysis may provide evidence that food hygiene regulations have been contravened or may provide the basis for additional inspection and/or examination. The single sample concept is likely to retain a role in assessing food safety in small-scale food production businesses which will have fewer resources for implementation of HACCP but will nevertheless have to take proper account of the risk to public health posed by each individual operation.

There are two types of sampling plans:

(1) Variables plans, when the microbial counts conform to a log-normal distribution (Section 8.4). These data would be known by a producer and are not applicable to an importer at the port of entry situation.
(2) Attribute plans, when no prior knowledge of the distribution of microorganisms in the food is known, i.e. at port of entry or the distribution of target organism is not log-normal (Sections 8.5 and 8.6).

Attributes sampling plans can be according to either a two-class plan or a three-class plan. The two class plan is used almost exclusively for pathogens, whereas a three class plan is frequently used to examine for hygiene indicators. The main advantage of using sampling plans is that they are statistically based and provide a uniform basis for acceptance against defined criteria. The type of sampling plan required can be decided using Fig. 8.1.

Attributes plans also involve the concept of 'choice of case', based on microbiological risk. 'Case' is a classification of sampling plans ranging from 1 (least stringent) to 15 (most stringent). The choice of case, and therefore the sampling plan, depends on:

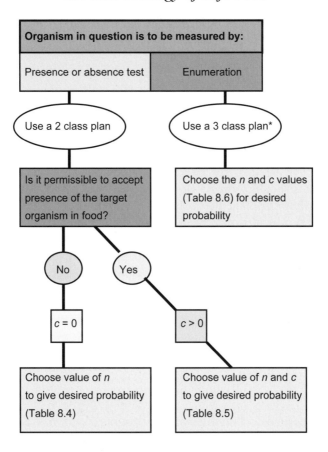

* A variable plan may be applicable if the organism is distributed in a log-normal fashion.

Fig. 8.1　Decision tree for choosing a sampling plan (adapted from ICMSF 1986).

• The relative severity of the hazard to food quality or consumer health on the basis of the microorganisms involved (see Chapter 5).
• The expectation of their destruction, survival or multiplication during normal handling of the food (see Chapter 2).

Table 8.1 and the decisions trees of Figs 8.1 and 8.2 should be referred to when deciding on the appropriate sampling plan. For example, cases 1 to 3 refer to utility applications, such as shelf life, whereas cases 13, 14 and 15 refer to severely hazardous foodborne pathogens. The severity of the

Table 8.1 Sampling plans in relation to degree of health hazard and conditions of use (ICMSF, 1986. Reprinted with permission of the University of Toronto Press).

Type of hazard	Conditions in which food is expected to be handled and consumed after sampling		
	Reduce degree of hazard	Cause no change in hazard	May increase hazard
No direct health hazard			
Utility, e.g. reduced shelf life, and spoilage	Case 1 3-class, $n = 5$, $c = 3$	Case 2 3-class, $n = 5$, $c = 2$	Case 3 3-class, $n = 5$, $c = 1$
Health hazard			
Low, indirect (indicator)	Case 4 3-class, $n = 5$, $c = 3$	Case 5 3-class, $n = 5$, $c = 2$	Case 6 3-class, $n = 5$, $c = 1$
Moderate, direct, limited spread	Case 7 3-class, $n = 5$, $c = 2$	Case 8 3-class, $n = 5$, $c = 1$	Case 9 3-class, $n = 10$, $c = 1$
Moderate, direct, potentially extensive spread	Case 10 2-class, $n = 5$, $c = 0$	Case 11 2-class, $n = 10$, $c = 0$	Case 12 2-class, $n = 20$, $c = 0$
Severe, direct	Case 13 2-class, $n = 15$, $c = 0$	Case 14 2-class, $n = 30$, $c = 0$	Case 15 2-class, $n = 60$, $c = 0$

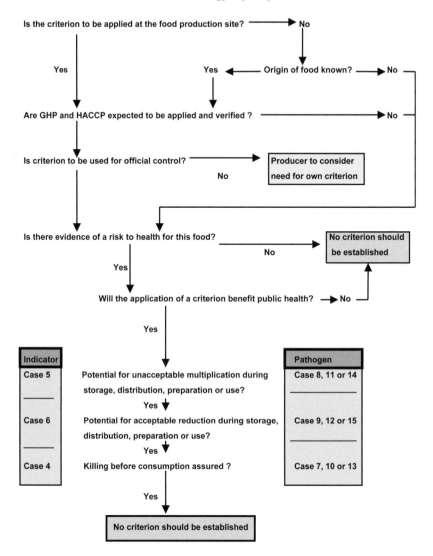

Fig. 8.2   Decision tree for choice of criteria for microbiological pathogens and indicator organisms (Source: EU microbiological criteria Web site).

microbiological hazard has been covered in Chapter 5 and the foodborne pathogens grouped (1986 version) to assist in referring to Table 8.1 (see Table 5.2). Note the ICMSF is altering its categorisation in a forthcoming publication.

Sampling plans and recommended microbiological limits were published by ICMSF (1986) for the following foods:

(1) Raw meats, processed meats, poultry and poultry products
(2) Pet foods
(3) Dried milk and cheese
(4) Pasteurised liquid, frozen and dried egg products
(5) Seafoods
(6) Vegetables, fruit, nuts and yeast
(7) Cereals and cereal products
(8) Peanut butter and other nut butters
(9) Cocoa, chocolate and confectionery
(10) Infant and certain categories of dietetic foods
(11) Bottled water.

8.4 Variables plans

Variables plans can be applied when the number of microorganisms in the food is distributed log-normally, that is the logarithms of the viable counts conform to a normal distribution (Fig. 8.3; Kilsby *et al.* 1979). This applies to certain foods which have been analysed over a period of time by the producer and therefore does not apply at port of entry.

If the microorganisms' distribution within a lot is log-normal then sampling plans can be used to develop acceptance sampling plans. The sample mean (x) and standard deviation (s) are determined from previous studies and are used to decide whether a 'lot' of food (Section 8.6.1) should be accepted or rejected. In addition:

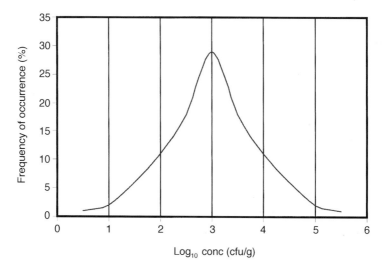

Fig. 8.3 Normal log distribution of a microorganism.

- The proportion (p_d) of units in a lot which can have a concentration above the limit value, V, must be decided.
- The desired probability P can be chosen where P is the probability of rejecting a lot which contains at least a proportion p_d above V.

The lot of food is rejected if

$$x + k_1 s > V$$

Where k_1 is obtained from reference tables (Table 8.2) according to the p_d and P values. It is therefore dependent upon the stringency of the sampling plan and number of sample units, n, analysed. V is the microbial count as a log-concentration which has been set as a safety limit.

Table 8.2 Safety and quality specification (reject if $x + k_1 s > V$) (adapted from ICMSF 1986 and reprinted with permission of the University of Toronto Press).

| Probability (P) of rejection | Proportion (p_d) exceeding V | Number of sample units | | | |
		3	5	7	10
0.95	0.05	7.7	4.2	3.4	2.9
	0.1	6.2	3.4	2.8	2.4
	0.3	3.3	1.9	1.5	1.3
0.90	0.1	4.3	2.7	2.3	2.1
	0.25	2.6	1.7	1.4	1.3

Deciding that a lot would be rejected if 10% ($p_d = 0.1$) of samples exceeded the value V, with a probability of 0.95 and taking five sample units (n) gives $k_1 s$ as 3.4. The more samples (n) are taken, the lower the chance of rejecting an acceptable lot of food.

The value of V is set by the microbiologist from previous experience. It can be similar to M in the three-class plan (Section 8.5.2). For example, the aerobic plate count for ice-cream from the Milk Products Directive 92/46/EEC (Section 8.9.2) gives $M = 500\,000$ cfu/g.

$500\,000 = \log 5$
therefore $V = 5$

Previous analysis gave a mean log value of 5.111 and a standard deviation of 0.201.
Therefore, deciding

(1) $p_d = 0.1$, the probability that a lot would be rejected if 10% of samples exceeded V

(2)   Probability of rejection = 0.95
(3)   Number of sample units = 3

gives $k_1 = 6.2$. Hence:

$$x + k_1 s = 5.111 + (6.2 \times 0.201) = 6.3572$$

Since $V = 5$, the lot would be rejected.

Variables plans can be applied to GMP standards using k_2 values in Table 8.3 (Kilsby 1982; ICMSF 1986). A similar formula is applied where the lot is accepted if $x + k_2 s < v$. The values of P and p_d are decided as before, v is similar to m in the three-class plan (Section 8.5.2) and the GMP values of IFST (1999) can be used. For example, GMP value for the aerobic plate count for raw poultry (IFST 1999) = $< 10^5$, which is less than log 5.0. Previous analysis had given $x = 4.3$ with a standard deviation (s) of 0.475.

Table 8.3   Determining the Good Manufacturing Practice limit (accept if $x + k_2 s < v$) (adapted from ICMSF 1986 and reprinted with permission of the University of Toronto Press).

Probability (P) of rejection	Proportion (p_d) exceeding v	Number of sample units			
		3	5	7	10
0.90	0.05	0.84	0.98	1.07	1.15
	0.1	0.53	0.68	0.75	0.83
	0.3	−0.26	−0.05	0.04	0.12
0.75	0.01	1.87	1.92	1.96	2.01
	0.1	0.91	0.97	1.01	1.04
	0.25	0.31	0.38	0.42	0.46
	0.5	−0.47	−0.33	−0.27	−0.22

Therefore, deciding:

(1)   Proportion of acceptance, $P = 0.9$
(2)   Proportion exceeding v, $p_d = 0.1$
(3)   Number of sample units = 7

gives $k_2 = 0.75$. Hence

$$x + k_2 s = 4.3 + (0.75 \times 0.475) = 4.65625$$

Therefore the lot of raw chicken is below the GMP value.

8.5 Attributes sampling plan

The attributes sampling plan(s) is applied when there is no previous microbiological knowledge of the distribution of microorganisms in the food or the microorganisms are not distributed log-normally. There are two types of attributes sampling plans as defined by ICMSF (1986):

- Two class plan; $n = 5$, $c = 0$ or $n = 10$, $c = 0$
- Three class plan; $n = 5$, $c = 1$, $m = 10^2$, $M = 10^3$

The two class plan is used almost exclusively for pathogens, whereas a three class plan is often applied for indicator organisms. The main advantage of using sampling plans is that they are statistically based and provide a uniform basis for acceptance against defined criteria.

8.5.1 Two class plan

A two class plan consists of the specifications n, c and m, where

n = number of sample units from a lot that must be examined.
c = maximum acceptable number of sample units that may exceed the value of m; the lot is rejected if this number is exceeded.
m = maximum number of relevant bacteria/g; values greater than this are either marginally acceptable or unacceptable.

For example:

$$n = 5, c = 0$$

This means that five sample units are analysed for a specific pathogen (for example *Salmonella*). If one unit contains *Salmonella* then the complete batch is unacceptable. Each sample unit analysed is normally $25\,g$ for *Salmonella* testing (Section 6.5.2).

8.5.2 Three class plan

The additional parameter in a three-class plan is M, a quantity that is used to separate marginally acceptable from unacceptable. A value at or above M in any sample is unacceptable. Hence the three class plan is where the food can be divided into three classes according to the concentration of microorganisms detected:

- 'Acceptable' if counts are below m.
- 'Marginally acceptable' if counts are above m but less than M.
- 'Unacceptable' (reject) if counts are greater than M.

For example, a sampling plan for coliforms could be:

$$n = 5, c = 2, m = 10, M = 100$$

This means that two units from a sample number of five can contain between 10 and 100 coliforms and be acceptable. However, if three units contain coliforms between 10 and 100, or just one sample has greater than 100 coliforms, then the batch is unacceptable and the lot is rejected. Hence the three class sampling plan includes a tolerance value for the random distribution of microbes in foods.

The stringency of the sampling plan can be decided using the ICMSF (1986) concept based on the hazard potential of the food and the conditions which a food is expected to be subject to before consumption (Table 8.1). Examples of microbiological criteria are given in Sections 8.8 and 8.9.

8.6   Principles

8.6.1   Defining a 'lot' of food

A lot is 'a quantity of food or food units produced and handled under uniform conditions'. This implies homogeneity within a lot. However, in most instances the distribution of microorganisms within a lot of food is heterogeneous. If a lot is in fact composed of different production batches then the producer's risk (i.e. the risk that an acceptable lot will be rejected, Section 8.6.3) can be high since the sample units analysed may by chance be those from a poor quality batch. In contrast, by defining individual production batches as lots a more precise identification of poor quality (reject) food can be made.

8.6.2   Sample unit number

The number of sample units, n, refers to the number of units that are chosen randomly. The samples should represent the composition of the lot from which it is taken. A sample unit can be an individual package or portions. The sample units must be taken in an unbiased fashion and must represent the food lot as well as possible. Microorganisms in food are often heterogeneously distributed and this makes the interpretation of sample unit results difficult. Random choice of samples is required to try and avoid biased sampling; however difficulties arise when the food is nonhomogeneous, for example a quiche.

The choice of n is usually a compromise between what is an ideal probability of assurance of consumer safety and the work load the laboratory can handle. It is important first to determine the nature of the hazard and then determine the appropriate probabilities of acceptance (Tables 8.4 to 8.6). It is uneconomical to test a large portion of a food lot. However, the stringency of a sampling plan for a hazardous

Table 8.4 Probability of accepting (P_a%) a food lot; two class plan, $c = 0$ (adapted from ICMSF 1986 and reprinted with permission from the University of Toronto Press).

Number of samples (n)	Probability of acceptance (P_a%)						
	Actual percentage of defective samples						
	2	5	10	20	30	40	50
3	94	86	73	51	34	22	13
5	90	77	59	33	17	8	3
10	82	60	35	11	3	1	(<0.5)
20	67	36	12	1	(<0.5)	(<0.5)	(<0.5)

Table 8.5 Probability of accepting (P_a%) a food lot; two class plan, $c = 1$ to 3 (adapted from ICMSF 1986 and reprinted with permission from the University of Toronto Press).

Number of samples (n)	Value of c	Probability of acceptance (P_a%)						
		Actual percentage of defective samples						
		2	5	10	20	30	40	50
5	1	100	98	92	74	53	34	19
	2	100	100	99	94	84	68	50
	3	100	100	100	99	97	91	81
10	1	98	91	74	38	15	5	1
	2	100	99	93	68	38	17	5
	3	100	100	99	88	65	38	17
15	1	96	83	55	17	4	1	<0.5
	2	100	96	82	40	13	3	<0.5
	4	100	100	99	84	52	22	6
20	1	94	74	39	7	1	<0.5	<0.5
	4	100	100	96	63	24	5	1
	9	100	100	100	100	95	76	41

Table 8.6 Probability of accepting (P_a%) a food lot; three class plan (adapted from ICMSF 1986 and reprinted with permission from the University of Toronto Press).

Percentage defective (P_d%)	Value of c	Percentage marginal (P_m%)					
		10	20	30	50	70	90
Number of samples (n) = 5							
50	3	3	3	2	< 0.5		
	2	3	2	1	< 0.5		
	1	2	1	< 0.5			
40	3	8	7	6	2	< 0.5	
	2	8	6	4	< 0.5		
	1	6	4	1	< 0.5		
30	3	17	16	15	7	< 0.5	
	2	16	14	11	2	< 0.5	
	1	14	9	5	2	< 0.5	
20	3	33	32	31	20	4	< 0.5
	2	32	29	24	9	1	< 0.5
	1	29	21	13	2	< 0.5	
10	3	59	58	56	43	18	<0.5
	2	58	55	47	23	5	< 0.5
	1	53	41	27	7	1	< 0.5
5	3	77	77	75	60	31	2
	2	77	72	63	35	9	< 0.5
	1	70	55	38	12	1	< 0.5
0	3	100	99	97	81	47	8
	2	99	94	84	50	16	1
	1	92	74	53	19	3	< 0.5
Number of samples (n) = 10							
40	3	1	< 0.5				
	2	< 0.5					
	1	< 0.5					
30	3	3	2	1	< 0.5		
	2	2	1	< 0.5			
	1	2	< 0.5				
20	3	10	8	5	< 0.5		
	2	9	6	2	< 0.5		
	1	7	3	1	< 0.5		

(Contd)

Table 8.6   *(Contd)*

Percentage defective (P_d%)	Value of c	Percentage marginal (P_m%)					
		10	20	30	50	70	90
10	3	34	29	20	3	< 0.5	
	2	32	21	10	1	< 0.5	
	1	24	11	4	1	< 0.5	
5	3	59	51	36	20	8	2
	2	55	39	20	8	2	< 0.5
	1	43	21	8	2	< 0.5	
0	3	99	88	65	17	1	< 0.5
	2	93	68	38	5	< 0.5	
	1	74	38	15	5	1	< 0.5

microorganism can be set using the relationship between the number of sample units analysed and the acceptance/rejection criteria (n and c values; Section 8.6.2).

8.6.3   Operating characteristic curve

It is possible, when using a sample plan, that a relatively poor lot of food will be accepted and a good lot is rejected. This is represented by the 'operating characteristic' curve. This is a plot of:

(1)   The probability of acceptance (P_a) on the y-axis, where P_a is the expected proportion of times a lot of this given quality is sampled for a decision.

(2)   The percentage of the defective sample units comprising a lot (p) on the x-axis. This is also known as a measure of lot quality.

Figure 8.4 gives the operating characteristic curve for the sampling plan $n = 5$, $c = 3$. The operating characteristic curve changes according to the values of n and c. Figure 8.5b is a selected area of Fig. 8.5a to emphasise the high chance of accepting lots with up to 30% defectives. If a producer sets a limit of 10% defectives (i.e. $p = 10\%$), using a two class plan of $n = 5$, $c = 2$, then the probability of acceptance (P_a) is 99%. This means that on 99 of every 100 occasions when a 10% defective lot is sampled, one may expect to have two or fewer of the five tests showing the presence of the organism and thus calling for 'acceptance', while on 1 of every 100 times there will be three or more positives, calling for nonacceptance. Therefore a sampling plan of $n = 5$, $c = 2$ will mean that 10% defective lots will

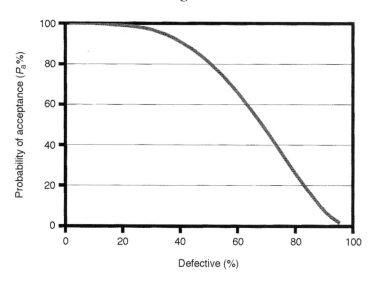

Fig. 8.4 Operating characteristic curve for sample plan $n = 5$, $c = 3$.

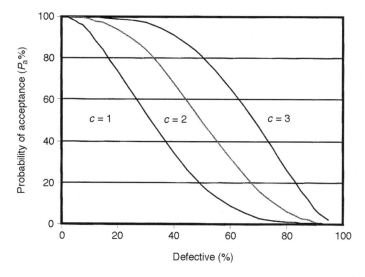

Fig. 8.5a Operating characteristic curve for $n = 5$, $c = 1$ to 3.

be accepted on most (99%) sampling occasions! Even increasing the number of samples to 10 ($n = 10$, $c = 2$) means that 10% defective batches will be accepted on 93% of occasions. Hence the need for the proactive approach of HACCP for assured food safety.

It is therefore apparent that no practical sampling plan can ensure the

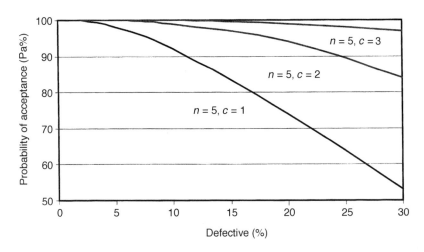

Fig. 8.5b   Enlarged area of Fig. 8.5a.

absence of the target microorganism and the concentration of the target organism may be greater than the set limit in parts of the food lot not sampled. The absence of a target organism in five randomly chosen samples only gives a 95% confidence that the food lot is less than 50% contaminated. If 30 samples had been analysed then the food lot is (with 95% confidence) contaminated at less than 10%. It requires 300 randomly taken samples giving the absence of the target microorganism a 95% confidence that the food lot is less than 1% contaminated. Therefore no sampling plan can guarantee the absence of a pathogen, unless every gram of the food was analysed, leaving nothing for consumption.

8.6.4   *Producer's risk and consumer's risk*

It follows from the operating characteristic curve than it is possible that a 'bad' lot of food will on occasions be accepted, and conversely a 'good' lot will be rejected. This is known as the 'consumer's risk' and 'producer's risk', respectively. The consumer's risk is considered to be the probability of accepting a lot whose actual microbial content is substandard as specified in the plan, even though the microbiological analysis of the sample units conforms to acceptance (P_a). The producer's risk is expressed by $1 - P_a$ (Fig. 8.6).

8.6.5   *Stringency of two and three class plans, setting* n *and* c

The stringency of the two class sampling plan depends upon the values chosen for n and c, where n is the number of sample units analysed and c

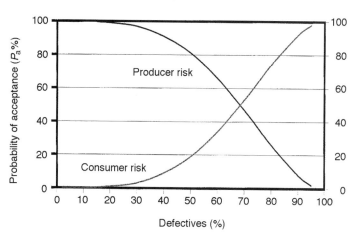

Fig. 8.6 Producer risk/consumer risk curve.

is the maximum acceptable number of sample units that may exceed the value of m, the maximum number of relevant bacteria/g. The lot is rejected if c is exceeded. If n is increased for a given value for c, then the better in microbiological terms the food lot must be to have the same chance of being passed. Conversely, for a given sample size n, if c is increased the sampling plan becomes more lenient as there is a higher probability of acceptance (P_a). This can be seen in Fig. 8.7a, where increasing $c = 0$ to 3 means the plan becomes more stringent, with less chance of accepting a defective lot. In Fig. 8.7b, however, where n is increased from 5 to 20 and c is fixed ($c = 1$), the plan becomes more stringent.

In a three-class plan it is the values of n and c which determine the probability of acceptance, P_a, for a food lot of given microbiological quality. The microbiological quality is given by determining the percentage of 'defective' proportions:

P_d = % 'defective'; above M
P_m = % 'marginally acceptable'; m to M
P_a = % 'acceptable'; equal to or less than m

Since the three terms must equal 100%, then only the first two terms need to be determined. Probability values for three class plans are given in Table 8.5.

Taking a lot of food of which 20% of the sample counts are marginally acceptable ($P_m = 20\%$) and 10% 'defective' ($P_d = 10\%$), the effect of n and c can be compared in Table 8.6. For $n = 5$, $c = 3$, the probability of

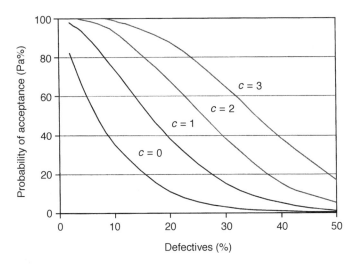

Fig. 8.7a   Stringency of sampling plans, illustrated by $n = 10$, $c = 0$ to 3.

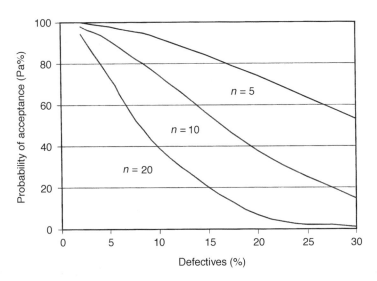

Fig. 8.7b   Stringency of sampling plans, illustrated by $n = 5$ to 20, $c = 1$.

acceptance (P_a) is 58% of occasions; if c is lowered to 1 then P_a decreases to 41%, whereas if n is increased to 10 ($c = 3$) then P_a is 29% of occasions. The most stringent is $n = 10$, $c = 3$, where the $P_a = 11\%$ of occasions.

This level of acceptance (11 in 100) for lots of which 10% are defective and 20% are marginally acceptable reinforces the fact that microbiological hazards must be controlled in food using the proactive HACCP approach rather than a retrospective end-product testing approach.

Therefore the setting of n and c varies with the desired stringency (probability of rejection). For stringent cases n is high and c is low; for lenient cases n is low and c is high. As n decreases the chance of acceptance of bad lots increases.

8.6.6 Setting the values for m and M

The level of the target organism which is acceptable and attainable in the food is m. It can be set from levels attained in GMP or, if the target organism is a pathogen, then m may be set at zero for a given volume of sample (e.g. 25 g).

M is only used in three class plans as the hazardous or unacceptable level of contamination caused by poor hygienic practice. There are three methods of setting the value of M:

(1) As a utility (spoilage or shelf life) index where it relates to levels of bacteria causing detectable spoilage (odour, flavour) or to a decrease in shelf life to an unacceptable short period (Section 4.2).
(2) As a general hygiene indicator, relating to levels of an indicator organism reflecting an unacceptable condition of hygiene.
(3) As a health hazard where it relates to infectious dose. This value can be set using epidemiological and laboratory data, and similar sources of information (see Table 7.7).

Therefore the values of m and M are independent of each other and have no set relationship.

8.7   Microbiological limits

8.7.1   Definitions

There are various terms used in reference to microbiological limits and these are defined as follows:

- 'Microbiological standards' refers to compulsory microbiological levels laid down in statute.
- 'Microbiological guidelines' refers to levels set out in guidance which does not have legal force.
- 'Microbiological criteria' may refer to either of the above or to levels in use by the food industry.

316 *The Microbiology of Safe Food*

- 'Microbiological specifications' are agreed within or between companies and generally have no direct legal implications.

Thus specifications can be prepared for raw materials supplied to a food processor, of foods at various stages of preparation and for final products. In the last case the microbiological specifications may be those agreed as reasonable and attainable by the company or they may be standards imposed by or jointly agreed with an external agency. Specifications may include standards for total numbers of microorganisms, food pathogens, indicator or spoilage organisms.

When compiling microbiological specifications for raw materials and final products it is desirable to start with as wide a range of relevant test methods as is practicable so that comprehensive data on the background microbiology can be built up. These methods should give the highest recovery of organisms and reproducible results. Ideally, test samples should be exchanged between laboratories and analysed by the agreed methods to ensure that similar results are being obtained. Greater attention should be paid to raw materials and foods where erratic or unexpected results are obtained. Specifications should reflect what is attainable under Good Manufacturing Practice, but should include tolerances to allow for sampling inaccuracies.

8.7.2 Limitations of microbiological testing

Using microbiological criteria for testing food does include a large number of problems to be considered:

(1) Cost of analysis with regard to trained personnel, equipment and consumables.
(2) Sampling problems: obtaining 'representative' samples is very difficult.
(3) Accepting a food lot which contains unacceptable levels of microorganisms/toxins, simply because of their presence at low levels and is a heterogeneous distribution (see Section 8.6.2).
(4) Variation in results, such as plate counts which have 95% confidence limits of ± 0.5 log cycles (see Table 6.1).
(5) Destructive testing means that samples cannot be retested.
(6) Length of time involved due to the need for prolonged incubation periods.
(7) Sensitivity and robustness of detection methods.

8.8   Examples of sampling plans

Egg products
Eggs are perishable since they are highly nutritious and hence subject to microbiological growth and spoilage. They are associated with certain foodborne pathogens, most notably *Salmonella* spp. Despite pasteurisation processes (liquid egg: 64.4°C, 2.5 min or 60°C, 3.5 min), eggs and egg products may be contaminated due to insufficient heat treatment or from post-pasteurisation contamination. Salmonella cells can subsequently multiply to an infectious dose due to temperature abuse after thawing or rehydration. *St. aureus* has been identified as a foodborne pathogen associated with pasta where it can grow and form toxic levels of enterotoxin. Sampling plans for eggs and egg products are given in Table 8.7.

Table 8.7   Sampling plan for egg products (adapted from ICMSF 1986 and reprinted with permission from the University of Toronto Press).

Target organism	Case	Plan cases	n	c	Limit (g^{-1})	
					m	M
Aerobic plate count	2	3	5	2	5×10^4	10^6
Coliforms	5	3	5	2	10	10^3
Salmonella spp.	10	2	5	0	0	—
(General population)	11	2	10	0	0	—
	12	2	20	0	0	—
Salmonella spp.	10	2	15	0	0	—
(High risk population)	11	2	30	0	0	—
	12	2	60	0	0	—

Milk and milk products
Dairy products are highly nutritious and have a neutral pH and water activity conducive to the growth of foodborne pathogens. Dairy products are divided into two groups:

(1)   The more perishable ('fresh') products such as milk, cream, flavoured milk and skimmed milk drinks, fresh cheese (cottage cheese) and fermented milks.

(2)   The relatively stable products having extended shelf life under appropriate conditions of storage, such as hard cheese, butter, dried milk products, ice cream mixes, evaporated (canned) milk and sterilised or ultra high temperature (UHT) milk (for fluid consumption).

Microbiological criteria cannot be applied effectively to group 1, since the products will probably have been consumed before the microbiological

The Microbiology of Safe Food

analysis has been completed. Milk products of group 2 which are associated with microbiological hazards (for example dried milk and ripened cheese) are usually microbiologically tested before distribution. Sampling plans proposed by ICMSF (1986) are given in Table 8.8.

Table 8.8 Sampling plan for dairy products (adapted from ICMSF 1986 and reprinted with permission from the University of Toronto Press).

Product	Target organism	Case	Plan Case	n	c	Limit (g^{-1})	
						m	M
Dried milk	Aerobic plate count	2	3	5	2	3×10^4	3×10^5
	Coliforms	5	3	5	1	10	10^2
	Salmonella spp.	10	2	5	0	0	—
	(Normal	11	2	10	0	0	—
	population)	12	2	20	0	0	—
	Salmonella spp.	10	2	15	0	0	—
	(High risk	11	2	30	0	0	—
	population)	12	2	60	0	0	—
Cheese, hard and semi-soft types	*St. aureus*	8	2	5	0	10^4	—

Processed meats

Processed meats include a range of meat products that have been processed by heat treatment, curing, drying and fermenting. There are a number of microbiological hazards associated with meat products. The ICMSF (1986) sampling plans are given in Table 8.9. The main target organisms are *St. aureus*, *Cl. perfringens* and salmonella.

Table 8.9 Sampling plan for processed meats (adapted from ICMSF 1986 and reprinted with permission from the University of Toronto Press).

Product	Target organism	Case	Plan class	n	c	Limit (g^{-1})	
						m	M
Dried blood, plasma and gelatin	*St. aureus*	8	3	5	1	10^2	10^4
	Cl. perfringens	8	3	5	1	10^2	10^4
	Salmonella spp.	11	2	10	0	0	—
Roast beef and pâté	*Salmonella* spp.	12	2	20	0	0	—

Cereals and cereal products
A range of bakery products is covered by this sampling plan (Table 8.10). Since the products are often dry (low water activity), moulds and the persistence of bacterial sporeformers are important.

Table 8.10 Sampling plan for cereals and bakery products (adapted from ICMSF 1986 and reprinted with permission from the University of Toronto Press).

Product	Target organism	Case	Plan case	n	c	Limit (g^{-1})	
						m	M
Cereals	Moulds	5	3	5	2	10^2-10^{4a}	10^5
Soya flour, concentrates and isolates	Moulds	5	3	5	2	10^2-10^4	10^5
	Salmonella spp.	10	2	5	0	0	—
Frozen bakery products (ready-to-eat) with low acid or high a_w fillings or toppings	*St. aureus*	9	3	5	1	10^2	10^4
	Salmonella spp.	12	2	20	0	0	—
Frozen bakery products (to be cooked) with low acid or high a_w fillings or toppings (e.g. meat pies, pizzas)	*St. aureus*	8	3	5	1	10^2	10^4
	Salmonella spp.	10	2	5	0	0	—
Frozen entrées containing rice or corn flour as a main ingredient	*B. cereus*	8	3	5	1	10^3	10^4
Frozen and dried products	*St. aureus*	8	3	5	1	10^2	10^4
	Salmonella spp.	10	2	5	0	0	—

[a] The exact value will vary with the type of grain.

Cook-chill and cook-freeze products
Guidelines from the Department of Health (UK) are given in Table 8.11. The guidelines target five foodborne pathogens and the aerobic plate count as an indicator of microbial load.

Seafoods
The Food and Drug Administration and Environmental Protection Agency

Table 8.11 Sampling plan for cook-chill and cook-freeze foods at point of consumption (adapted from Department of Health (UK) guidelines).

Target organism	Limit
Aerobic plate count	$< 10^5 \, g^{-1}$
E. coli	$< 10 \, g^{-1}$
St. aureus	$< 100 \, g^{-1}$
Cl. perfringens	$< 100 \, g^{-1}$
Salmonella spp.	Absent in 25 g
L. monocytogenes	Absent in 25 g

have given guidance levels for seafood microbiological hazards (Table 8.12).

8.9 Implemented microbiological criteria

8.9.1 Microbiological criteria in the European Union

Most countries in the EU have implemented the Food Hygiene Directive (93/43/EEC; Section 10.4.5) into national law, although most have restricted the implementation to the first five principles of HACCP. There are differences in implementation within Europe; for example, the Netherlands have a high emphasis on sampling programmes within inspections. Many businesses use microbiological criteria in contractual agreements with suppliers and customers and as a means of monitoring the hygiene of the production environment. These criteria are laid out in guidelines developed by individual companies of the relevant industry sector, and are generally based on GMP. Under the EC Food Hygiene Directive, it will be possible to include microbiological guidelines in industry guides to GHP. Currently the EU microbiological criteria are under revision and proposed amendments are given in Tables 8.13 to 18.17.

The Food Hygiene Directive (93/34/EEC) provides:

Without prejudice to more specific Community rules, microbiological criteria and temperature control criteria for certain classes of foodstuffs may be adopted in accordance with the procedure laid down in Article 14 and after consulting the Scientific Committee for Food set up by Decision 74/234/EEC.

An EU survey revealed that sampling plans have only limited application for food inspection in Europe, probably because they are too expensive

Table 8.12 Food and Drug Administration and Environmental Protection Agency guidelines for seafood microbiology hazards.

Product	Guideline/tolerance
Ready to eat fishery products (minimal cooking by consumer)	Enterotoxigenic *Escherichia coli* (ETEC) - 1×10^3 ETEC/g, LT or ST positive
Ready to eat fishery products (minimal cooking by consumer)	*Listeria monocytogenes* - presence of organism
All fish	*Salmonella* species - presence of organism
All fish	*Staphylococcus aureus* - (1) positive for staphylococcal enterotoxin, or (2) *Staphylococcus aureus* level is equal to or greater than 10^4/g (MPN)
Ready to eat fishery products (minimal cooking by consumer)	*Vibrio cholerae* - presence of toxigenic 01 or non-01
Ready to eat fishery products (minimal cooking by consumer)	*Vibrio parahaemolyticus* - levels equal to or greater than 1×10^4/g (Kanagawa positive or negative)
Ready to eat fishery products (minimal cooking by consumer)	*Vibrio vulnificus* - presence of pathogenic organism
All fish	*Clostridium botulinum* - (1) Presence of viable spores or vegetative cells in products that will support their growth; or (2) presence of toxin
Clams and oysters, and mussels fresh or frozen - imports	Microbiological - (1) *E. coli* - MPN of 230/100 g (average of subs or 3 or more of 5 subs); or (2) APC - 500 000/g (average of subs or 3 or more of 5 subs)
Clams, oysters, and mussels, fresh or frozen - domestic	Microbiological - (1) *E. coli* or fecal coliform - 1 or more of 5 subs exceeding MPN of 330/100 grams or 2 or more exceeding 230/100 grams; or (2) APC - 1 or more of 5 subs exceeding 1 500 000/g or 2 or more exceeding 500 000/g
Salt-cured, air-dried uneviscerated fish	Not permitted in commerce (*Note:* small fish exemption)
Tuna, mai mahi, and related fish	Histamine - 500 ppm set based on toxicity, 50 ppm set as defect action level, because histamine is generally not uniformly distributed in a decomposed fish. Therefore, if 50 ppm is found in one section, there is the possibility that other units may exceed 500 ppm

Table 8.13 Sampling plans for minced meat and meat preparations (94/65/EEC) and proposed amendments.

Target organism	Limits	Sampling plan				Comment[a]
		n	c	m	M	
Minced meat						
Aerobic mesophilic bacteria		5	2	5×10^5/g	5×10^6/g	
E. coli		5	2	50/g	500/g	Change sample size to 25 g
Salmonella spp.	Absent in 10 g	5	0			Delete
St. aureus		5	2	100/g	5000/g	
Meat preparations						
E. coli		5	2	500/g	5000/g	
St. aureus		5	1	500/g	5000/g	
Salmonella spp.	Absent in 1 g	5	0			Change sample size to 25 g

[a] Proposed amendment (Anon 1999d).

Table 8.14 Sampling plans for egg products (89/437/EEC).

Target organism	Limits	Sampling plan				Comment[a]
		n	c	m	M	
Salmonella spp.	Absent in 25 g or ml					Consider sampling plans (minimum $n = 5$)
Aerobic mesophilic bacteria					10^5 in 1 g or ml	Consider sampling plans, e.g. $n = 5$, $c = 2$
Enterobacteriaceae					10^2 in 1 g or ml	Consider sampling plans
St. aureus	Absent in 1 g or ml					Delete

[a] Proposed amendment (Anon 1999d).

Table 8.15 Sampling plans for dairy products (92/46/EEC).

Target organism	Limits		Sampling plan				Comment[a]
			n	c	m	M	
Raw cow's drinking milk							
Salmonella spp.	Absent in 25 g		5	0			Deletion
St. aureus			5	2	100/g	500/g	Replace with *E. coli*
Aerobic microorganisms (30°C)	5×10^4						
Pasteurised drinking milk							
L. monocytogenes	Absent in 25 g		5	0			Deletion
Salmonella spp.	Absent in 25 g		5	0			Deletion
Coliforms (30°C)			5	1	0/g or ml	5/g or ml	Replace with Enterobacteriaceae
Aerobic microorganisms (21°C)			5	1	5×10^4/g	5×10^5/g	
Sterilised and UHT drinking milk							
Aerobic microorganisms (30°C)	10 per 0.1 ml						Delete
Hard cheese made from heat-treated milk							
L. monocytogenes	Absent in 1 g						Delete
Salmonella spp.	Absent in 25 g		5	0			Delete
Hard cheese made from raw or thermised milk							
L. monocytogenes	Absent in 1 g						Delete
Salmonella spp.	Absent in 25 g		5	0			Delete
St. aureus			5	2	1000/g	1×10^4/g	
E. coli			5	2	1×10^4/g	1×10^5/g	

(Contd)

Table 8.15 *(Contd)*

Target organism	Limits	Sampling plan				Comment[a]
		n	c	m	M	
Fresh cheese						
L. monocytogenes	Absent in 25 g	5	0			Retain if made with thermised milk
Salmonella spp.	Absent in 25 g	5	0			Retain if made with thermised milk
St. aureus		5	2	10/g	100/g	Delete if produced by fermentation
Cheese other than hard or fresh made from heat-treated milk						
L. monocytogenes	Absent in 25 g	5	0			
Salmonella spp.	Absent in 25 g	5	0			Delete
St. aureus		5	2	100/g	1000/g	Delete
E. coli		5	2	100/g	1000/g	Delete
Coliforms (30°C)		5	2	1×10^4/g	1×10^5/g	Replace with Enterobacteriaceae
Cheese other than hard or fresh made from raw or thermised milk						
L. monocytogenes	Absent in 25 g	5	0			
Salmonella spp.	Absent in 25 g	5	0			
St. aureus		5	2	1000/g	1×10^4/g	
E. coli		5	2	1×10^4/g	1×10^5/g	Possibly *m* and *M* too high
Pasteurised butter						
L. monocytogenes	Absent in 1 g	5	0			Delete
Salmonella spp.	Absent in 25 g	5	2	0	10/g	Delete
Coliforms (30°C)						Delete

(Contd)

Table 8.15 *(Contd)*

Target organism	Limits	Sampling plan				Comment[a]
		n	*c*	*m*	*M*	
Milk powder						
Salmonella spp.	Absent in 25 g	10	0			Delete
L. monocytogenes	Absent in 1 g	5	0			Delete
St. aureus		5	2	10/g	100/g	
Coliforms (30°C)		5	0	0	10/g	Replace with Enterobacteriaceae
Liquid dairy products						
L. monocytogenes	Absent in 1 g	5	0			Retain if made with raw/ thermised milk
Salmonella spp.	Absent in 25 g	5	0			Retain if made with raw/ thermised milk
Coliforms (30°C)		5	2	0	5/g	Replace with Enterobacteriaceae
Aerobic microorganisms (21°C)		5	2	$5 \times 10^4/g$	$1 \times 10^5/g$	
Frozen dairy products including ice cream						
L. monocytogenes	Absent in 1 g	5	0			Delete
Salmonella spp.	Absent in 25 g	5	0			Delete
Coliforms (30°C)		5	2	10	100	Replace with Enterobacteriaceae
Plate counts (30°C)		5	2	$1 \times 10^5/g$	$5 \times 10^5/g$	Delete
Other milk products						
L. monocytogenes	Absent in 1 g	5	0			Retain if made with raw or thermised milk
Salmonella spp.	Absent in 25 g	5	0			Retain if made with raw or thermised milk

[a] Proposed amendment (Anon 1999b).

Table 8.16 Sampling plan for cooked shellfish and molluscs (93/51/EEC).

Target organism	Limits	Sampling plan				Comment[a]
		n	c	m	M	
Salmonella spp.	Absent in 25 g or ml	5				Deletion
St. aureus		5	2	100/g	1000/g	
Thermotolerant coliforms (44°C)		5	2	10/g	100/g	Delete
E. coli		5	1	10/g	100/g	
Aerobic mesophile bacteria (30°C)		5	2	1×10^4/g	1×10^5/g	Whole product
Aerobic mesophile bacteria (30°C)		5	2	5×10^4/g	5×10^5/g	Peeled or shelled products except crab flesh. Delete
Aerobic mesophile bacteria (30°C)		5	1	1×10^4/g	1×10^6/g	Crab flesh

[a] Proposed amendment (Anon 1999d).

Table 8.17 Sampling plans for live bivalve products (91/492/EEC).

Target organism	Limits	Sampling plan	Comment[a]
Salmonella spp.	Absent in 25 g or ml	None	Sampling plan required
Faecal coliforms	<300 per 100 g flesh		Delete
E. coli	<230 per 100 g flesh		

[a] Proposed amendment (Anon 1999d).
From Anon (1999d): the main hazard is viruses (Norwalk-like). Criteria should be linked to management and intended use (raw or cooked). Algal biotoxins ASP, DSP and PSP were not considered.

and time consuming for routine use (Table 8.18; SCOOP 1998). However, some larger companies who have yet to fully establish HACCP-based systems may have to rely on the use of microbiological criteria with stated sampling plans to assess the effectiveness of processing and safety of their products. The SCOOP survey obtained results from all member states except Germany, Greece and Ireland. The survey was completed in November 1995, although it was not published until 1998, hence the current situation may be different. There were large differences between the legal status of microbiological criteria and a clear distinction needed to be drawn between microbiological standards which are included in regulations (mandatory) and microbiological guidelines which arise from other sources (see Section 8.7.1).

In France and Spain there were 81 and 61 microbiological standards, respectively, in contrast to the UK with one (dairy products). French producers are required to send samples regularly (monthly or weekly) to approved laboratories. The costs of regular sampling and testing by producers are high. In March 1993, the Netherlands introduced new standards applying at the point of sale to foods that are to receive no further treatment before consumption. The same level for six pathogenic microorganisms is applied to all ready-to-eat foods. Levels were set on the basis of industry-wide data on microbiological loads, covering all ready-to-eat foods produced under hygienic conditions. The levels set are therefore intended to be readily achievable. The situation in Denmark and Norway is different to other participating countries in that microbiological guidelines have been issued by the National Food Control Authorities. For all other countries, the most common source was the advice of national (for example the PHLS in the UK) or international (for example ICMSF) groups of experts, recommendations of Codex Alimentarius and codes issued by professional organisations. For the official control of foods, consideration is usually given to such microbiological guidelines where no microbiological standards are laid down in national regulations. However, the status and use of such guidelines vary among participating countries.

A large diversity exists in the types of microorganisms specified in different countries. Where foodborne disease bacteria are concerned, 12 countries which answered the questionnaire have statutory requirements concerning *Salmonella* and *St. aureus* in specified food commodities. *Cl. perfringens* and *B. cereus* were referred to in nine and eight countries, respectively. *P. aeruginosa* was cited five times with specific reference to mineral water. Reference to other pathogens, such as *Shigella, Y. enterocolitica, Campylobacter, V. parahaemolyticus, Cl. botulinum* (spores) and β-haemolytic streptococci, was made by only one or two countries. Regarding indicator organisms, all 13 countries answering the questionnaire used the aerobic mesophilic plate count. The coliform group

Table 8.18 Summary of sampling plans for dairy products in European countries (SCOOP 1998).

Designation	OBLIGATORY CRITERIA: pathogenic bacteria		ANALYTICAL CRITERIA: bacteria indicating a hygiene defect		INDICATOR BACTERIA: GUIDELINES		
	Listeria monocytogenes (1)	Salmonella spp. (in 25)	Staphylococcus aureus (per g or per ml)	Escherichia coli (per g)	Coliforms 30°C (per g)	Aerobic microorganisms (per g) at 21°C	at 30°C
DRINKING MILK Raw cow's milk			$m = 100$, $M = 500$, $n = 5$, $c = 2$				5×10^4 (3)
Pasteurised milk	absence in 25 g $n = 5, c = 0$	absence $n = 5, c = 0$			$m = 0$, $M = 5$, $n = 5, c = 1$	$m = 5 \times 10^4$ $M = 5 \times 10^5$ $n = 5, c = 1$	
Sterilised and U.H.T. milk							10 per 0.1 ml (4)
CHEESE (1) Hard • made from heat treated milk	absence in 1 g						
• made from raw milk or from thermised milk	absence in 1 g		$m = 1000$, $M = 10\,000$, $n = 5$, $c = 2$ (5)	$m = 10\,000$ $M = 100\,000$, $n = 5, c = 2$ (5)			
(2) Fresh	absence in 25 g $n = 5, c = 0$	absence $n = 5, c = 0$	$m = 10$, $M = 100$, $n = 5, c = 2$				

(Contd)

Table 8.18 (Contd)

Designation	OBLIGATORY CRITERIA: pathogenic bacteria		ANALYTICAL CRITERIA: bacteria indicating a hygiene defect		INDICATOR BACTERIA: GUIDELINES		
	Listeria monocytogenes (1)	Salmonella spp. (in 25)	Staphylococcus aureus (per g or per ml)	Escherichia coli (per g)	Coliforms 30°C (per g)	Aerobic micro-organisms (per g) at 21°C	at 30°C
(3) Other than hard or fresh							
• made from heat-treated milk	absence in 25 g $n=5$, $c=0$ (6)	absence $n=5$, $c=0$	$m=100$, $M=1000$, $n=5$, $c=2$ (7), (5)	$m=100$, $M=1000$, $n=5$, $c=2$ (7), (5)	$m=10000$ (7), (5) $M=100000$ $n=5$, $c=2$		
• made from raw milk or from thermised milk	absence in 25 g $n=5$, $c=0$ (6)	absence $n=5$, $c=0$	$m=1000$, $M=10000$, $n=5$, $c=2$ (5)	$m=10000$ $M=100000$ $n=5$, $c=2$ (5)			
PASTEURISED BUTTER	absence in 1 g $n=5$, $c=0$	absence $n=5$, $c=0$					
MILK POWDER	absence in 25 g $n=5$, $c=0$	absence $n=10$, $c=0$	$m=10$, $M=100$, $n=5$, $c=2$		$m=0$, $M=10$, $n=5$, $c=0$		
LIQUID DAIRY PRODUCTS	absence in 1 g $n=5$, $c=0$	absence $n=5$, $c=0$			$m=0$, $M=5$, $n=5$, $c=2$	$m=50000$ $M=10^5$ $n=5$, $c=2$ (8)	
OTHER MILK PRODUCTS	absence in 1 g $n=5$, $c=0$	absence $n=5$, $c=0$					

(1) This test is not obligatory for sterilised milk, preserved milk or dairy products heat-treated after wrapping or packaging.

(2) After incubation at 6°C for 5 days.

(3) After incubation at 30°C for 15 days.

(4) Geometric mean measured over a 2-month period, with at least two samples per month.

(5) Whenever the standard M is exceeded, testing must be carried out for the possible presence of enterotoxigenic $S.$ $aureus$ or strains of $E.$ $coli$ presumed to be pathogenic and if necessary, for the possible presence of staphylococcal toxins by methods to be established in accordance with the procedure laid down in Article 13.

(6) The 25 g are obtained by taking 5 aliquots of 5 g from the same sample, at different points.

(7) Specific criterion for soft cheese made from heat-treated milk.

(8) For non-fermented thermically treated liquid dairy products.

was used by 12 countries, *E. coli* was quoted five times, sulphite-reducing bacteria five times and yeast and/or mould count ten times. Other indicators were mentioned only occasionally.

The status of regulatory provisions for six pathogens (*L. monocytogenes, Salmonella, B. cereus, Cl. perfringens, St. aureus* and *Campylobacter*) showed two trends; for similar products, safety expectations (as expressed by the numerical values for pathogens) were mostly comparable. However, there is a marked difference in the approach to sampling and use of sampling plans.

The survey conducted on the present status and use of microbiological criteria in Europe revealed that, in several participating countries, food inspection services rely on the examination of single food samples. To a large extent the microbiological limits which are used as part of these examinations differed between countries.

Because of the variation in microbiological criteria in the EU Directives, for example the volume of material tested for the presence of salmonella varies from 1 to 25 g, the criteria are undergoing revisions at the present time (see Tables 8.13 to 8.17).

8.9.2   EU Directives specifying microbiological standards for foods

Some microbiological criteria are standards which are specified in EC product-specific food hygiene Directives. Some of these 'vertical' Directives make provision for them to be introduced in the future, or, where standards have been set, there is scope for them to be revised or added to. The EU may also lay down suitable laboratory methods. The main vertical EU Directives specifying microbiological criteria are given below and a summary of their microbiological requirement is given in Tables 8.13 to 8.17. Such Directives relate to food primarily during manufacturing, transport and wholesale storage, but not usually at retail or catering level (except where small producers sell direct to the consumer). These criteria were set about 5 to 10 years ago and are currently undergoing revision for consolidation and simplification. Therefore the following section should be regarded as an indication of microbiological status in the EU and exact details should be checked with the appropriate regulatory authority. The recommended revisions, which have *not* been adopted, are indicated where appropriate (Anon 1999d).

Food safety in the EU is partially controlled through:

(1)   Vertical Directives, which deal with products of animal origin (fresh meat, poultry, milk, fish, eggs). These apply at manufacture, storage and during transport.

(2) Horizontal Directives, which apply safety measures to all foodstuffs not covered by the vertical Directives and when all foods enter the retail market.

See Section 10.4 for a fuller account of EU regulations.
The five Directives which include microbiological criteria are:

(1) Minced Meat and Meat Preparation Directive (94/65/EEC).
(2) Milk and Milk based Products Directive (92/46/EEC).
(3) Fishery Products Directive (91/493/EEC) and the Commission decision on the microbiological criteria applicable to the production of cooked crustaceans and molluscan shellfish (93/51/EEC).
(4) Live Bivalve Molluscs Directive (91/492/EEC).
(5) Egg Products Directive (89/437/EEC).

Obviously foodborne pathogens which have recently been recognised, such as campylobacter, STEC, and Norwalk-like virus, are not referred to in the Directives. Also there is a wide diversity of criteria, such as the amount of food sampled for the absence of *Salmonella* spp. ranges from 1 to 25 g and *St. aureus* should be absent in a 1 g portion in one Directive, yet has a microbiological limit of 15 000 cfu/g in another. It has also been proposed that many of the microbiological criteria are used as 'microbiological guidelines' as opposed to mandatory 'microbiological standards'. Where appropriate these proposed amendments are indicated in the relevant tables.

Meat Products Directive 94/65/EEC
Council Directive 94/65/EEC lays down the requirements for the production of, and trade in, minced meat, meat and meat preparations and amends Directives 64/433/EEC, 71/118/EEC, 72/462/EEC and 88/657/EEC. Standards are laid down for four groups of bacteria: aerobic mesophiles, *E. coli*, *St. aureus* and *Salmonella* spp. (Table 8.13).

Egg Products Directive 89/437/EEC
The Directive contains criteria for aerobic plate counts, *Salmonella* spp., Enterobacteriaceae and *St. aureus* (Table 8.14).

Milk and Milk Products Directive 92/46/EEC
Council Directive 92/46/EEC (which superseded Council Directive 85/397/EEC) is the major legislation of health rules for the production and placing on the market of raw milk, heat-treated milk and milk-based

Table 8.19 Aerobic plate count categories for different types of ready-to-eat foods (adapted from Anon 1996).

Food group	Product	Category
Meat	Beefburgers and pork pies	1
	Poultry (unsliced)	2
	Pate, sliced meat (except ham and tongue)	3
	Tripe and other offal	4
	Salami and fermented meat products	5
Seafood	Herring and other pickled fish	1
	Crustaceans, seafood meals	3
	Shellfish (cooked), smoked fish, taramasalata	4
	Oysters (raw)	5
Dessert	Mousse/dessert	1
	Cakes, pastries, slices and desserts – without dairy cream	2
	Trifle	3
	Cakes, pastries, slices and desserts – with dairy cream	4
	Cheesecake	5
Savoury	Bhaji	1
	Cheese-based bakery products	2
	Spring rolls, satay	3
	Houmus, tzatziki and other dips	4
	Fermented foods, bean curd	5
Vegetable	Vegetables and vegetable meals (cooked)	2
	Fruit and vegetables (dried)	3
	Prepared mixed salads	4
	Fruit and vegetables (fresh)	5
Dairy	Ice cream (dairy and nondairy)	2
	Cheese, yoghurt	5
Ready-to-eat meals		2
Sandwiches and filled rolls	Without salad	3
	With salad	4

products. Products covered are cheeses, milk powder and frozen milk-based products including ice cream. *L. monocytogenes* and *Salmonella* spp. must be absent from all 25 g samples analysed. *St. aureus* and *E. coli*, coliforms and total viable counts (after incubation at $6°C$ for 5 days) have prescribed limits (Table 8.15).

Table 8.20 Guidelines for ready-to-eat foods (adapted from Anon 1996).

Food category (see Table 8.19)	Criterion			
	Satisfactory	Borderline – limit of acceptability	Unsatisfactory	Unacceptable/ potentially hazardous
Aerobic colony count[a] (30°C; 48 ± 2 h)				
1	$<10^3$	$10^3-<10^4$	$\geq 10^4$	N/A
2	$<10^4$	$10^4-<10^5$	$\geq 10^5$	N/A
3	$<10^5$	$10^5-<10^6$	$\geq 10^6$	N/A
4	$<10^6$	$10^6-<10^7$	$\geq 10^7$	N/A
5	—*	—*	—*	—*
Indicator organisms				
E. coli (total) 1-5	<20	$20-<100$	$100-<10^4$	$\geq 10^4$
Enterobacteriaceae[a] 1-5	<100	10^2-10^4	$>10^4$	N/A
Listeria spp. (not L. monocytogenes) 1-5	not detected in 25 g $(<20)^a$	Present in 25 g–<200/g $(20-<100)^a$	$200-<10^4$ $(100-<10^4)^a$	$\geq 10^4$
Pathogens				
Salmonella serovars 1-5	not detected in 25 g			present in 25 g
Campylobacter spp. 1-5	not detected in 25 g			present in 25 g
E. coli O157:H7 & other STEC 1-5	not detected in 25 g			present in 25 g
V. parahaemolyticus – seafoods 1-5	not detected in 25 g	present in 25 g–<200/g	$200-<10^3$	$\geq 10^3$
L. monocytogenes 1-5	not detected in 25 g	present in 25 g–<200/g	$200-<10^3$	$\geq 10^3$
S. aureus 1-5	<20	$20-100$	$100-<10^4$	$\geq 10^4$
C. perfringens 1-5	$<20^a$	$10-<100$	$100-<10^4$	$\geq 10^4$
B. cereus and B. subtilis group 1-5	$<10^3$	$10^3-<10^4$	$10^4-<10^5$	$\geq 10^5$

*Guidelines for aerobic plate counts may not apply to certain fermented foods, e.g. salami, soft cheese and unpasteurised yoghurt. These foods fall into category 5. Acceptability is based on appearance, smell, texture and the levels or absence of pathogens.
N/A denotes not applicable.
[a] Proposed revision in 2000. (See Comm. Dis. Pub. Health 3, 163–7 (2000)).

Fishery Products Directive 91/493/EEC and Commission Decision on crustaceans and shellfish 93/51/EEC
The microbiological criteria for the Fishery Directive (91/493/EEC) and Decision 93/51/EEC are given in Table 8.16. They include a range of different aerobic plate count criteria (at 30°C) for different products.

Live bivalve products (91/492/EEC)
This Directive include criteria for toxins causing paralytic and diarrhoeic shellfish poisoning (Table 8.17). It does not (as yet) include criteria for Norwalk-like viruses and algal toxins (Sections 5.3 and 5.4).

8.10   Public Health (UK) Guidelines for Ready-To-Eat Foods

The Public Health guidelines for ready-to-eat foods were first published in 1992 and a revised version was published in 1996 (Gilbert 1992; Anon 1996). They were collated to assist in the implementation of the UK Food Safety Act 1990 (Section 30) which provides for food examiners to examine samples of food submitted to them and issue certificates specifying the results of their examination. The guidelines are for application during the shelf life of the product and are not at point of production. They are not mandatory government standards. Foods are divided into five categories (Table 8.19) for the aerobic plate count analysis, whereas the same criteria are used for all indicator organisms and foodborne pathogens. The Guidelines will be revised in 2000 to include Enterobacteriaceae and *E. coli* criteria (Table 8.20).

Apart from setting proscriptive limits for certain pathogens, the guidelines recommend ranges of bacterial counts for a number of different types of food which allow the division of results into 'satisfactory', 'borderline – limit of acceptability', 'unsatisfactory' or 'unacceptable/potentially hazardous'. Although the guidelines have no formal status and refer only to ready-to-eat food, they do reflect the opinions of experienced workers with access to a wealth of unpublished data collected over 50 years by the PHLS. They are applied to single samples and therefore do not have the statistical validity of the ICMSF sampling plans.

9

MICROBIOLOGICAL RISK ASSESSMENT

Changes in food processing techniques, food distribution and the emergence of new foodborne pathogens will change the epidemiology of foodborne diseases. Therefore new strategies are required for evaluating and managing food safety risks. Microbiological Risk Assessment (MRA) generates models which will enable the changes in food processing, distribution and consumption to be assessed with regard to their influence on food poisoning potential.

Food safety must be ensured by the proper design of the food product and the production process. This means that the optimal interaction needs to be assured between intrinsic and extrinsic parameters (which are appropriate for the product's shelf life) as well as conditions for handling, storage, preparation and use. As previously outlined the assured method of controlling hazards is HACCP (Section 7.3). This should not be confused with MRA which is principally a regulatory activity. MRA is the stepwise analysis of hazards that may be associated with a particular type of food product, permitting an estimation of the probability of occurrence of adverse effects on health from consuming the product in question (Notermans & Mead 1996). It is sometimes referred to as 'Quantitative Microbial Risk Assessment' (Haas *et al.* 1999).

MRA is not designed to be performed by food companies, although they will be able to contribute data and experience to some of the steps. The primary role of the food companies is to manage the manufacture of safe foods by applying the appropriate control measures. Safety concepts need to be built into the development of food products, for example through HACCP implementation. These in turn must be incorporated into Good Manufacturing Practices, Good Hygienic Practices and Total Quality Management (Sections 7.9 and 7.11). MRA should provide better information for the development of HACCP schemes.

Microbiological Risk Analysis (MRA) has been defined by the Codex Alimentarius Commission (1999a). It is a management tool for governmental bodies to define an appropriate level of protection and establish guidelines to ensure the supply of safe foods. Within this concept 'risk' is defined as 'a health effect caused by a hazard in a food and the likelihood of its occurrence'. Increasingly, government bodies at national and international levels are addressing food safety risk analysis associated with biological hazards. MRA is a deliberate, structured and formalised approach to understanding and, where necessary, reducing risk. MRA information is useful in determining what hazards are of such a nature that their prevention, elimination or reduction to acceptable levels is necessary. It is generally recognised that MRA consists of three components (Fig. 9.1):

(1) Risk assessment
(2) Risk management
(3) Risk communication

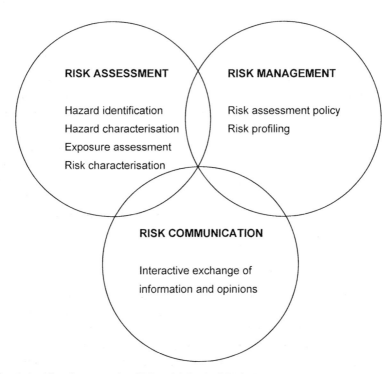

Fig. 9.1   The framework of Microbiological Risk Assessment.

9.1 Risk assessment (RA)

Risk assessment is a scientific approach to estimating a risk and to understanding the factors that influence it. Starting with a statement of purpose or problem formulation, the process as defined by Codex is composed of elements:

• Statement of purpose of risk assessment
• Hazard identification
• Exposure assessment
• Hazard characterisation (including a dose–response assessment)
• Risk characterisation
• Production of a formal report

Figure 9.2 shows the sequence of steps. The knowledge in each step is combined to represent a cause-and-effect chain from the prevalence and concentration of the pathogen to the probability and magnitude of health effects (Lammerding & Paoli 1997). In risk assessment, 'risk' consists of both the probability and impact of disease. Therefore risk reduction can be achieved either by reducing the probability of disease or by reducing its severity.

9.1.1 Statement of purpose

The specific purpose of the risk assessment should be clearly stated. The output form and possible output alternatives should be defined. This stage refers to problem formulation. During this stage, the cause of concern and the goals, breadth and focus of the risk assessment should be defined. The statement may also include data requirements, as they may vary depending on the focus and use of the risk assessment and the questions relating to uncertainties that need resolving. Output might, for example, take the form of an estimate of an annual occurrence of illness, or an estimate of annual rate of illness per 100 000 population, or an estimate of the rate of human illness per eating occurrence.

9.1.2 Hazard identification

Hazard identification consists of the identification of biological, chemical and physical agents (microorganisms and toxins) capable of causing adverse health effects which may be present in a particular food or group of foods. The purpose of hazard identification is to identify the microorganism or microbial toxin of concern and to evaluate whether the microorganism or the toxin is a potential hazard when present in food.

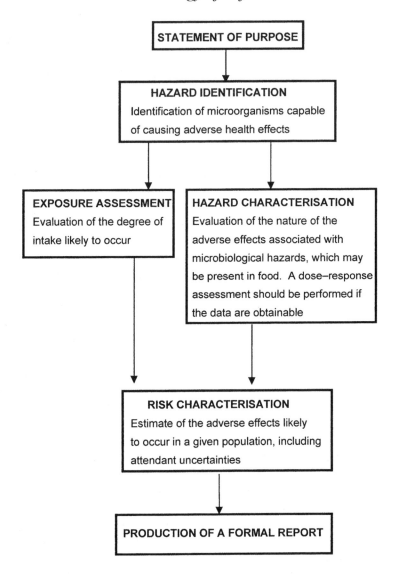

Fig. 9.2   Risk assessment flowchart (adapted from Notermans *et al.* 1996).

Information on potentially hazardous organisms can be obtained from government surveillance studies and similar sources (for example ICMSF publications). The information may describe growth and death conditions (pH, a_w, D values, etc.). Microorganisms causing foodborne illness are given in Chapter 5 and their relevance in particular foods in Table 7.6. The key to hazard identification is the availability of public health data and a preliminary estimate of the sources, frequency and amount of the agent(s) under consideration. The information collected is later used in exposure assessment, where the effect of food processing, storing and distributing (covering from processing to consumption) on the number of foodborne pathogens is assessed.

9.1.3 Exposure assessment

Exposure assessment is the qualitative and/or quantitative evaluation of the likely intake of biological, chemical and physical agents via food as well as exposure from other sources if relevant. It describes the pathways through which a pathogen population enters the food chain and is subsequently distributed and challenged in the production, distribution and consumption of food.

The ultimate goal of exposure assessment is to evaluate the level of microorganisms or microbial toxins in the food at the time of consumption. This may include an assessment of actual or anticipated human exposure. For foodborne microbiological hazards, exposure assessment might be based on the possible extent of food contamination by a particular hazard and on consumption patterns and habits.

Depending upon the scope of the risk assessment, exposure assessment can begin with either (a) pathogen prevalence in raw materials or (b) the description of the pathogen population at subsequent steps.

Topics of interest include:

- The microbial ecology of the food.
- The initial contamination of the raw materials.
- The effect of the production, processing, handling, distribution steps and preparation by the final consumer on the microbial agent (i.e. the impact of each step on the level of the pathogenic agent of concern).
- The variability in processes involved and the level of process control.
- The level of sanitation.
- The potential for (re)contamination (e.g. cross-contamination from other foods; recontamination after a heat treatment).
- The methods or conditions of packaging, distribution and storage of the food (e.g. temperature of storage, relative humidity of the environment, gaseous composition of the atmosphere).
- The characteristics of the food that may influence the potential for growth of the pathogen (and/or toxin production) in the food under various conditions, including abuse (e.g. pH, moisture content or water activity, nutrient content, presence of antimicrobial substances, competitive flora).

Data on microbial survival and growth in foods can be obtained from food poisoning outbreaks, storage tests, historical performance data of a food process, microbiological challenge tests and predictive microbiology (Sections 2.8 and 4.2). These tests provide information on the likely numbers of organisms (or quantity of toxin) present in a food at the point of consumption. Exposure assessment includes various levels of uncertainty. Therefore this uncertainty should be estimated, for example by using event tree analysis, fault tree analysis, Hazard Analysis and Operability Study (HAZOP) and probabilistic scenario analysis.

Figure 9.3a and b gives an example of *B. cereus* in pasteurised milk (Notermans & Mead 1996).

Information on consumption patterns and habits may include:

- Socio-economic and cultural background, ethnicity.
- Consumer preferences and behaviour as they influence the choice and the amount of the food intake (e.g. frequent consumption of high risk foods).
- Average serving size and distribution of sizes.
- Amount of food consumed over a year considering seasonality and regional differences.
- Food preparation practices (e.g. cooking habits and/or cooking time, temperature used, extent of home storage and conditions, including abuse).
- Demographics and size of exposed population(s) (e.g. age distribution, susceptible groups).

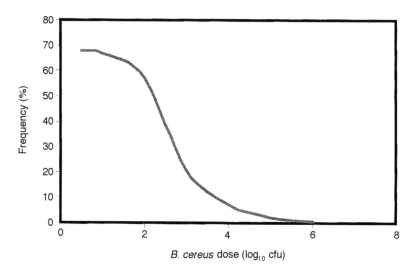

Fig. 9.3a *B. cereus* dose–response curve.

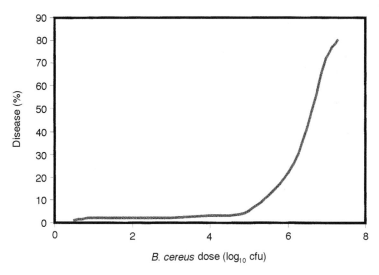

Fig. 9.3b *B. cereus* disease response curve.

9.1.4 Hazard characterisation

Hazard characterisation is the qualitative and/or quantitative evaluation of the nature of the adverse effects associated with biological, chemical and physical agents that may be present in food. It may include a dose-response assessment if data are available. The purpose of hazard characterisation is to provide an estimate of the nature, severity and duration of the adverse effects associated with harmful agents in food. Factors important to consider relate to the microorganisms, the dynamics of infection and the sensitivity of the host.

A dose-response assessment is used to translate the final exposure to a pathogen population into a health response in the population of consumers. Important factors are the physiology and the pathogenicity/virulence of the microorganism, the dynamics of infection and the host susceptibility. There is currently a paucity of data concerning pathogen-specific responses and the effect of the host's immunocompetence on the pathogen-specific responses. When data are obtainable, a dose-response assessment can be performed.

Factors relating to the microorganism may include:

- Microbial replication; generation time
- Virulence factors; toxin production, attachment factors, antigenic properties, immune evasion properties
- Microbial adaptation
- Antigenic variation

- Tolerance to adverse conditions
- Acquisition of new traits through DNA uptake (i.e. antibiotic resistance)

Factors related to the microorganism's infectivity include:

- Rate of infection
- Latency
- Disease pattern; incubation period, severity, persistence

Host-related factors include:

- Immune status
- Susceptibility; use of antacids and antibiotics

Interactions between host, microorganism and food matrix include:

- Stomach pH; increased due to age and use of antacids
- Stomach residence time; short residence time for liquids
- Acid tolerance; pre-exposure to stress conditions inducing stress response and increased acid resistance, entrapment in fatty material

A dose–response assessment determines the relationship between the magnitude of exposure (dose) to the pathogen and the severity and/or frequency of adverse health effects (response). Sources of information include:

(1) Foodborne disease analysis
(2) Population characteristic surveys
(3) Animal trials
(4) Human volunteer studies

The dose–response relationship can be described using beta-Poisson distribution and exponential distribution. The beta-Poisson distribution is described by the equation:

$$P_i = 1 - (1 + N/\beta)^{-\alpha}$$

where: P_i is the probability of infection, N is the exposure (pathogen level in colony-forming units) and α and β are coefficients specific to the pathogen.

This empirical model is particularly effective for describing dose–response relations when assessing low levels of bacterial pathogens. It generates a sigmoid dose–response relationship that assumes no threshold value for infection (Figs 9.3b and 9.4). Instead it assumes that there is a small but finite risk that an individual can become infected after exposure to a single cell of a bacterial pathogen. These values are different from the minimum infectious dose values commonly given (Table 7.7).

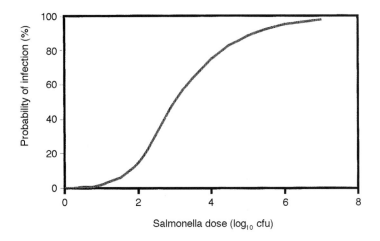

Fig. 9.4 Salmonella probability of infection curve.

From the human volunteer studies probability of infection (P_i) values have been determined for a number of foodborne pathogens and related organisms (Table 9.1, cf Table 7.7).

Table 9.1 gives the P_i for various foodborne pathogens and other infectious organisms. For example there is a 1 in 2000 chance of an individual becoming infected from a single salmonella cell compared to a 1 in 7 million chance from *V. cholerae*. The problem with human volunteer studies is that they are almost always conducted with healthy, young adults, usually men. These are not the most vulnerable members of society (elderly and very young). Since the minimum infective dose (MID) may vary widely from person to person, the MID concept may not be appropriate for risk assessment in a population. See also Section 9.5.3 concerning salmonella infectious dose estimates.

9.1.5 Risk characterisation

Risk characterisation is the quantitative and/or qualitative estimation including attendant uncertainties of the probability of occurrence and severity of known or potential adverse health effects in a given population

Table 9.1   Probability of infection (P_i).

Enteric pathogen	P_i
Campylobacter jejuni	7×10^{-3}
Salmonella spp.	2×10^{-3}
Shigella spp.	1×10^{-3}
V. cholerae (classical)	7×10^{-6}
V. cholerae El Tor	1.5×10^{-5}
Cl. perfringens	1×10^{5a}
B. cereus (diarrhoeal type)	1×10^{5a}
Cl. botulinum	0.5-5 ng[b]
St. aureus	0.5-5 µg[b]
Rotavirus	3×10^{-1}
Giardia spp.	2×10^{-2}

[a] Organisms producing toxin, numbers required to cause disease.
[b] Quantity of toxin causing symptoms.
Sources: Notermans & Mead 1996; Buchanan & Whiting 1996.

based on hazard identification, hazard characterisation and exposure assessment. It involves integrating the information gathered in the previous steps to estimate the risk to a population, or a particular type of consumer. The degree of confidence in the final estimation of risk depends on the variability, uncertainty and assumptions identified in the previous steps. Risk characterisation is the last step in risk assessment from which a risk management strategy can be formulated. These estimates can be assessed by comparison with independent epidemiological data that relate hazards to disease prevalence.

Using probabilistic models, computer spreadsheet risk assessments can be used to modify previous assumptions and values to ascertain their relative importance.

9.1.6   Production of a formal report

The risk assessment should be fully and systematically documented. To ensure transparency, the final report should indicate in particular any constraints and assumptions relative to the risk assessment. The report should be made available to independent parties on request.

9.2   Risk management

Risk management is the complex analysis and decisions which aim to reduce the probability of occurrence of unacceptable risks. It must be separated from Risk Assessment. Risk assessment of microbiological hazards is a scientific process aimed at identifying and characterising a microbiological hazard and estimating the risk of that hazard to a population. Risk Management is a separate process aimed at identifying options for action(s) needed to manage that risk and it has a policy function. If required, it may select and implement appropriate control options, including regulatory measures. It is required when epidemiological and surveillance data demonstrate that specific foods are a possible hazard factor towards consumer health due to the presence of hazardous microorganisms or toxic compounds from microbial origin. Governmental risk managers must decide on appropriate control options to manage this risk. In order to understand the risk for consumers more explicitly, the risk managers may ask expert food scientists to perform an MRA.

9.2.1 Risk assessment policy

Guidelines for value judgement and policy choices which may need to be applied at specific decision points in the risk assessment process are known as risk assessment policy. Risk assessment policy setting is risk management responsibility, which should be carried out in full collaboration with risk assessors, and which serves to protect the scientific integrity of the risk assessment. The guidelines should be documented so as to ensure consistency and transparency. Examples of risk assessment policy setting are establishing the population(s) at risk, establishing criteria for ranking hazards and guidelines for application of safety factors.

9.2.2 Risk profiling

Risk profiling is the process of describing a food safety problem and its context, in order to identify those elements of the hazard or risk relevant to various risk management decisions. The risk profile would include identifying aspects of hazards relevant to prioritising and setting the risk assessment policy and aspects of the risk relevant to the choice of safety standards and management options. A typical risk profile might include the following: a brief description of the situation, product or commodity involved; the values expected to be placed at risk, e.g. human health, economic concerns; potential consequences; consumer perception of the risks and the distribution of risks and benefits.

9.3 Risk communication

Risk communication is the interactive exchange of information and opinions concerning risk among risk assessors, risk managers, consumers and other interested parties. It also deals with communicating the outcome of the decision making process to the consumers using appropriate tools and channels.

9.4   Food Safety Objectives

MRA is primarily the prerogative of government bodies that have access to all necessary data and research findings. Although it is recognised that, with regard to food microbiology, the formalised approach to risk analysis is in its infancy, it is likely that in the near future MRA have a greater importance in the determination of the level of consumer protection that a government considers necessary and achievable. For practical implementation in specific sectors of the food chain, it is the responsibility of governmental authorities to translate the expected level of protection into *food safety objectives*. Such objectives delineate the specific target(s) that any food operator concerned should endeavour to achieve through appropriate interventions.

Food safety objectives are a statement of the maximum level of a microbiological hazard in a food considered acceptable for human consumption. Whenever possible, food safety objectives should be quantitative and verifiable. Food safety objectives as defined by governmental authorities represent the minimum target on which food operators base their own approach (see Fig. 1.4). The government's food safety objectives may be adopted as such in the form of a company's food safety requirements. Alternatively, depending on commercial factors, a company may wish to establish more demanding food safety requirements. Food safety requirements provide input to the food safety programme. They direct product and process planning, design and implementation of GMP, GHP, HACCP and quality assurance systems with the aim of fulfilling the food safety requirements.

There are several factors that strongly affect consumer exposure but which are poorly understood. These factors are:

- Incidence of different pathogens in raw materials.
- Effect of processing conditions (including alternative technologies).
- Effect of distribution conditions (e.g. chilled chain).
- Recontamination during handling.
- Abuse by the consumer.
- Heterogeneous distribution of microorganisms within a food.
- Person-to-person transmission.
- Host effects: age, pregnancy, nutritional status, concurrent or recent infections, use of medication, immunological status.
- Food vehicle effects.

The most thoroughly documented MRA is the shelled eggs MRA by the Food Safety and Inspection Service (1998), which is summarised in Section 9.5.3.

9.5   Application of MRA

There are a few detailed examples of complete MRA in the literature (see Appendix for Web site of references). The following are abridged examples to give an indication of the process.

9.5.1   Salmonella *spp. in cooked chicken (Buchanan & Whiting 1996)*

A three-stage example concerning *Salmonella* spp. in cooked chicken is described. In this process raw chicken is stored at 10°C for 48 hours before being cooked at 60°C for 3 minutes and then stored at 10°C for 72 hours before consumption. The 10°C stored temperature is in the 'Danger Zone' and represents mild temperature abuse (Section 7.6.2).

(1)   *Number of* Salmonella *spp. in raw chicken before cooking.* The number of salmonella cells on raw chicken will vary, however an expected level of contamination is given in Fig. 9.5. The contamination range varies from no salmonella cells in 75% of samples to

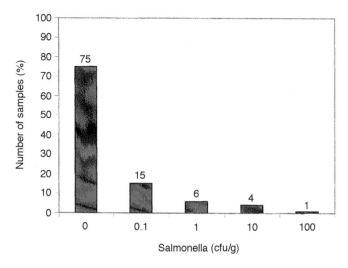

Fig. 9.5   Distribution of salmonella in raw foods (Buchanan & Whiting 1996).

1% containing 100 cells per gram of meat. The amount of salmonella growth at 10°C for 48 hours before cooking can be determined using growth models (Section 2.8) by assuming the meat pH is pH 7.0 and the sodium chloride level is 0.5%.

(2) *Effect of cooking (60°C, 3 minutes) on salmonella numbers in chicken.* The decimal reduction time (D value) at 60°C is 0.4 minutes. The effect of heat treatment on salmonella numbers can be calculated using the equation:

$$\log (N) = \log (N_0) - (t/D)$$

where N is the number of microorganisms (cfu/g) after the heat treatment, N_0 is the initial number of bacteria (cfu/g), D is the D value, log (cfu/g)/min, and t is the duration of the heat treatment (min). Note, for simplicity, no effect on salmonella numbers is taken into consideration for the time period during warming the food to 60°C and cooling afterwards. This equation gives the number of surviving salmonella after the cooking process and is designed to give a 7 D kill (Table 9.2).

(3) *Salmonella cell numbers following storage at 10°C, 72 h before consumption.* As before in stage (1) a growth curve for salmonella can be generated to estimate the number of salmonella in cooked chicken after storage, before consumption (Table 9.2).

Determination of the survival number and subsequent growth for each initial population level of salmonella gives an estimate of the numbers of salmonella that a population of consumers is likely to ingest. In this

Table 9.2 Hypothetical risk assessment for *Salmonella* spp. in food at various processing steps (adapted from Buchanan & Whiting 1996).

Process step	Salmonella cell count ($\log_{10}$ cfu/g) Initial population distribution levels (%)[a]				
	75	15	6	4	1
1	—	−1.0	0.0	1.0	2.0
2	—	−0.8	0.2	1.1	2.1
3	—	−8.3	−7.4	−6.4	−5.4
4	—	−7.5	−6.6	−5.7	−4.8
P_i^b	0	8.8×10^{-11}	6.5×10^{-10}	5.1×10^{-9}	4.1×10^{-8}

[a] See Fig. 9.5 for distribution of *Salmonella* spp. in raw food.
[b] Probability of infection per gram of food consumed.
Process step: 1 Before initial storage
2 After initial storage at 10°C for 48 hours
3 After cooking at 60°C for 3 minutes
4 After final storage at 10°C for 72 hours

example, 1% of the chicken samples contained 100 salmonella cells per gram, giving a probability of infection (P_i) of 4.1×10^{-8}/g food consumed (see Section 9.1.4 for an explanation of P_i). This means there was less than one cell surviving for every 10 000 g of food. Hence the salmonella risk associated with cooked chicken under these conditions of storage is minimal.

The above example can be used as a template to determine the effect of changing the cooking regime and storage conditions. For example, raising the initial storage temperature to 15°C and reducing the cooking time to 2 minutes (5 D kill) causes the probability of infectivity (P_i) to be unacceptably high (Table 9.3).

Table 9.3 Hypothetical risk assessment for *Salmonella* spp. in food at various processing steps after temperature abuse (adapted from Buchanan & Whiting 1996).

Process step	Salmonella cell count ($\log_{10}$ cfu/g) Initial population distribution levels (%)[a]				
	75	15	6	4	1
1	—	-1.0	0.0	1.0	2.0
2	—	1.6	2.6	3.6	4.6
3	—	-3.4	-2.4	-1.4	-0.4
4	—	1.8	2.7	3.7	4.7
P_i[b]	0	1.1×10^{-1}	4.1×10^{-1}	7.0×10^{-1}	8.6×10^{-1}

[a] See Fig. 9.5 for distribution of *Salmonella* spp. in raw food.
[b] Probability of infection per gram of food consumed.
Process step: 1 Before initial storage
2 After initial storage at 15°C for 48 hours
3 After cooking at 60°C for 2 minutes
4 After final storage at 15°C for 72 hours

9.5.2 E. coli O157:H7 in ground beef (Cassin et al. 1998)

A model of *E.coli* O157:H7 in ground beef was constructed to assess the impact of different control strategies (Fig. 9.6). The model describes the pathogen population from carcass processing through to consumer cooking and consumption. To generate a representative distribution of risk, the model is simulated many times with different values selected from the probability distributions. This technique is known as Monte Carlo simulation (McNab 1998).

The immediate output of the model is the distribution of health risk from eating ground beef hamburger patties. Additionally, the model enables changes in health risk associated with changes in particular parameters (temperature abuse, infectious dose, etc.) to be predicted. The

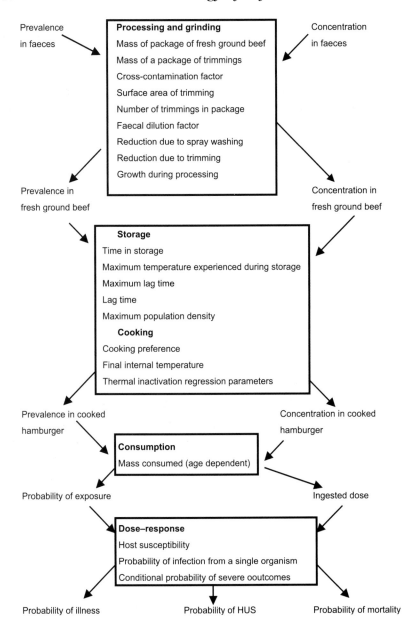

Fig. 9.6 Flow diagram for determining the exposure assessment and dose-response for *E. coli* in hamburgers (adapted from Cassin *et al.* 1998).

model predicted a probability of haemolytic uraemic syndrome of 3.7×10^{-6} and a probability of mortality of 1.9×10^{-7} per meal for the very young. The average probability of illness was predicted to be reduced by 80% under a strategy of reducing microbial growth during the retail storage through a reduction in storage temperature (Table 9.4). This strategy was predicted to be more effective than the reduction in the concentration of the pathogen in cattle faeces.

Table 9.4 Risk mitigation strategy, per cent reduction per meal illness from *E. coli* O157:H7 following assumed compliance (Cassin *et al.* 1998).

Strategy	Control variable	Predicted reduction in illness (%)
(1) Storage temperature	Maximum storage temperature	80
(2) Pre-slaughter screening	Concentration of *E. coli* O157:H7 in faeces	46
(3) Hamburger cooking Consumer information programme on cooking hamburgers	Cooking temperature	16

9.5.3 S. enteritidis *in shell eggs and egg products (FSIS)*

The Food Safety and Inspection Service (FSIS) have completed a 2-year comprehensive risk assessment of *S. enteritidis* in shell eggs which can be accessed from the Web (see Web pages Appendix for the URL). The model can also be downloaded from this site. It requires EXCEL (Microsoft™) and @RISK (Palisade Europe) to run.

The objectives were to:

• Identify and evaluate potential risk reduction strategies
• Identify data needs
• Prioritise future data collection efforts

The risk assessment model consists of five modules (Fig. 9.7):

• *Egg production module:* this estimates the number of eggs produced that are infected (or internally contaminated) with *S. enteritidis.*
• *Shell egg processing and distribution module:* this module follows the shell eggs from collection on the farm through processing, transportation and storage. The eggs remain intact throughout this module. Therefore the primary factors affecting the *S. enteritidis* are the

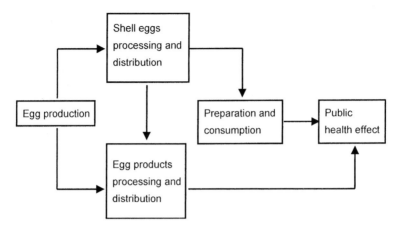

Fig. 9.7   Farm to table risk assessment model for eggs and egg products (Food Safety Inspection Service 1998).

cumulative temperatures and times of the various processing, transportation and storage stages. The two important modelling components are the time until the yolk membrane loses its integrity and the growth rate of *S. enteritidis* in eggs after breakdown of the yolk membrane.

* *Egg products processing and distribution module:* this module tracks the change in numbers of *S. enteritidis* in egg processing plants from receiving through pasteurisation (Figs 9.7 and 9.8). Figures 9.9 and 9.10 summarise the death rate (and D value; Section 2.4.2) of *S. enteritidis* in

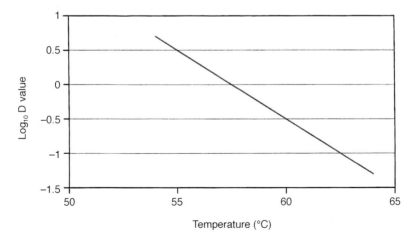

Fig. 9.8   Death rate of *S. enteritidis* in liquid whole eggs.

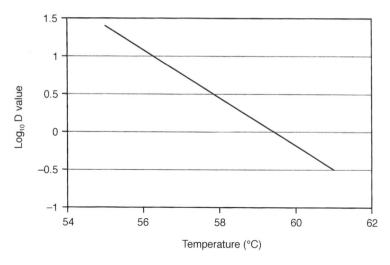

Fig. 9.9 Death rate of *S. enteritidis* in liquid egg yolk.

whole eggs and yolk during pasteurisation (liquid egg pasteurisation: 64.4°C, 2.5 min or 60°C, 3.5 min). There are two sources of *S. enteritidis* in egg products: from the internal contents of eggs and from cross-contamination during breaking.

- *Preparation and consumption module:* this estimates the increase or decrease in the numbers of *S. enteritidis* organisms in eggs or egg products as they pass through storage, transportation, processing and preparation.
- *The public health module:* this calculates the incidences of illnesses and four clinical outcomes (recovery without treatment, recovery after treatment by a physician, hospitalisation and mortality) as well as the cases of reactive arthritis associated with consuming *S. enteritidis* positive eggs (Table 9.5).

The baseline model simulates:

- Average production of 46.8 billion shell eggs per year (in the US)
- 2.3 million eggs contain *S. enteritidis*
- This results in 661 633 human illnesses per year, of which:
    94% of illnesses recover without medical care
    5% visit a physican
    0.5% are hospitalised
    0.05% of cases result in death
- 20% of the population are considered to be at a higher risk: infants,

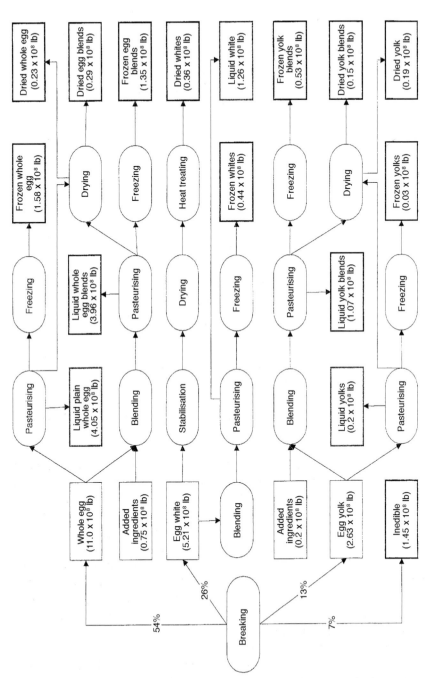

Fig. 9.10 Liquid egg product flow for year 1996 (Food Safety Inspection Service 1998).

Table 9.5 Public health module results.

	Category	Mean
Normal population	Exposed	1 889 200
	Ill	448 803
	Recover w/ no treatment	425 389
	Physician visit and recovery	21 717
	Hospitalised and recovered	1 574
	Death	123
	Reactive arthritis	13 578
Susceptible population	Exposed	521 705
	Ill	212 830
	Recover w/ no treatment	196 295
	Physician visit and recovery	14 491
	Hospitalised and recovered	1 776
	Death	269
	Reactive arthritis	6 416
Total population	Exposed	2 410 904
	Ill	661 633
	Recover w/ no treatment	621 684
	Physician visit and recovery	36 208
	Hospitalised and recovered	3 350
	Death	391
	Reactive arthritis	19 994

Source: Food Safety and Inspection Service (1998).

elderly, transplant patients, pregnant women, individuals with certain diseases.

The beta-Poisson model (Section 9.1.4) from *Salmonella* human volunteer feeding trials (1930–1973) estimates a probability of infection of 0.2 from ingesting 10^4 salmonella cells. Because an infectious dose does not necessarily lead to illness, the probability of infection is greater than the probability of illness. These data were obtained using serotypes other than *S. enteritidis* and hence may not be totally appropriate.

The baseline egg products model predicts that the probability is low that any cases of *S. enteritidis* will result from the consumption of pasteurised egg products. However, the current FSIS time and temperature regulations do not provide sufficient guidance to the egg products industry for the large range of products the industry produces (Table 9.6). Time and temperature standards based on the amount of bacteria in the raw product, how the raw product will be processed and the intended use of the final product will provide greater protection to the consumers of egg products.

Table 9.6   USDA minimum time and temperature requirements for three egg products.

Liquid egg product	Minimum temperature requirements (°F)	Minimum holding time requirements (minutes)
Albumen	134	3.5
	132	6.2
Whole egg	140	3.5
Plain yolk	142	3.5
	140	6.2

Source: Regulations Governing the Inspection of Eggs and Egg Products (7 CFR Part 59). May 1, 1991, USDA, FSIS, Washington, D.C. 20250.

The per cent reduction for total human illnesses was calculated for two scenarios differing from current practice within the shell egg processing and distribution module. The first scenario was that if all eggs were immediately cooled after lay to an internal temperature of 7.2°C (45°F), then maintained at this temperature, then a 12% reduction in human illnesses would be the result. Similarly, an 8% reduction in human illnesses would be the result if eggs were maintained at an ambient (i.e. air) temperature of 7.2°C (45°F) throughout shell egg processing and distribution.

9.5.4   L. monocytogenes *(EC 1999)*

The absence of agreed reference values for *L. monocytogenes* (except for dairy products) has led to controversy, especially intra-community trade of the EU where the lack of microbiological reference values has led to food products being declared unfit for human consumption because of non-quantified demonstration of *L. monocytogenes* contamination. Therefore MRA leading to the setting of food safety objectives was required and has been reported (EC 1999).

Although human listeriosis is mainly caused by a few serovars (4b and 1/2 a, b) it was concluded that a wide range of strains might cause serious disease. Additionally, since none of the typing methods discriminates pathogenic from nonpathogenic or less virulent strains, all *L. monocytogenes* should be regarded as potentially pathogenic.

The risk assessment of *L. monocytogenes* facilitates the classification of six food groupings relative to the organism's control (Table 9.7). Examples of products are:

• Groups B and D: meat products such as cooked ham, wiener sausages or hot smoked fish, soft cheese made from pasteurised milk.

Table 9.7 Grouping of ready-to-eat food commodities relative to the control potential for *Listeria monocytogenes*.

A	Foods heat-treated to a listericidal level in the final package
B	Heat-treated products that are handled after heat treatment. The products support growth of *L. monocytogenes* during the shelf life at the stipulated storage temperature
C	Lightly preserved products, not heat-treated. The products support growth of *L. monocytogenes* during the shelf life at the stipulated storage temperature
D	Heat-treated products that are handled after heat treatment. The products are stabilised against growth of *L. monocytogenes* during the shelf life at the stipulated storage temperature
E	Lightly preserved products, not heat-treated. The products are stabilised against growth of *L. monocytogenes* during the shelf life at the stipulated storage temperature
F	Raw, ready-to-eat foods

- Groups C and E: cold smoked or gravid fish and meat, cheese made from unpasteurised milk.
- Group F: tartar, sliced vegetables and sprouts.

Groups B and D, and C and E are separated according to the technology used.

A concentration of 100 *L. monocytogenes* cells/g of food at the point of consumption is considered a low risk to consumers. However, due to the uncertainties related to this risk, levels lower than 100 cells/g may be required for those foods in which listeria growth may occur. *L. monocytogenes* levels above 100 cfu/g can be achieved after in-food growth. Therefore, risk management should be focused on those foods which support *L. monocytogenes* growth.

Suggested levels of *L. monocytogenes* are:

(1) Food groups D, E and F: < 100 cfu/g throughout the shelf life and at point of consumption.
(2) Food groups A, B and C: not detectable in 25 g at time of production.

The food safety objective should be to keep the concentration of *L. monocytogenes* in food below 100 cfu/g and to reduce the fraction of foods with a concentration above 100 *L. monocytogenes* per gram significantly. In risk communication special attention should be addressed to consumer groups at increased risk (immunocompromised) which represent a considerable and growing section of the total population.

9.5.5 Salmonella *spp. in cooked patty (Whiting 1997)*

The steps in the MRA are (Fig. 9.11a and b):

(1) *Initial populations:* the initial microbial load was taken from published data (Surkiewicz *et al.* 1969), where 3.5% of the samples have salmonella present at levels greater than 0.44 cfu/g (Fig. 9.11a).

(2) *Storage:* conditions chosen were 21°C, 5 h and the growth rate predicted from a salmonella model (Gibson *et al.* 1988).

(3) *Cooking:* published D values were used to determine the extent of heat treatment at 60°C for 6 minutes.

(4) *Consumption:* a typical serving of 100 g was assumed.

(5) *Infectious dose.*

The model calculates that one salmonella cell has the mean probability of $10^{-4.6}$ of being an infectious dose (Fig. 9.11b). However, 3% of the predictions gave risks greater than 10^{-3} due to the small number of initial samples with high salmonella contamination. The usefulness of the model is in determining the effect of altering the variables, such as cooking temperature to 61°C which reduces the median probability of $10^{-7.4}$.

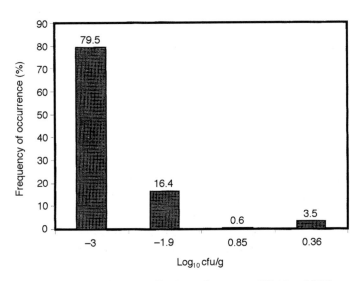

Fig. 9.11a Frequency of salmonella in poultry patty (Whiting 1997).

STAGE	Statistics
Initial distribution	−2.7 log cfu/g
Storage (21°C for 5 hours)	0.17 log cfu/g
Cooking (60°C for 6 minutes)	−4.42 log cfu/g
Consumption (100 g)	−2.42 log cfu/g
Infectious dose (probability)	$10^{-4.6}$ ($P = 1 - e^{-0.00752N}$)

Fig. 9.11b Risk-assessment model for salmonella in a cooked poultry patty (adapted from Whiting 1997).

9.5.6 Cl. botulinum *in ready-to-eat chill stored food* *(Baird-Parker 1994)*

Given the equation:

$$P_s = P_e + P_a + P_{pa} + P_n$$

where:

P_s = probability of contamination with psychrotrophic spores
P_e = probability of contamination from the environment
P_a = probability of contamination during assembly
P_{pa} = probability of contamination from packaging
P_n = probability of contamination from product ingredients

the probability of contamination with psychrotrophic spores is:

$$P_s = {}<10^{-8} + {}<10^{-7} + {}<10^{-9} + 0.8 \times 10^{-6}$$
$$P_s = 10^{-6}$$

In other words, 1 in 1 million.

This example showed that contamination of the ingredients will probably determine the contamination level of the final product. Therefore, if an unacceptable level of end-product contamination is found then the microbiological quality of the ingredients should be re-evaluated.

10

REGULATIONS AND AUTHORITIES

10.1 Regulations in international trade of food

The globalisation of food trade and increasing problems worldwide with emerging and re-emerging foodborne pathogens have increased the risk of cross-border transmission of infectious agents. Because of the global nature of food production, manufacturing and marketing, infectious agents can be disseminated from the original point of processing and packaging to locations thousands of miles away.

Food safety requires enhanced levels of international cooperation in setting standards and regulations. Food safety measures are not uniform around the world and such differences can lead to trade disagreements among countries. This is particularly true if microbiological requirements are not justified scientifically. A useful Web site giving food and agricultural import regulations and standards can be found in the listing of Web addresses in the Appendix.

The standards, guidelines and recommendations adopted by the Codex Alimentarius Commission (CAC, Section 10.2) and International trade agreements, such as those administered by the World Trade Organisation (WTO, Section 10.3), are playing an increasingly important role in protecting the health of consumers and ensuring fair practices in trade. In 1962, the Joint Food Administration Organisation/World Health Organisation Food Standards Programme was created with the CAC as its executive organ. The Codex Alimentarius, or the food code, is a collection of international food standards that have been adopted by the CAC. Codex standards cover all the main foods, whether processed, semi-processed or raw. The principal objectives of CAC are to protect the health of consumers and ensure fair practices in the food trade.

In the case of microbiological hazards, Codex has elaborated standards, guidelines and recommendations that describe processes and procedures for the safe preparation of food. The application of these standards, guidelines and recommendations is intended to prevent or eliminate hazards in foods or reduce them to acceptable levels.

The WTO SPS Agreement entered into force in 1995 and applies to all sanitary and phytosanitary measures, which may, directly or indirectly, affect international trade. It provides basic rights and obligations for WTO members and directs them to harmonise sanitary guidelines and recommendations. For food safety, the standards, guidelines and recommendations established by the CAC relating to food additives, veterinary drug and pesticide residues, contaminants, sampling and methods of analysis, and codes and guidelines of hygienic practice are recognised as the basis for harmonisation of sanitary measures.

WTO members may introduce or maintain measures that result in a higher level of sanitary or phytosanitary protection than would be achieved by measures based on international standards, guidelines and recommendations. In this regard, WTO members are required to ensure that their sanitary and phytosanitary measures are based on an assessment, as appropriate to the circumstances, of the risks to human, animal or plant life or health, taking into account the risk assessment techniques developed by the relevant international organisations. Article 5 of the SPS Agreement provides an impetus for the development of microbiological risk assessment to support the elaboration of standards, guidelines and recommendations related to food safety.

The first Joint FAO/WHO Expert Consultation on the Application of Risk Analysis to Food Standards Issues was held in 1995. It delineated the basic terminology and principles of risk assessment and concluded that the analysis of risks associated with microbiological hazards presents unique challenges. The report of the Joint FAO/WHO Expert Consultation on Risk Management and Safety held in 1997 identified a risk management framework and the elements of risk management for food safety (FAO 1997). The Joint FAO/WHO Expert Consultation on the Application of Risk Communication to Food Standards and Safety Matter held in 1998 identified elements and guiding principles of risk communication and strategies for effective risk communication (FAO 1998). In addition to the foundation provided by the series of Joint FAO/WHO Expert Consultations, the Codex Committee on Food Hygiene has elaborated principles and guidelines for microbiological risk assessment. The 'Draft Principles and Guidelines for the Conduct of Microbiological Risk Assessment' were adopted by the 23rd Session of the Codex Commission in June 1999 (Codex Alimentarius Commission 1999b).

10.2 Codex Alimentarius Commission

The Codex Alimentarius Commission was established in 1962 to implement the Joint Food and Agricultural Organisation (FAP) and World Health Organisation (WHO) Food Standards Programme. The aim of the programme is to set minimum standards to protect the consumer and provide a framework to ensure fair trade across international borders. Codex also plays a role in promoting coordination on all food standards work between international governmental and nongovernmental organisations, which includes helping to set priorities for work in this area (Table 10.1).

Table 10.1 Codex committees.

Subject	Host country
Food labelling	Canada
Food additives and contaminants	The Netherlands
Food hygiene	USA
Pesticide residues	The Netherlands
Veterinary drugs in foods	USA
Methods of analysis and sampling	Hungary
Food import and export inspection and certification systems	Australia
General principles	France
Nutrition and foods for special dietary uses	Germany

10.2.1 Codex Alimentarius

The Codex Alimentarius (Latin for food law or code) is a collection of internationally accepted food standards presented in a uniform manner. The aim of these food standards is to protect consumers' health and to ensure fair practices in the food trade. Codex also publishes advisory tests in the form of codes of practice, guidelines and other recommended measures to assist in achieving its objective.

10.2.2 Codex committees and structure

The Codex machine operates through a labyrinthine structure of committees which feed recommendations up to the main Codex Alimentarius Commission. There are a number of commodity type committees dealing with hygiene and quality issues; examples are: Codex Committee for Food Hygiene, Codex Committee for Milk and Milk Products, and Codex Committee for Nutrition and Foods for Special Dietary Uses. There are also a number of other committees dealing with executive, procedural and regional issues, for example Codex Committee on General Principles and Codex Coordinating Committee for Europe.

10.3   Sanitary and Phytosanitary measures (SPS), Technical Barriers to Trade (TBT) and the World Health Organisation (WHO)

The World Trade Organisation (WTO) agreements on Sanitary and Phytosanitary measures (SPS) and Technical Barriers to Trade (TBT) seek to remove barriers and harmonise trade at an international level (Section 7.1). Paragraph 3 of Article 2 (Basic Rights and Obligations) of the SPS Agreement states that

> Members shall ensure that any sanitary and phytosanitary measure is applied only to the extent necessary to protect human, animal or plant life or health, is based on scientific principles and is not maintained without sufficient scientific evidence

Article 3 encourages the harmonisation of food safety measures on as wide a basis as possible. It implies that food safety measures which conform to Codex standards and related texts are deemed necessary to protect human life or health and are presumed to be consistent with the relevant provisions of the SPS Agreement.

All countries need to develop the necessary capacities to conduct risk assessments and implement appropriate risk management activities, particularly regarding emerging biological or chemical problems. Article 4 *inter alia* refers to the conclusion of bilateral or multilateral agreements on the recognition of the equivalence of the level of protection of specified food safety measures. Failure to conclude equivalence agreements with major trading partners can have serious economic consequences. The development of such agreements is, however, facilitated through the use of Codex standards, guidelines and recommendations as a basis for the food control legislation of individual countries. Such countries, however, need to develop the capacities to ensure adherence to the harmonised legislation. To assist with this aim the WHO has recognised Codex Standards as the international reference for settling disputes between governments on food safety related issues. Codex, mindful of the increased importance of their standards and likely effects on world trade, has since embarked on the process of revisiting a number of their standards to ensure that they are soundly based. National controls in variance with Codex Standards which give rise to trade disputes have to be justified on the basis of scientific risk assessment. While controls may differ, they must achieve an equivalent level of protection. The increased importance of Codex will be reflected in future years.

10.4 Food authorities in the United States

10.4.1 US Department of Health and Human Services

This includes the Food and Drug Administration (FDA) and the Centers for Disease Control and Prevention (CDC).

The FDA's responsibilities are for:

- All domestic and imported food sold in interstate commerce, including shell eggs, but not meat and poultry
- Bottled water
- Wine beverages with less than 7% alcohol

Its role is to enforce food safety laws concerning domestic and imported food, except meat and poultry. The FDA Food Code is endorsed by USDA's Food Safety and Inspection Service (FSIS) and Centers for Disease Control and Prevention. It provides a model by which US state regulatory authorities may develop or update their own food safety rules concerning food safety in restaurants, grocery stores, nursing homes and other institutional and retail settings. The Food Code is neither federal law nor regulation. However, it is used by more than 3000 state and local regulatory agencies across the US and aims to achieve consistency across the various regulatory jurisdictions. The Food Code is updated every 2 years and the Web site is given in the Appendix section.

The CDC has responsibility for all foods. Their role is to:

- Investigate sources of foodborne disease outbreaks.
- Maintain a nationwide system of foodborne disease surveillance.
- Develop and advocate public health policies to prevent foodborne diseases.
- Conduct research to help prevent foodborne illness.
- Train local and state safety personnel.

10.4.2 US Department of Agriculture (USDA)

The USDA is composed of the FSIS, Co-operative State Research, Education, and Extension Service and the National Agricultural Library USDA/FDA Foodborne Illness Education Information Center.

The FSIS enforces food safety laws governing domestic and imported meat and poultry products. It has responsibility for:

• Domestic and imported meat and poultry and related products, such as meat- or poultry-containing stews, pizzas and frozen foods
• Processed egg products (generally liquid, frozen and dried pasteurised egg products)

The Co-operative State Research, Education, and Extension Service is responsible for all domestic foods and some imported foods. Its food safety role is with US colleges and universities to develop research and education programmes on food safety for farmers and consumers.

The National Agricultural Library (NAL) USDA/FDA Foodborne Illness Education Information Center maintains a database of computer software, audiovisuals, posters, games, teachers' guides and other educational materials on preventing foodborne illness. It also helps educators, food service trainers and consumers locate educational materials on preventing foodborne illness.

10.4.3   US Environmental Protection Agency (EPA)

The US EPA oversees drinking water and has a food safety role with regard to foods made from plants, seafood, meat and poultry. It:

• Establishes safe drinking water standards.
• Regulates toxic substances and wastes to prevent their entry into the environment and food chain.
• Assists states in monitoring the quality of drinking water and finding ways to prevent contamination of drinking water.
• Determines the safety of new pesticides, sets tolerance levels for pesticide residues in foods and publishes directions on safe use of pesticides.

10.4.4   Food Outbreak Response Coordinating Group (FORC-G)

The FORC-G was formed in 1997 to:

• Increase coordination and communication among federal, state and local food safety agencies.
• Guide efficient use of resources and expertise during an outbreak.
• Prepare for new and emerging threats to the US food supply.

The group is formed from the Department of Health and Human Services (which includes the FDA), the US Department of Agriculture and the EPA.

10.5 European Union legislation

There are three types of legislation within the EU:

10.5.1 Regulation

A legal act which has general applications and is binding in its entirety and directly applicable to the citizens, courts and governments of all Member States. Regulations do not, therefore, have to be transferred into domestic laws and are chiefly designed to ensure uniformity of law across the Community.

10.5.2 Directive

A binding law directed to one or more Member States. The law states objectives which the Member States(s) are required to confirm within a specified time. A Directive has to be implemented by Member States by amendment of their domestic laws to comply with the stated objectives. This process is known as 'approximation of laws' or 'harmonisation' since it involves the alignment of domestic policy throughout the Community.

10.5.3 Decision

An act which is directed at specific individuals, companies or Member States which is binding in its entirety. Decisions addressed to Member States are directly applicable in the same way as directives.

10.5.4 Single Act Measures

Food hygiene law and technical legislation such as additives or labelling are mostly 'Single Act Measures', this means they are part of the progress towards the single market. Single Act Measures are those which are essential for the free flow of goods and services in a truly common market; they are subject to majority voting.

The aim of EU food legislation is to ensure a high standard of public health protection and that the consumer is adequately informed of the nature and, where appropriate, the origin of the product. In the context of a developing internal market, the Commission has adopted a particular approach to the food sector in order to establish a large market without barriers, and at the same time ensuring consumer safety with the widest possible choice. The Commission has combined the principles of mutual recognition of national standards and rules, contained within Articles 30 to 36 of the Treaty of Rome – this approach works alongside 100a which resulted from the amendment to the treaty by the Single European Act. The primary source for European Union information is the 'Official Journal'.

The 1985 European Commission White Paper 'Completing the Internal Market' catalogued the measures necessary to allow for the free movement of goods (including food services, capital and labour) which would lead to the removal of all physical, technical and fiscal barriers between Member States. Since 1 January 1993, food has moved freely within the EU with the minimum of inspection at land or sea frontiers. Harmonised rules have been adopted, applicable to all food produced in the EU, underpinned by the principle of mutual recognition of national standards and regulations for matters that do not require EC legislation. Specific directives were in place for meat and meat products, live bivalve molluscs, fishery products, milk and milk products and eggs (Section 8.9). Foods entering the EC from countries outside the EU will be subject to EU hygiene standards.

10.5.5 Food Hygiene Directive (93/43/EEC)

One of the most significant EU directives for the food industry was the adoption of the Food Hygiene Directive (93/43/EEC) on the hygiene of foodstuffs (EEC 1993). It is extremely significant in the development of EU food law and will form the basis for food hygiene control across Europe for years to come. The Directive deals with general rules of hygiene for foodstuffs and the procedures for verification of compliance with the rules. 'Food hygiene' means all measures necessary to ensure the safety and wholesomeness of foodstuffs. 'Wholesome food' means food which is fit for human consumption. 'Food business' means any undertaking whether public or private, and whether operating for profit or not. The measures cover all stages after primary production (harvesting, slaughter and milking), preparation, processing, manufacturing, packaging, storing, transportation, distribution, handling and offering for sale or supply to the consumer. Food business operators must identify any step in their activities which is critical to ensuring food safety and ensure that adequate safety procedures are identified, implemented, maintained and reviewed on the basis of the principles used to develop HACCP (Section

7.3). The Directive is a horizontal Directive and therefore applies across the whole of the food industry. It covers producers, manufacturers, distributors, wholesalers, retailers and caterers. In essence this Directive combines the proactive approach of food safety by HACCP implementation and codes of Good Hygienic Practice into food law.

10.6 Food safety agencies

There are numerous national food safety agencies established around the world; the Web sites for a few of them are given in the Appendix. The Danish National Food Agency takes a 'plough to plate' approach and has a broad remit. The Swedish National Food Administration (SMFA) also has a broad remit, has a proactive health role and has powers to legislate. In Germany the enforcement responsibilities are within the regional governments, the German Federal Institute for Consumer Health and Veterinary Medicine (BgVV) and the Ministry of Health. The Australian and New Zealand Food Authority (ANZFA) has a narrower remit and currently focuses on the development of food standards and codes of practice to protect public health and promote fair trade. Its role is to make recommendations and it has industry representation. This is in contrast to the Food Authority in Ireland where commercial interests have been deliberately excluded in order to establish public confidence in its independence. The Canadian Food Inspection Agency (CFIA) has

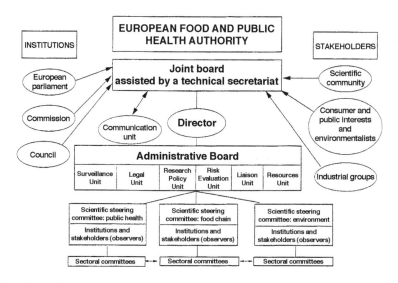

Fig. 10.1 The proposed structure of the new EU institution.

responsibility for human, plant and animal health and has enforcement powers. There is a proposed European Public Health and Food Agency (see Web site for details) which will have extensive powers to investigate and regulate farms, shops and restaurants. This Agency would be similar to the Food and Drug Administration (FDA) of the United States. The FDA, however, does not have responsibility for meat, this is the responsibility of the US Department of Agriculture.

Many agencies are in the process of restructuring in response to the considerable public concern over food safety, partially due to both the BSE/nvCJD situation in the UK and dioxin contamination in Belgium. The structure of the proposed European Agency is given in Fig. 10.1. In order to be implemented the Maastricht Treaty will have to be amended and receive unanimous approval from all EU countries. According to the EU Web site (see Appendix for URL address), the Scientific Committee on Food responsibilities will be: scientific and technical questions concerning consumer health and food safety associated with the consumption of food products and in particular questions relating to toxicology and hygiene in the entire food production chain, nutrition, and applications of agrifood technologies, as well as those relating to materials coming into contact with foodstuffs, such as packaging.

GLOSSARY OF TERMS

CCP: Critical Control Point.

CCP Decision Tree: a sequence of questions to assist in determining whether a control point is a Critical Control Point.

Cleaning: the removal of soil, food residue, dirt, grease or other objectionable matter.

Contaminant: any biological or chemical agent, foreign matter, or other substances not intentionally added to food which may compromise food safety or suitability.

Contamination: the introduction or occurrence of a contaminant in food or food environment.

Control (verb): to take all necessary actions to ensure and maintain compliance with criteria established in the HACCP plan.

Control (noun): the state wherein correct procedures are being followed and criteria are being met.

Control Measure: any action or activity that can be used to prevent, eliminate or reduce a food safety hazard to an acceptable level.

Control Point: any step at which biological, chemical or physical factors can be controlled.

Corrective action: any action to be taken when the results of monitoring at the CCP indicate a loss of control.

Criterion: a requirement on which a judgement or decision can be based.

Critical Control Point (CCP): a step at which control can be applied and is essential to prevent or eliminate a food safety hazard or reduce it to an acceptable level.

Critical Limit: a maximum and/or minimum value to which a biological, chemical or physical parameter must be controlled at a CCP to prevent, eliminate or reduce to an acceptable level the occurrence of a food safety hazard. A criterion which separates acceptability from unacceptability.

Deviation: failure to meet a critical limit.

Disinfection: the reduction, by means of chemical agents and/or physical methods, of the number of microorganisms in the environment, to a level that does not compromise food safety or suitability.

Dose–response assessment: the determination of the relationship between the magnitude of exposure (dose) to a biological, chemical or physical agent and the severity and/or frequency of associated adverse health effects (response).

Enterotoxins: substances that are toxic to the intestinal tract causing vomiting, diarrhoea, etc.; most common enterotoxins are produced by bacteria.

Establishment: any building or area in which food is handled and the surroundings under the control of the same management.

Exposure assessment: the qualitative and/or quantitative evaluation of the likely intake of biological, chemical and physical agents via food as well as exposures from other sources, if relevant.

Flow diagram: a systematic representation of the sequence of steps or operations used in the production or manufacture of a particular food item.

Food handler: any person who directly handles packaged or unpackaged food, food equipment and utensils, or food contact surfaces and is therefore expected to comply with food hygiene requirements.

Food hygiene: all conditions and measures necessary to ensure the safety and suitability of food at all stages of the food chain.

Food safety: assurance that food will not cause harm to the consumer when it is prepared and/or eaten according to its intended use.

Food safety objective: a government target considered necessary to protect the health of consumers (this may apply to raw materials, a process or finished products).

Food safety requirement: a company-defined target considered necessary to comply with a food safety objective.

Food suitability: assurance that food is acceptable for human consumption according to its intended use.

HACCP: a systematic approach to the identification, evaluation and control of food safety hazards.

HACCP Plan: the document which is based upon the principles of HACCP and which delineates the procedures to be followed.

HACCP System: the result of the implementation of the HACCP plan.

HACCP Team: the group of people responsible for developing, implementing and maintaining the HACCP system.

Hazard: a biological, chemical or physical agent in a food, or condition of a food, with the potential to cause an adverse health effect.

Hazard Analysis: the process of collecting and evaluating information on hazards associated with the food under consideration to decide which are significant and must be addressed in the HACCP plan.

Hazard characterisation: the qualitative and/or quantitative evaluation of the nature of the adverse health effects associated with biological, chemical or physical agents which may be present in food. For chemical agents, a dose–response assessment should be performed. For biological or chemical agents, a dose–response assessment should be performed if the data are obtainable.

Hazard identification: the identification of biological, chemical and physical agents capable of causing adverse health effects which may be present in a particular food or group of foods.

Lot: a quantity of food or food units produced and handled under uniform conditions.

Monitoring: the act of conducting a planned sequence of observations or measurements of control parameters to assess whether a CCP is under control.

Prerequisite Programmes: procedures, including Good Manufacturing Practices, that address operational conditions providing the foundation for the HACCP system.

Primary production: those steps in the food chain up to and including, for example, harvesting, slaughter, milking, fishing.

Proteome: the protein complement of a genome.

Quality: the totality of characteristics of an entity that bear on its ability to satisfy stated or implied needs.

Quality assurance: all the planned and systematic activities implemented within the quality system, and demonstrated as needed, to provide adequate confidence that an entity will fulfil requirements for quality.

Quality control: the operational techniques and activities used to fulfil requirements of quality.

Quality management: all activities of the overall management function that determine the quality policy objectives and responsibilities that implement them by means such as quality planning, quality control, quality assurance and quality improvement with the quality system.

Quality system: the organisational structure, procedures, processes and resources needed to implement quality management.

Risk: a health effect caused by a hazard in a food and the likelihood of its occurrence.

Risk analysis: a process consisting of three components: risk assessment, risk management and risk communication.

Risk assessment: a scientifically based process consisting of the following steps: (a) hazard identification, (b) hazard characterisation, (c) exposure assessment and (d) risk characterisation.

Risk characterisation: the qualitative and/or quantitative estimation, including attendant uncertainties, of the probability of occurrence and severity of known or potential adverse health effects in a given population based on hazard characterisation and exposure assessment.

Risk communication: the interactive exchange of information and opinions concerning risk among risk assessors, risk managers, consumers and other interested parties.

Risk management: the process of weighing policy alternatives in light of the results of risk assessment and, if required, selecting and implementing appropriate control options, including regulatory measures.

Safety policy: the overall intentions and direction of an organisation with regard to safety as formally expressed by top management.

Severity: the seriousness of the effect(s) of a hazard.

Step: a point, procedure, operation or stage in the food chain including raw materials, from primary production to final consumption.

Total Quality Management: an organisation's management approach centred on quality, based on the participation of all its members and aimed at long-term success through customer satisfaction and benefits to the members of the organisation and to society.

Validation: that element of verification focused on collecting and evaluating scientific and technical information to determine whether the HACCP plan, when properly implemented, will effectively control the hazards.

Verification: those activities, other than monitoring, that determine the validity of the HACCP plan and that the system is operating according to the plan.

Appendix

FOOD SAFETY RESOURCES ON THE
WORLD WIDE WEB

Since Web sites have a habit of moving, it may be necessary to use a search engine with the name of the organisation or topic to relocate the URL. The prefix 'http://' has been omitted from all addresses.

Organisation or topic	URL
Author's Food Microbiology Information Centre	science.ntu.ac.uk/external/fhc
Web page up-dates for the book	www.blackwell-science.com/MicroSafeFoods
European antimicrobial resistance surveillance system	www.earss.rivm.nl
Codex – Food Hygiene	www.fao.org/waicent/faoinfo/economic/esn/ CODEX/STANDARD/standard.htm
EU microbiological criteria	www.europa.eu.int/comm/dg24/health/sc/scv/out26_en.html
FDA Bad Bug Book	vm.cfsan.fda.gov/~mow/intro.html
Food and agricultural import regulations and standards	www.fas.usda.gov/itp/ofsts/fairs-country.html
Food irradiation	www.nal.usda.gov/fnic/foodborne/fbindex/035.htm
Food safety education (Aust)	www.safefood.net.au/
Food safety education (USA)	www.nal.usda.gov/fnic/foodborne/foodborn.htm
FSIS *Salmonella* Enteritidis Risk Assessment	www.europa.eu.int/comm/dg24/health/sc/scv/out26_en.html.
Hazard Analysis Critical Control Point (HACCP):	
Generic HACCP models (USA)	www.fsis.usda.gov/OA/haccp/models.htm
Generic HACCP models (Canada)	www.cfia-acia.agr.ca/english/ppc/haccp/haccp.html
General HACCP (various foods)	vm.cfsan.fda.gov/~lrd/haccp.html
HACCP95	www.cvm.uiuc.edu/announcements/haccp95/haccp95.html
HACCP resources (USDA)	www.nal.usda.gov/foodborne/haccp/resource/resource.html
HACCP-TQM in retail	www.hi-tm.com
International HACCP Alliance	aceis.agr.ca
NACMCF HACCP plan	www.fst.vt.edu/haccp97
Seafood HACCP	seafood.ucdavis.edu/haccp/compendium/compend.htm
International Commission on Microbiological Specifications for Foods (ICMSF)	www.dfst.csiro.au/icmssf.htm
International Life Science Institute (ILSI)	www.ilsi.org/europe.html
Microbial growth media and detection kits (OXOID Ltd.)	www.oxoid.co.uk
Microbiological Risk Assessment	www.who.org/fsf/mbriskassess/index
Risk Analysis software (Palisade Europe)	www.palisade-europe.com
Risk assessment and risk communication	www.nal.usda.gov/fnic/foodborne/risk.htm

National Governmental Food Departments and Agencies (or equivalents)

Austria. Rechtsinformationssystem	www.ris.bka.gv.at
Austria. Federal Environment Agency	www.ubavie.gv.at
Australia. Department of Agric., Fish. and Forestry	www.affa.gov.au
Australia & New Zealand Food Authority	www.anzfa.gov.au
Belgium. Centre for Agricultural Research (CLO)	www.clo.fgov.be
Canada . Food Inspection Agency	www.cfia-acia.agr.ca/english/overvew2.html
Denmark. Fødevareministeriet	www.fvm.dk
England. Ministry of Agriculture Fisheries and Food	www.maff.gov.uk
European Union:	
DG VI – Agriculture	europa.eu.int/comm/dg06/index_en.htm
DG XIV – Fisheries	europa.eu.int/comm/dg14/dg14.htm
European Union	www.europa.eu.int
European Union	www.eu.int/comm/dg24/health
EU public health and food issues agency	europa.eu.int/comm/dg24/library/press/press37_en.html
Finland. Maa- ja Metsätalousministeriö	www.mmm.fi.htm
France. Ministry of Agriculture	www.agriculture.gouv.fr
Germany. Bundes.Ernah., Landw. Forsten	www.bml.de
Greece. Hellenic Republic Ministry of Agriculture	www.minagric.gr
India. Ministry of Food Processing	www.allindia.com/gov/ministry/fpi/policy.htm
Ireland. Dept. of Agriculture and Food	www.irlgov.ie/daff
Ireland. Food Safety Authority	www.fsai.ie/
Italy. Istituto Nazionale di Economia Agraria	www.inea.it
Japan. Ministry of Agriculture Fisheries and Food	www.maff.go.jp/eindex.html
Korea. Food and Drug Administration	www.kfda.go.kr/english/index.html
Netherlands. Ministry of Agric., Nat. Man. Fish.	www.minlnv.nl/international/
Portugal. Minist. Agric.Desen. rural e das Pescas	www.min-agrcultura.pt
Portugal. Minist. Equip.Plane. Admin.	www.min-plan.pt
Russia. Ministry of Agriculture and Food	www.aris.ru/N/WIN_R/PARTNER
Scotland	www.scotland.gov.uk/food/

Spain. Agritel – Minist. de Agric. Pesc. Aliment. www.sederu.es
Sweden. Jordbruksdepartementet www.sb.gov.se/info_rosenbad/department/jordbruk.html
United Nations:
  Codex Alimentarius Commission www.fao.org/es/esn/codex/Default.htm
  Codex Standards www.fao.org/es/esn/codex/STANDARD/standard.htm
  Food and Agriculture Organization www.fao.org
USA:
  Centres for Disease Control ftp.cdc.gov/pub/mmwr/MMWRweekly
  Centres for Disease Control and Prevention www.cdc.gov
  Co-operative State Research, Education and Extension Service www.reeusada.gov
  Department of Agriculture www.usda.gov
  Food and Drug Administration (FDA) vm.cfsan.fda.gov/list.html
  Food Code (FDA) vm.cfsan.fda.gov/~dms/foodcode.html
  Food Safety and Inspection Service www.fsis.usda.gov
  National Agricultural Library USDA/FDA Foodborne Illness www.nal.usda.gov/fnic
  World Health Organization www.who.int/fsf/

REFERENCES

Abee, T. & Wouters, J.A. (1999) Microbial stress response in minimal processing. *Int. J. Food Microbiol.*, **50**, 65-91.

Adams, M.R. & Marteau, P. (1995) On the safety of lactic acid bacteria from food. *Int. J. Food Microbiol.*, **27**, 263-4.

Adams, M.R., Little, C.L. & Easter, M.C. (1991) Modelling the effect of pH, acidulant and temperature on the growth rate of *Yersinia enterocolitica. J. Appl. Bacteriol.*, **71**, 65.

Allos, B.M. (1998) *Campylobacter jejuni* infection as a cause of the Guillain-Barré syndrome. *Emerg. Infect. Dis.*, **12**, 173-84.

Amann, R.I., Ludwig, W. & Scheifer, K.-H. (1995) Phylogenetic identification and *in situ* detection of individual microbial cells without cultivation. *Microbiol. Rev.*, **59**, 143-69.

Anon (1992) HACCP and Total Quality Management winning concepts for the 90's a review. *J. Food Protect.*, **55**, 459-62.

Anon (1993a) Generic HACCP for raw beef. *Food Microbiol. (Lond.)*, **10**, 449-88.

Anon (1993b) HACCP implementation: A generic model for chilled foods. *J. Food Protect.*, **56**, 1077-84.

Anon (1993c) *Listing of Codes of Practice Applicable to Foods*. Institute of Food Science and Technology, London.

Anon (1995a) Antibiotic resistance in salmonellas from humans in England and Wales: the situation in 1994. *PHLS Microbiol. Digest*, **12**, 131-3.

Anon (1996) Microbiological guidelines for some ready-to-eat foods sampled at the point of sale: an expert opinion from the Public Health Laboratory Service (PHLS). *PHLS Microbiol. Digest*, **13**, 41-3.

Anon (1997a) Preliminary report of an outbreak of *Salmonella anatum* infection linked to infant formula milk. *Eurosurv.*, **2**, 22-4.

Anon (1997b) Surveillance of enterohaemorrhagic *E. coli* (EHEC) infections and haemolytic uraemic syndrome (HUS) in Europe. *Euroserv.*, **2**, 91-6.

Anon (1999a) Management of the investigation by Enter-Net of international foodborne outbreaks of gastrointestinal organisms. *Eurosurv.*, **4**, 58-62.

Anon (1999b) Opinion of the scientific committee on veterinary measures relating to public health on the evaluation of microbiological criteria for food products of animal origin for human consumption.

Anon (1999c) AOAC International qualitative and quantitiatve microbiology guidelines for methods validation. *J. AOAC Int.*, **82**, 402-16.

Anon (1999d) Where have all the gastrointestinal infections gone? *EuroSurveil. Weekly Rep.*, **3**, 990114 (http://www.eurosurv.org).

AOAC (1998) Bacteriological Analytical Manual, 8th edn. Association of Official Analytical Chemists Food and Drug Administration.

Atlas, R.M. (1999) Probiotics – snake oil for the new millennium? *Environ. Microbiol.*, **1**, 377–80.

Baig, B.H., Wachsmuth, I.K., Morris, G.K. & Ill, W.E. (1986) Probing of *Campylobacter jejuni* with DNA coding for *Escherichia coli* heat-labile enterotoxin. *J. Infect. Dis.*, **154**, 542.

Baik, H.S., Bearson, S., Dunbar, S. & Foster, J.W. (1996) The acid tolerance response of *Salmonella typhimurium* provides protection against organic acids. *Microbiology*, **142**, 3195–200.

Baird-Parker, A.C. (1994) Foods and microbiological risks. *Microbiology*, **140**, 687–95.

Baker, D.A. (1995) Application of modelling in HACCP plan development. *Int. J. Food Microbiol.*, **25**, 251–61.

Baker, D.A. & Genigeorgis, C. (1990) Predicting the safe storage of fresh fish under modified atmospheres with respect to *Clostridium botulinum* toxigenesis by modeling length of the lag phase of growth. *J. Food Protect.*, **53**, 131–40.

Baranyi, J. & Roberts, T.A. (1994) A dynamic approach to predicting bacterial growth in food. *Int. J. Food Microbiol.*, **23**, 277–4.

Baranyi, J. & Roberts, T.A. (1995) Mathematics of predictive food microbiology. *Int. J. Food Microbiol.*, **26**, 199–218.

Barer, M.R. (1997) Viable but non-culturable and dormant bacteria: time to resolve an oxymoron and a misnomer? *J. Med. Microbiol.*, **46**, 629–31.

Barer, M.R., Kaprelyants, A.S., Weichart, D.H., Harwood, C.R. & Kell, D.B. (1998) Microbial stress and culturability: conceptual and operational domains. *Microbiology*, **144**, 2009–10.

Bauman, H.E. (1974) The HACCP concept and microbiological hazard categories. *Food Technol.*, **28**, 30–4, 74.

Bearson, S., Bearson, B. & Foster, J.W. (1997) Acid stress responses in enterobacteria. *FEMS Microbiol. Lett.*, **147**, 173–80.

Bell, C. & Kyriakides, A. (1998) Listeria – a practical approach to the organism and its control in foods. Blackie Academic and Professional, London.

Berg, R.D. (1998) Probiotics, prebiotics or 'conbiotics'? *Trends Microbiol.*, **6**, 89–92.

Berke, T., Golding, B., Jiang, K., *et al.* (1997) *J. Med. Microbiol.*

Beuchat, L.R. (1996) Pathogenic microorganisms associated with fresh produce. *J. Food Protect.*, **59**, 204–16.

Biller, J.A, Katz, A.J., Flores, A.F., Buie, T.M. & Gorbach, S.L. (1995) Treatment of recurrent *Clostridium difficile* colitis with *Lactobacillus* GG. *J. Pediatr. Gastroenterol.*, **21**, 224–6.

Blackstock, W.P. & Weir, M.P. (1999) Proteomics: quantitative and physical mapping of cellular proteins. *TIBTECH*, **17**, 121–7.

Blocher, J.C. & Busta, F.F. (1985) Multiple modes of inhibition of spore germination and outgrowth by reduced pH and sorbate. *J. Appl. Bacteriol.*, **59**, 467–78.

Bloomfield, S.F., Stewart, G.S.A.B., Dodd, C.E.R., Booth, I.R. & Power, E.G.M. (1998) The viable but non-culturable phenomenon explained? *Microbiology*, **144**, 1–3.

de Boer, E., van Herwaarden, C. & Tilburg, J. (1993) Isolation of *Campylobacter* from chicken products using a selective semi-solid medium. *Acta Gastro-Enterol. Belg.*, **56**, 22.

de Boer, E., Tilburg, J.J.H.C., Woodward, D.L., Lior, H. & Johnson, W.M. (1996) A selective medium for the isolation of *Arcobacter* from meats. *Lett. Appl. Microbiol.*, **23**, 64-6.

Booth, I.R. & Kroll, R.G. (1989) The preservation of foods by low pH. In: *Mechanisms of Action of Food Preservation Procedures* (ed. G.W. Gould), pp. 119-60. Elsevier, London.

Boquet, P., Munro, P., Fiorentini, C. & Just, I. (1998) Toxins from anaerobic bacteria: specifically and molecular mechanisms of action. *Curr. Opin. Microbiol.*, **1**, 66-74.

Borch E. & Wallentin, C. (1993) Conductance measurement for data generation in predictive modelling. *J. Ind. Microbiol.*, **12**, 286.

Booth, I.R., Pourkomailian, B., McLaggan, D. & Koo, S.-P. (1994) Mechanisms controlling compatible solute accumulation: a consideration of the genetics and physiology of bacterial osmoregulation. *J. Food Eng.*, **22**, 381-97.

Borsch, E., Nesbakken, T. & Christensen, H. (1996) Hazard identification in swine slaughter with respect to foodborne bacteria. *Int. J. Food Microbiol.*, **30**, 9-25.

Bower, C.K. & Daeschel, M.A. (1999) Resistance responses of microorganisms in food environments. *Int. J. Food Microbiol*, **50**, 33-44.

Boyd, E.F., Wang, F.S., Whitham, T.S., & Selander, R.K. (1996) Molecular relationship of the salmonellae. *Appl. Env. Microbiol.*, **62**, 804-808.

Bracey, D., Holyoak, C.D. & Coote, P.J. (1998) Comparison of the inhibitory effect of sorbic acid and amphotericin B on *Saccharomyces cerevisiae*: is growth inhibition dependent on reduced intracellular pH? *J. Appl. Microbiol.*, **85**, 1056-66.

Brenner, D.J. (1984) Facultatively anaerobic Gram-negative rods. In: *Bergey's Manual of Systematic Bacteriology*, Vol. 1 (eds N.R. Krieg & J.C. Holt), pp. 408-516. Williams & Wilkins, Baltimore.

Broughall, J.M. & Brown, C. (1984) Hazard analysis applied to microbial growth in foods: development and application of three-dimensional models to predict bacterial growth. *Food Microbiol.*, **1**, 12-22.

Broughall, J.W., Anslow, P.A. & Kilsby, D.C. (1983) Hazard analysis applied to microbial growth in foods: development of mathematical models describing the effect of water activity. *J. Appl. Bacteriol.*, **55**, 101-10.

Brown, M.H., Davies, K.W., Billon, C.M.P., Adair, C. & McClure, P.J. (1998) Quantitative Microbiological Risk Assessment: Principles applied to determining the comparative risk of salmonellosis from chicken products. *J. Food Protect.*, **61**, 1446-53.

Brown, W.L. (1991) Designing *Listeria monocytogenes* thermal inactivation studies for extended shelf-life refrigerated foods. *Food Technol.*, **45**, 152-3.

Bruce, M.E., Will, R.G., Ironside, J.W., *et al.* (1997) Transmissions to mice indicate that 'new variant' CJD is caused by the BSE agent. *Nature*, **389**, 498-501.

Brul, S. & Klis, F.M. (1999) Review: mechanistic and mathematical inactivation studies of food spoilage fungi. *Fungal Gen. Biol.*, **27**, 199-208.

Buchanan, R.L. (1995). The role of microbiological criteria and risk assessment in HACCP. *Food Microbiol. (Lond.)*, **12**, 421-4.

Buchanan, R.L. (1997) National Advisory Committee on Microbiological Criteria for Foods 'Principles of Risk Assessment for illness caused by foodborne biological agents'. *J. Food Protect.*, **60**, 1417-19.

Buchanan, R.L., Damert, W.G., Whiting, R.C. & Van Schothorst, M. (1997) Use of epidemiological and food survey data to estimate a purposefully conservative dose-response relationship for *Listeria monocytogenes* levels and incidence of listeriosis. *J. Food Protect.*, **60**, 918-22.

Buchanan, R.L. & Edelson, S.G. (1999) Effect of pH-dependent, stationary phase acid resistance on the thermal tolerance of *Escherichia coli* O157:H7. *Food Microbiol.*, **16**, 447-58.

Buchanan, R.L. & Whiting, R. (1996) Risk assessment and predictive microbiology. *J. Food Protect.*, **Suppl.**, 31-6.

Buchanan, R.L. & Whiting, R.C. (1998) Risk Assessment: a means for linking HACCP plans and public health. *J. Food Protect.*, **61**, 1531-4.

Bückenhuskes, H.J. (1997) In: *Fermented vegetables. Food Microbiology: Fundamentals and Frontiers* (eds M.P. Doyle, L.R. Beuchat & T.J. Montville), pp. 595-609. ASM Press, Washington DC.

Büllte, M., Klien, G. & Reuter, G. (1992) Pig slaughter. Is the meat contaminated by *Yersinia enterocolitica* strains pathogenic to man? *Fleischwirtschaft*, **72**, 1267-70.

Butzler, J.P., Dekeyser, P., Detrain, M. & Dehaen, F. (1973) Related vibrio in stools. *J. Pediatr.*, **82**, 493-5.

Calvert, R.M., Hopkins, H.C., Reilly, M.J. & Forsythe, S.J. (2000) Caged ATP - an internal calibration method for ATP biluminescence assays. *Lett. Appl. Microbiol.*, **30**, 223-7.

Caplice, E. & Fitzgerald, G.F. (1999) Food fermentations: role of microorganisms in food production and preservation. *Int. J. Food Microbiol.*, **50**, 131-49.

Cassin, M.H., Lammerding, A.M., Todd, E.C.D, Ross, W. & McColl, R.S. (1998) Quantitative risk assessment for *Escherichia coli* O157:H7 in ground beef hamburgers. *Int. J. Food Microbiol.*, **41**, 21-44.

Centers for Disease Control (CDC) (1993) Multistate outbreak of *Escherichia coli* O157:H7 infections from hamburgers - Western United States, 1992-1993. *MMWR*, **42**, 258-63.

Çetinkaya, B., Egan, K. & Morgan, K.L. (1996) A practice-based survey of the frequency of Johne's disease in south west England. *Vet. Record*, **134**, 494-7.

Chapman, P.A. & Siddons, C.A. (1996) A comparison of immunomagnetic separation and direct culture for the isolation of verocytotoxin-producing *Escherichia coli* O157 from cases of bloody diarrhoea, non-bloody diarrhoea and asymptomatic contact. *J. Med. Microbiol.*, **44**, 267-71.

Chiodini, R.J. & Hermon-Taylor, J. (1993) The thermal resistance of *Mycobacterium paratuberculosis* in raw milk under conditions simulating pasteurisation. *J. Vet. Diagn. Invest.*, **5**, 629-31.

Chirife, J. & del Pilarbuera, M. (1996) Water activity, water glass and dynamics, and the control of microbiological growth in foods. *Crit. Rev. Food Sci. Nutr.*, **36**, 465-513.

Claesson, B.E.B., Holmlund, D.E.W, Linghagen, C.A. & Matzsch, T.W. (1994) *Plesiomonas shigelloides* in acute cholecystitis: a case report. *J. Clin. Microbiol.*, **20**, 985-7.

Clark, D.S. & Takács, J. (1980) Gases as preservatives. In: *Microbial Ecology of Foods, Factors Affecting Life and Death of Microorganisms*, Vol. I (eds J.H. Silliker & R.P. Elliot), pp. 170-92. Academic Press, New York.

Codex Alimentarius Commission (1993) *Codex Guidelines for the Application of the Hazard Analysis Critical Control Point (HACCP) System.* Joint FAO/WHO Codex Committee on Food Hygiene. WHO/FNU/FOS/93.3 Annex II.

Codex Alimentarius Commission (1995) *Hazard Analysis Critical Control Point (HACCP) System and Guidelines for its Application.* Alinorm 97/13, Annex to Appendix II.

Codex Alimentarius Commission (1997a) *Hazard Analysis Critical Control Point*

(HACCP) System and Guidelines for its Application. Annex to CAC/RCP 1-1969, Rev. 3. 1997.

Codex Alimentarius Commission (1997b) *Principles for the Establishment and Application of Microbiological Criteria for Foods*. CAC/GL 21 - 1997.

Codex Alimentarius Commission (1997c) *Principles for the Development of Microbiological Criteria for Animal Products and Products of Animal Origin Intended for Human Consumption*. European Commission, Luxembourg.

Codex Alimentarius Commission (1999a) *Draft Principles and Guidelines for the Conduct of Microbiological Risk Assessment*. Alinorm 99/13A, Appendix A.

Codex Alimentarius Commission (1999b) *Principle and Guidelines for the Conduct of Microbiological Risk Assessment*. CAC/GL 30.

Coghlan, A. (1998) Deadly *E. coli* strains may have come from South America. *N. Scientist*, 10 January, 12.

Collinge, J., Sidle, K.C.L., Meads, J., Ironside, J. & Hill, A.F. (1996) Molecular analysis of prion strain variation and the aetiology of 'new variant' CJD. *Nature*, **383**, 685-90.

Colwell, R.R., Brayton, P., Herrington, D., *et al.* (1996) Viable but nonculturable *Vibrio cholerae* O1 revert to a cultivable state in the human intestine. *World J. Microbiol. Biotechnol.*, **12**, 28-31.

Commission Decision, 93/51/EEC (1993) On the microbiological criteria applicable to the production of cooked crustasceans and molluscan shellfish. *Off. J. Euro. Comm.*, **L13**, 11.

Cone, L.A., Voodard, D.R., Schievert, P.M. & Tomory, G.S. (1987) Clinical and bacteriological observations of a toxic-shock-like syndrome due to *Streptococcus pyogenes*. *N. Engl. J. Med.*, **317**, 146-9.

Corlett, D.A. (1998) *HACCP User's Manual*. A Chapman & Hall Food Science Title. An Aspen Publication. Aspen Publishers, Gaithersburg, Maryland.

Corthier, G., Delorme, C., Ehrlich, S.D. & Renault, P. (1998) Use of luciferase genes as biosensors to study bacterial physiology in the digestive tract. *Appl. Env. Microbiol.*, **64**, 2721-2.

Council Decision, 91/180/EEC (1991) Methods for analysis of raw and heat-treated milk. *Off. J. Euro. Comm.*, **L93**, 1.

Council Directive, 92/46/EEC (1992) Laying down the health rules for the production and placing on the market of raw milk, heat-treated milk and milk-based products. *Off. J. Euro. Comm.*, **L268**, 1-28.

Council Directive, 93/43/EEC (1993) On the hygiene of foodstuffs. *Off. J. Euro. Comm.*, **L175**, 1-11.

Cousens, S.N., Vynnycky, E., Zeidler, M., Will, R.G. & Smith, P.G. (1997) Predicting the CJD epidemic in humans. *Nature*, **385**, 197-8.

Cowden, J.M., Wall, P.G., Adak, G., *et al.* (1995) Outbreaks of foodborne infectious intestinal disease in England and Wales: 1992 and 1993. *CDR Review*, **5**, R109-R117.

Dainty, R.H. (1996) Chemical/biochemical detection of spoilage. *Int. J. Food Microbiol.*, **33**, 19-33.

Dalgaard, P., Gram, L. & Huss, H.H. (1993) Spoilage and shelf-life of cod fillets packed in vacuum or modified atmospheres. *Int. J. Food Microbiol.*, **19**, 283-94.

D'Aoust, J.-Y. (1994) Salmonella and the international food trade. *Int. J. Food Microbiol.*, **24**, 11-31.

Davidson, P.M. (1997) Chemical preservatives and natural antimicrobial compounds. In: *Food Microbiology: Fundamental and Frontiers* (eds M.P.

Doyle, L.R. Beuchat & T.J. Montville), pp. 520–56. ASM Press, Washington DC.

Denyer, S.P., Gorman, S.P. & Sussman, M. (1993) *Microbial Biofilms*. The Society for Applied Bacteriology Technical Series No 30. Blackwell Scientific Publications, London.

Diplock, A.T., Aggett, P.J., Ashwell, M., *et al.* (1999) Scientific concepts of functional foods in Europe: consensus document. *Br. J. Nutr.*, **81**, S1–S27.

Dodd, C.E.R., Sharman, R.L., Bloomfield, S.F., Booth, I.R. & Stewart, G.S.A.B. (1997) Inimical processes: bacterial self-destruction and sub-lethal injury. *Trends Food Sci. Technol.*, **8**, 238–41.

Druggan, P., Forsythe, S.J. & Silley, P. (1993) Indirect impedance for microbial screening in the food and beverage industries. In: *New Techniques in Food and Beverage Microbiology* (eds R.G. Kroll, *et al.*). Society for Applied Bacteriology, Technical Series No 31. Blackwell Science, Oxford.

Earnshaw, R.G., Appleyard, J. & Hurst, R.M. (1995) Understanding physical inactivation processes: combined preservation opportunities using heat, ultrasound and pressure. *Int. J. Food Microbiol.*, **28**, 197–219.

EC (1999) Opinion of the Scientific Committee on Veterinary Measures relating to Public Health on *Listeria monocytogenes*. European Commission. Health & Consumer Protection Directorate-General. Web site.

EEC (1993) Council Directive 93/43/EEC on the hygiene of foodstuffs. *Offic. J. Euro. Comm.* No L 175/1.

Eklund, T. (1985) The effect of sorbic acid and esters of para-hydroxybenzoic acid on the proton motive force in *Escherichia coli* membrane vesicle. *J. Gen. Microbiol.*, **131**, 73–6.

Elliott, P.H. (1996) Predictive microbiology and HACCP. *J. Food Protect.*, **Suppl.**, 48–53.

Ellis, M.J. (1994) The methodology of shelf-life determination. In: *Shelf-life Evaluation of Foods* (eds C.M.D. Man & A.A. Jones), pp. 40–51. Blackie Academic & Professional, London.

Entis, P. & Lerner, I. (1997) 24-hour presumptive enumeration of *Escherichia coli* O157:H7 in foods by using the ISO-GRID(R) method with SD-39 agar. *J. Food Protect.*, **60**, 883–90.

Ewing, W.H. (1986) The taxonomy of Enterobacteriaceae, isolation of Enterobacteriaceae and preliminary identification. The genus *Salmonella*. In: *Identification of Enterobacteriaceae* (eds P. Edwards & W.H. Ewing), 4th edn, pp. 1–91, 181–318. Elsevier, New York.

Falik, E., Aharoni, Y., Grinberg, S., Copel, A. and Klein (1994) Postharvest hydrogen peroxide treatment inhibits decay in eggplant and sweet red pepper. *Crop Protect.*, **13**, 451–4.

FAO (1996) *Biotechnology and Food Safety*. Report of a joint FAO/WHO consultation. UN Food and Agriculture Organisation, Rome.

FAO (1997) *Risk Management and Food Safety*. Report of a Joint FAO/WHO Expert Consultation, Rome, Italy, 1997. FAO Food and Nutrition Paper, No 65.

FAO (1998) *The Application of Risk Communication to Food Standards and Safety Matters*. Report of a Joint FAO/WHO Expert Consultation, Rome, Italy, 1998. FAO Food and Nutrition Paper, No 70.

Farber, J.M., Ross, W.H. & Harwig, J. (1996) Health risk assessment of *Listeria monocytogenes* in Canada. *Int. J. Food Microbiol.*, **31**, 145–56.

Ferguson, N.M., Donnelly, C.A., Ghani, A.C. & Anderson, R.M. (1999) Predicting the size of the epidemic of the new variant of Creutzfeldt-Jakob disease. *Br. Food J.*, **101**, 86-98.

Fischetti, V.A., Medaglini, D., Oggioni, M. & Pozzi, G. (1993) Expression of foreign proteins on Gram-positive commensal bacteria for mucosal vaccine delivery. *Curr. Opin. Biotechnol.*, **4**, 603-10.

Fisher, I.S.T. (1997a) *Salmonella enteritidis* and *Salmonella typhimurium* in Western Europe for 1993-1995, a surveillance report from Salm-Net. *Eurosurv.*, **2**, 4-6.

Fisher, I.S.T. (1997b) Salm/Enter-Net records a resurgence in *Salmonella enteritidis* infection throughout the European Union. *Eurosurv. Weekly*, June, 26.

Fisher, I.S.T. (1999) *S. enteritidis* in Western Europe 1995-1998 - a surveillance report from Enter-Net. *Eurosurv.*, **4**, 56.

Foegeding, P.M. & Busta, F.F. (1991) Chemical food preservatives. In: *Disinfection, Sterilization and Preservation* (ed. S. Block), pp. 802-32. Lea & Febiger, Philadelphia.

Food Safety and Inspection Service (1998) *Salmonella enteritidis* Risk Assessment. Shell Eggs and Egg Products. http://www.europa.eu.int/comm/dg24/health/sc/scv/out26_en.html.

Forsythe, S.J. & Hayes, P.R. (1998) *Food Hygiene, Microbiology and HACCP*. A Chapman & Hall Food Science Book. Aspen Publishers, Gaithersburg.

Fuller, R. (1989) Probiotics in man and animals. *J. Appl. Bacteriol.*, **66**, 365-78.

Gale, P., Young, C., Stanfield, G. & Oakes, D. (1998) A review: development of a risk assessment for BSE in the aquatic environment. *J. Appl. Microbiol.*, **84**, 467-77.

German, B., Schiffrin, E.J., Reniero, R., *et al.* (1999) The development of functional foods: lessons from the gut. *Trends Biotechnol.*, **17**, 492-9.

Gibson, A.M. & Roberts, T.A. (1996) The effect of pH, sodium chloride, sodium nitrite and storage temperature on the growth of *Clostridium perfringens* and fecal streptococci in laboratory media. *Int. J. Food Microbiol.*, **3**, 195-210.

Gibson, A.M., Bratchell, N. & Roberts, T.A. (1988) Predicting microbial growth: growth responses of salmonellae in a laboratory medium as affected by pH, sodium chloride, and storage temperature. *Int. J. Food Microbiol.*, **6**, 155-78.

Gilbert, R.J. (1992) Provisional microbiological guidelines for some ready-to-eat foods sampled at point of sale: notes for PHLS Food Examiners. *PHLS Microbiol. Digest*, **9**, 98-9.

Gill, C.O. & Phillips, D.M. (1985) The effect of media composition on the relationship between temperature and growth rate of *Escherichia coli*. *Food Microbiol.*, **2**, 285.

Gold, L.S., Slone, T.H., Stern, B.R., Manley, N.B. & Ames, B.N. (1992) Rodent carcinogens - setting priorities. *Science*, **258**, 261-5.

Gould, G.W. (ed.) (1995) *New Methods of Food Preservation*. Chapman and Hall, London.

Gould, G.W. (1996) Methods for preservation and extension of shelf life. *Int. J. Food Microbiol.*, **33**, 51-64.

Graham, A.F. & Lund, B.M. (1986) The effect of citric acid on growth of proteolytic strains of *Clostridium botulinum*. *J. Appl. Bacteriol.*, **61**, 39-49.

Grant, I.R., Ball, H.J., Neill, S.D. & Rowe, M.T. (1996) Inactivation of *Mycobacterium paratuberculosis* in cow's milk at pasteurisation temperatures. *Appl. Environ. Microbiol.*, **62**, 631-6.

Granum, P.E. & Lund, T. (1997) MiniReview. *Bacillus cereus* and its food poisoning toxins. *FEMS Microbiol. Lett.*, **157**, 223-8.

Grau, F.H. & Vanderlinde, P.B. (1992) Aerobic growth of *Listeria monocytogenes* on beef lean and fatty tissue: equations describing the effects of temperature and pH. *J. Food Protect.*, **55**, 4.

Graves, D.J. (1999) Powerful tools for genetic analysis come of age. *TIBTECH*, **17**, 127-34.

Green, D.H., Wakeley, P.R., Page, A., et al. (1999) Characterization of two *Bacillus* probiotics. *Appl. Environ. Microbiol.*, **65**, 4288-91.

Guarner, F. & Schaafsma, G.J. (1998) Probiotics. *Int. J. Food Microbiol.*, **39**, 237-8.

Haas, C.H., Rose, J.B. & Gerba, C.P. (1999) *Quantitative Microbial Risk Assessment*. John Wiley & Sons, New York.

Haire, D.L., Chen, G.M., Janzen, E.G., Fraser, L. & Lynch, J.A. (1997) Identification of irradiated foodstuffs: a review of the recent literature. *Food Res. Int.*, **30**, 249-64.

Haldenwang, W.G. (1995) The sigma factors of *Bacillus subtilis. Microbiol. Rev.*, **59**, 1-30.

Harrigan, W.F. & Park, R.A. (1991) *Making Safe Food. A Management Guide for Microbiolgoical Quality*. Academic Press, London.

Hathaway, S.C. (1997) Development of Food Safety Risk Assessment Guidelines for foods of animal origin in international trade. *J. Food Protect.*, **60**, 1432-8.

Hauschild, A.H.W., Hilsheimer, R., Jarvis, G. & Raymond, D.P. (1982) Contribution of nitrite to the control of *Clostridium botulinum* in liver sausage. *J. Food Protect.*, **45**, 500-506.

Havenaar, R. & Huis in't Veld, J.H. (1992) Probiotics: a general view. In: *The Lactic Acid Bacteria in Health and Disease, The Lactic Acid Bacteria*, Vol. 1 (ed. B.J. Wood), pp. 209-24. Chapman and Hall, New York.

Heitzer, A., Kohler, H.E., Reichert, P. & Hamer, G. (1991) Utility of phenomenological models for describing temperature dependence of bacterial growth. *Appl. Environ. Microbiol.*, **57**, 2656.

Heldman, D.R. & Lund, D.B. (1992) *Handbook of Food Engineering*. Marcel Dekker, New York.

Henderson, B., Wilson, W., McNab, R. & Lax, A. (1999) *Cellular Microbiology. Bacteria-Host Interactions in Health and Disease*. John Wiley & Sons, Chichester.

Hendrickx, M., Ludikhuyze, L., Vanden Broeck, I. & Weemaes, C. (1998) Effects of high pressure on enzymes related to food quality. *Trends Food Sci. Technol.*, **9**, 197-203.

Hill, A.F., Desbruslais, M., Joiner, S., et al. (1997) The same prion strain causes vCJD and BSE. *Nature*, **389**, 448-50.

Hirasa, K. & Takemasa, M. (1998) Antimicrobial and antioxidant properties of spices. In: *Spice Science and Technology*, pp. 163-200. Marcel Dekker, New York.

Hitchins, A.D., Feng, P., Watkins, W.D., Rippey, S.R. & Chandler, L.A. (1998) *Escherichia coli* and the coliform bacteria. In: *Food and Drug Administration Bacteriological Analytical Manual* (ed. R.L. Merker), 8th edn, revision A, Chapter 4. AOAC International, Gaithersburg, MD.

Holmes, C.J. & Evans, R.C. (1989) Resistance of bacterial biofilms to antibiotics. *J. Antimicrobiol. Chemother.*, **24**, 84.

Holyoak, C.D., Stratford, M., McMullin, Z., et al. (1996) Activity of the plasma-membrane H^+-ATPase and optimal glycolytic flux are required for rapid adaption and growth in the presence of the weak acid preservative sorbic acid. *Appl. Env. Microbiol.*, **62**, 3158-64.

Hugenholtz, J. & Kleerebezem, M. (1999) Metabolic engineering of lactic acid bacteria: overview of the approaches and results of pathway rerouting involved in food fermentations. *Curr. Opin. Biotechnol.*, **10**, 492–7.

Huis in't Veld, J.H.J. (1996) Microbial and biochemical spoilage of foods: an overview. *Int. J. Food Microbiol.*, **33**, 1–18.

Huisman, G.W. & Kolter, R. (1994) Sensing starvation: a homoserine lactone-dependent signalling pathway in *Escherichia coli*. *Science*, **265**, 537–9.

IFBC (1990) Biotechnologies and Food: Assuring the safety of foods produced by genetic modification. *Reg. Tox. Pharm.*, **12**, 3.

IFST (1999) *Development and Use of Microbiological Criteria for Foods*. Institute of Food Science & Technology (UK), London.

International Commission on Microbiological Specifications for Foods (ICMSF) (1986) *Microorganisms in Foods. Vol. 2. Sampling for Microbiological Analysis: Principles and Specific Applications*. University of Toronto Press, Toronto.

International Commission on Microbiological Specifications for Foods (ICMSF) (1988) *Microorganisms in Foods 4. Applications of the Hazard Analysis Critical Control Point (HACCP) System to Ensure Microbiological Safety and Quality*. Blackwell Science, Oxford.

International Commission on Microbiological Specifications for Foods (ICMSF) (1996) *Microorganisms in Foods 5. Characteristics of Microbial Pathogens*. Blackie Academic & Professional, London.

International Commission on Microbiological Specification for Foods (ICMSF) (1997) Establishment of microbiological safety criteria for foods in international trade. *Wld Hlth Statist. Quart.*, **50**, 119–23.

International Commission on Microbiological Specification for Foods (ICMSF) (1998) *Microorganisms in Foods 6. Microbial Ecology of Food Commodities*. Blackie Academic & Professional, London.

International Commission on Microbiological Specification for Foods (ICMSF) (1998) Potential application of risk assessment techniques to microbiological issues related to international trade in food and food products. *J. Food Protect.*, **61**, 1075–86.

ISO (1994) *ISO 9000 Series of Standards:* ISO 9000: Quality Management and Quality Assurance Standards, Part 1: Guidelines for selection and use. ISO 9001: Quality Systems – Model for Quality Assurance in design/development, production, installation and servicing. ISO 9002: Quality Systems – Model for Quality Assurance in production and installation. ISO 9004-1: Quality Management and Quality System elements, Part 1: guidelines. ISO 8402 Standard: Quality Management and Quality Assurance Standards – Guidelines for selection and use – vocabulary. International Standardisation Organisation.

Isolauri, E., Juntunen, M., Rautanen, T., Sillanaukee, P. & Koivula, T. (1991) A *Lactobacillus* strain (*Lactobacillus GG*) promotes recovery from acute diarrhoea in children. *Paediat.*, **88**, 90–7.

Jablonski, J.R. (1991) TQM implementation. In: *Implementing Total Quality Management: An Overview*. Pfeiffer and Co, San Diego.

Jacobs-Reitsma, W., Kan, C. & Bolder, N. (1994) The induction of quinolone resistance in *Campylobacter* bacteria in broilers by quinolone treatment. *Lett. Appl. Microbiol.*, **19**, 228–31.

Jessen, B. (1995) Start cultures for meat fermentation. In: *Fermented Meats* (eds G. Campbell-Platt & P.E. Cook), pp. 130–59. Blackie, Glasgow.

Johnson, R.P., Clarke, R.C., Wilson, J.B., *et al.* (1996) Growing concerns and recent outbreaks involving non-O157:H7 serotypes of verotoxigenic *Escherichia coli. J. Food Protect.*, **59**, 1112-2.

Jonas, D.A., Antignac, E., Antoine, J.-M., *et al.* (1996) The safety assessment of novel foods. *Food Chem. Toxicol.*, **34**, 931-40.

Jones, J.E. (1993) A real-time database/models base/expert system in predictive microbiology. *J. Ind. Microbiol.*, **12**, 268.

Jouve, J.L., Stringer M.F. & Baird-Parker, A.C. (1998) *Food Safety Management Tools.* International Life Sciences Institute. ILSI Europe, Brussels.

Juneja, V.K., Marmer, B.S., Philips, J.G. & Palumbo, S.A. (1996) Interactive effects of temperature, initial pH, sodium chloride, and sodium pyrophosphate on the growth kinetics of *Clostridium perfringens. J. Food Protect.*, **59**, 963-8.

Juven, B.J. & Pierson, M.D. (1996) Antibacterial effects of hydrogen peroxide and methods for its detection and quantification. *J. Food Protect.*, **59**, 1233-41.

Kaila, M., Isolauri, E., Soppi, E., *et al.* (1992) Enhancement of circulating antibody secreting cell response in human diarrhoea by a human *Lactobacillus* strain. *Pediatr. Res.*, **32**, 141-4.

Kaila, M., Isolauri, E., Saaxelin, M., Arvilommi & Vesikari, T. (1995) Viable versus inactivated *Lactobacillus* strain GG in acute rotavirus diarrhoea. *Arch. Dis. Child.*, **72**, 51-3.

Kalchayanand, N., Sikes, A., Dunne, C.P. & Ray, B. (1998) Factors influencing death and injury of foodborne pathogens by hydrostatic pressure-pasteurization. *Food Microbiol.*, **15**, 207-14.

Kapikian, A.Z., Wyatt, R.G., Dolin, R., *et al.* (1972) Visualisation by immune electron microscopy of a 27 nm particle associated with acute infectious nonbacterial gastroenteritis. *J. Virology*, **10**, 1075-81.

Kaur, P. (1986) Survival and growth of *Bacillus cereus* in bread. *J. Appl. Bacteriol.*, **60**, 513-16.

Kilsby, D. (1982) Sampling schemes and limits. In: *Meat Microbiology* (ed. M.H. Brown), pp. 387-421. Applied Science Publishers, London.

Kilsby, D.C., Aspinall, L.J. & Baird-Parker, A.C. (1979) A system for setting numerical microbiological specifications for foods. *J. Appl. Bacteriol.*, **46**, 591-9.

Klaenhammer, T.R. (1993) Genetics of bacteriocins produced by lactic acid bacteria. *FEMS Microbiol. Rev.*, **12**, 39-86.

Klaenhammer, T.R. & Kullen, M.J. (1999) Selection and design of probiotics. *Int. J. Food Microbiol.*, **50**, 45-57.

Kleerebezem, M., Quadri, L.E.N., Kuipers, O.P. & De Vos, W.M. (1997) Quorum sensing by peptide pheromones and two-component signal-transduction systems in Gram-positive bacteria. *Mol. Microbiol.*, **24**, 895-904.

Knorr, D. (1993) Effects of high-hydrostatic pressure processes on food safety and quality. *Food Technol.*, **47**, 156-62.

Kramer, J.M. & Gilbert, R.J. (1989) *Bacillus cereus* and other *Bacillus* species. In: *Foodborne Bacterial Pathogens* (ed. M.P. Doyle), pp. 21-70. Marcel Dekker, New York.

Krysinski, E.P., Brown, L.J. & Marchisello, T.J. (1992) Effect of cleaners and sanitizers on *Listeria monocytogenes* attached to product contact surfaces. *J. Food Protect.*, **55**, 246-51.

Kuipers, O.P., de Ruyter, P.G.G.A., Kleerebezem, M. & de Vos, W.M. (1997) Controlled overproduction of proteins by lactic acid bacteria. *TIBTECH*, **15**, 135-40.

Kuipers, O.P. (1999) Genomics for food biotechnology: prospects of the use of high-throughput technologies for the improvement of food microorganisms. *Curr. Opin. Microbiol.*, **10**, 511-16.

Kumar, C.G. & Anand, S.K. (1998) Significance of microbial biofilms in food industry: a review. *Int. J. Food Microbiol.*, **42**, 9-27.

Kwon, Y.M. & Ricke, S.C. (1998) Induction of acid resistance of *Salmonella typhimurium* by exposure to short-chain fatty acids. *Appl. Env. Microbiol.*, **64**, 3458-63.

Kyriakides, A. (1992) ATP bioluminescence applications for microbiological quality control in the dairy industry. *J. Soc. Dairy Technol.*, **45**, 91-3.

Lammerding, A.M. & Paoli, G.M. (1997) Quantitative Risk Assessment: an emerging tool for emerging foodborne pathogens. *Emerg. Infect. Dis.*, **3**.

Le Chevallier, M.W., Cawthon, C.D. & Lee, R.G. (1988) Inactivation of biofilm bacteria. *Appl. Environ. Microbiol.*, **54**, 2492-9.

Le Clercq-Perlat & Lalande (1994) Cleanability in relation to surface chemical composition and surface finishing of some materials commonly used in food industries. *J. Food Eng.*, **23**, 501-17.

Le Minor, L. (1988) Typing *Salmonella* species. *Euro. J. Clin. Microbiol. Infect. Dis.*, **7**, 214-18.

Lieb, S. (1983) *Outbreak of Shellfish Associated Gastroenteritis.* Bay County Department of Health and Rehabilitative Services, Tallahasse.

Lindgren, S.E. & Dobrogosz, W.J. (1990) Antagonistic activities of lactic acid bacteria in food and feed fermentations. *FEMS Microbiol. Rev.*, **87**, 149-3.

Lindroth, S.E. & Genigeorgis, C.A. (1986) Probability of growth and toxin production by nonproteolytic *Clostridium botulinum* in rockfish stored under modified atmospheres. *Int. J. Food Microbiol.*, **3**, 167-81.

Linkous, D.A. & Oliver, J.D. (1999) Pathogenesis of *Vibrio vulnificus*. *FEMS Microbiol. Lett.*, **174**, 207-14.

Linton, R.H., Carter, W.H., Pierson, M.D., Hackney, C.R. & Eifert, J.D. (1996) Use of a modified Gompertz equation to predict the effects of temperature, pH, and NaCl on the inactivation of *Listeria monocytogenes* Scott A heated in infant formula. *J. Food Protect.*, **59**, 16-23.

Lotong, N. (1998) Koji. In: *Microbiology of Fermented Foods* (ed. B.J.B. Wood), pp. 659-5. Blackie Academic & Professional, London.

Lund, B.M., Graham, A.F., George, S.M. & Brown, D. (1990) The combined effect of inoculation temperatuure, pH and sorbic acid on the probability of growth of non-proteolytic type B *Clostridium botulinum*. *J. Appl. Bacteriol.*, **69**, 481-92.

Luo, Y., Han, Z., Chin, S.M. & Linn, S. (1994) Three chemically distinct types of oxidants formed by iron mediated Fenton reactions in the presence of DNA. *Proc. Natl Acad. Sci.*, **91**, 12438-42.

McClure, P.J., Boogard, E., Kelly, T.M., Baranyi, J. & Roberts, T.A. (1993) A predictive model for the combined effects of pH, sodium chloride and temperature, on the growth of *Brochothrix thermosphacta*. *Int. J. Food Microbiol.*, **19**, 161-78.

McClure, P.J., Blackburn, C. de W., Cole, M.B., *et al.* (1994) Review paper. Modelling the growth, survival and death of microorganisms in foods: the UK Food MicroModel approach. *Int. J. Food Microbiol.*, **23**, 265-75.

McClure, P.J., Beaumont, A.L., Sutherland, J.P. & Roberts, T.A. (1997) Predictive modelling of growth of *Listeria monocytogenes*. The effects on growth of NaCl, pH, storage temperature and $NaNO_2$. *Int. J. Food Microbiol.*, **34**, 221-32.

McDonald, K. & Sun, D-W. (1999) Predictive food microbiology for the meat industry: a review. *Int. J. Food Microbiol.*, **52**, 1-27.

McMeekin, T.A., Olley, J.N., Ross, T. & Ratkowsky, D.A. (1993) *Predictive Microbiology*. John Wiley & Sons, Chichester.

McNab, W.B. (1998) Review - a general framework illustrating an approach to quantitative microbial food safety risk assessment. *J. Food Protect.*, **61**, 1216-28.

Mack, D.R., Michail, S., Wei, S., McDougall, L. & Hollingsworth, M.A. (1999) Probiotics inhibit enteropathogenic *E. coli* adherence *in vitro* by inducing intestinal mucin gene expression. *Am. J. Physiol.*, **276**, G941-950.

Malcolm, S. (1984) A note on the use of the non-central *t*-distribution in setting numerical microbiological specifications for foods. *J. Appl. Bacteriol.*, **46**, 591-9.

Man, C.M.D. & Jones, A.A. (eds) (1994) *Shelf Life Evaluation of Foods*. Blackie Academic & Professional, London.

Mansfield, L.P. & Forsythe, S.J. (1996) Collaborative ring-trial of Dynabeads[R] anti-Salmonella for immunomagnetic separation of stressed *Salmonella* cells from herbs and spices. *Int. J. Food Microbiol.*, **29**, 41-7.

Mansfield, L.P. & Forsythe, S.J. (2000a) Detection of salmonellae in food. *Rev. Med. Microbiol.*, **11**, 37-46.

Mansfield, L.P. & Forsythe, S.J. (2000b) Arcobacters, newly emergent human pathogens. *Rev. Med. Microbiol.*, **11**, 161-70.

Marie, D., Brussard, C.P.D., Thyhaug, R., Bratbak, G. & Vaulot, D. (1999) Enumeration of marine viruses in culture and natural samples by flow cytometry. *Appl. Environ. Microbiol.*, **65**, 45-52.

Marks, H. & Colman, M. (1998) Estimating distributions of numbers of organisms in food products. *J. Food Protect.*, **61**, 1535-40.

Marshall, D.L. & Schmidt, R.H. (1988) Growth of *Listeria monocytogenes* at 10°C in milk preincubated with selected pseudomonads. *J. Food Protect.*, **51**, 277.

Marteau, P., Flourie, B., Pochart, P., *et al.* (1990) Effect of the microbial lactase activity in yogurt on the intestinal absorption of lactose: an *in vivo* study in lactase-deficient humans. *Br. J. Nutr.*, **64**, 71-9.

Marteau, P., Vaerman, J.P., Dehennin, J.P., *et al.* (1997) Effects of intracellular perfusion and chronic ingestion of *Lactobacillus johnsonii* strain La1 on serum concentrations and jejunal secretions of immunoglobulins and serum proteins in healthy humans. *Gastroenterol. Clin. Biol.*, **21**, 293-8.

Matossian, M.K. (1981) Mould poisoning: an unrecognised English health problem, 1550-1800. *Med. History*, **25**, 73-84.

Matsuzaki, T. (1998) Immunomodulation by treatment with *Lactobacillus casei* strain Shirota. *Int. J. Food Microbiol.*, **41**, 133-40.

Mead, P.S., Slutsker, L., Dietz, V., *et al.* (1999) Food-related illness and death in the United States. *Emerg. Infect. Dis.*, **5**, 607-25.

Medema, G.J., Teunis, P.F.M., Havelaar, A.H. & Haas, C.N. (1996) Assessment of the dose-response relationship of *Campylobacter jejuni*. *Int. J. Food Microbiol.*, **30**, 101-11.

Meng, J. & Genigeorgis, C.A. (1993) Model lag phase of nonproteolytic *Clostridium botulinum* toxigenesis in cooked turkey and chicken breast as affected by temperature, sodium lactate, sodium chloride and spore inoculum. *Int. J. Food Microbiol.*, **19**, 109-22.

Meng, J. & Genigeorgis, C.A. (1994) Delaying toxigenesis of *Clostridum botulinum* by sodium lactate in 'sous-vide' products. *Lett. Appl. Microbiol.*, **19**, 20-23.

Mierau, I., Kunji, E.R.S., Leenhouts, K.J., *et al.* (1996) Multiple-peptidase mutants of *Lactococcus lactis* subsp. *cremoris* SK110 and its nisin-immune transconjugant in relation to flavour development in cheese. *Appl. Env. Microbiol.*, **64**, 1950–53.

Mortimore, S. & Wallace, C. (1994) *HACCP - A Practical Approach.* Practical Approaches to Food Control and Food Quality Series No.1. Chapman & Hall, London.

Moseley, B.E.B. (1999) The safety and social acceptance of novel foods. *Int. J. Food Microbiol.*, **50**, 25–31.

Mossel, D.A.A., Corry, J.E.L., Struijk, C.B. & Baird, R.M. (1995) In: *Essentials of the Microbiology of Foods. A Textbook for Advanced Studies.* John Wiley & Sons, Chichester.

Murata, M., Legrand, A.M., Ishibashi, Y., Fukui, M. & Yasumoto, Y. (1990) Structures and configurations of ciguatoxin from the moray eel *Gymnothora javanicus* and its likely precursor from the dinoflagellate *Gambierdiscus toxicus. J. Am. Chem. Soc.*, **112**, 4380–86.

Muyzer, G. (1999) DGGE/TGGE a method for identifying genes from natural ecosystems. *Curr. Opin. Microbiol.*, **2**, 317–22.

NACMCF (1992) Hazard analysis and critical control point system. *Int. J. Food Microbiol.*, **16**, 1–23.

NACMCF (1997) Hazard analysis and critical control point principles and application guidelines. *J. Food Protect.*, **61**, 1246–59.

NararowecWhite, M. & Farber, J.M. (1997a) *Enterobacter sakazakii*: a review. *Int. J. Food Microbiol.*, **34**, 103–13.

NararowecWhite, M. & Farber, J.M. (1997b) Incidence, survival and growth of *Enterobacter sakazakii* in infant formula. *J. Food Protect.*, **60**, 226–30.

NararowecWhite, M., McKellar, R.C. & Piyasena, P. (1999) Predictive modelling of *Enterobacter sakazakii* inactivation in bovine milk during high-temperature short-time pasteurization. *Food Res. Int.*, **32**, 375–9.

Nataro, J.P. & Kaper, J.B. (1998) Diarrheagenic *Escherichia coli. Clin. Microbiol. Rev.*, **11**, 142–201.

Nes, I.F., Diep, D.B., Havarstein, L.S., *et al.* (1996) Biosynthesis of bacteriocins in lactic acid bacteria. *Ant. van Leeuw.*, **70**, 113–28.

Notermans, S., Dufreene, J., Teunis, P., *et al.* (1997) A risk assessment study of *Bacillus cereus* present in pasteurised milk. *Food Microbiol.*, **14**, 143–51.

Notermans, S., Dufreene, J., Teunis, P. & Chackraborty, T. (1998) Studies on the risk assessment of *Listeria monocytogenes. J. Food Protect.*, **61**, 244–8.

Notermans, S., Gallhoff, G., Zwietering, M.H. & Mead, G.C. (1995a) The HACCP concept: specification of criteria using quantitative risk assessment. *Food Microbiol. (Lond.)*, **12**, 81–90.

Notermans, S., Gallhoff, G., Zwietering, M.H. & Mead, G.C. (1995b) Identification of critical control points in the HACCP system with a quantitative effect on the safety of food products. *Food Microbiol. (Lond.)*, **12**, 93–8.

Notermans, S. & van der Giessen, A. (1993) Foodborne diseases in the 1980's and 1990's: the Dutch experience. *Food Contam.*, **4**, 122–4.

Notermans, S. & Jouve, J.L. (1995) Quantitative risk analysis and HACCP: some remarks. *Food Microbiol. (Lond.)*, **12**, 425–9.

Notermans, S. & Mead, G.C. (1996) Incorporation of elements of quantitative risk analysis in the HACCP system. *Int. J. Food Microbiol.*, **30**, 157–73.

Notermans, S., Mead, G.C. & Jouve, J.L. (1996) Food products and consumer protection: a conceptual approach and a glossary of terms. *Int. J. Food Microbiol.*, **30**, 175-83.

Notermans, S., Nauta, M.J., Jansen, J., Jouve, J.L. & Mead, G.C. (1998). A risk assessment approach to evaluating food safety based on product surveillance. *Food Control*, **9**, 217-23.

Notermans, S. & Teunis, P. (1996) Quantitative risk analysis and the production of microbiologically safe food: an introduction. *Int. J. Food Microbiol.*, **30**, 9-25.

Notermans, S., Zwietering, M.H. & Mead, G.C. (1994) The HACCP concept: identification of potentially hazardous microorganisms. *Food Microbiol. (Lond.)*, **11**, 203-14.

Oberman, H. & Libudzisz, Z. (1998) Fermented milks. In: *Microbiology of Fermented Foods* (ed. B.J.B. Wood), pp. 308-50. Blackie Academic & Professional, London.

O'Donnell-Maloney, M.J., Smith, C.L. & Contor, C.R.E. (1996) The development of microfabricated arrays for DNA sequencing and analysis. *Trends Biotechnol.*, **14**, 401-407.

OECD (1993) *Safety Evaluation of Foods Produced by Modern Biotechnology – Concepts and Principles*. Organisation for Economic Cooperation and Development, Paris.

Olsvik, O., Popovic, T., Skjerve, E., *et al.* (1994) Magnetic separations techniques in diagnostic microbiology. *Clin. Microbiol. Rev.*, **7**, 43-54.

Ouwehand, A.C., Kirjavainen, P.V., Shortt, C. & Salminen, S. (1999) Probiotics: mechanisms and established effects. *Int. Dairy J.*, **9**, 43-52.

Palumbo, S.A. (1986) Is refrigeration enough to restrain foodborne pathogens? *J. Food Protect.*, **49**, 1003-1009.

Panisello, P.J. & Quantick, P.C. (1998) Application of Food MicroModel predictive software in the development of Hazard Analysis Critical Control Point (HACCP) systems. *Food Microbiol.*, **15**, 425-39.

Parry, R.T. (ed.) (1993) *Principles and Applications of Modified Atmosphere Packaging of Foods*. Blackie Academic & Professional, Glasgow.

Paton, J.C. & Paton, A.W. (1998) Pathogenesis and diagnosis of shiga toxin-producing *Escherichia coli* infections. *Clin. Microbiol. Rev.*, **11**, 450-79.

Peck, M.W. (1997) *Clostridium botulinum* and the safety of refrigerated processed foods of extended durability. *Trends Food Sci. Technol.*, **8**, 186-92.

Pickett, C.A. & Whitehouse, C.A. (1999) The cytolethal distending toxin family. *Trends Microbiol.*, **7**, 292-7.

Piper, P., Mahe, Y., Thompson, S., *et al.* (1998) The Pdr12 ATP-binding cassette ABC is required for the development of weak acid resistance in *Saccharomyces cerevisiae*. *EMBO J.*, **17**, 4257-65.

Post, D.E. Series of 6 monographs: Salmonella, Listeria, Campylobacter, *Cl. perfringens, B. cereus, E. coli*, Shigella, *St. aureus*. Oxoid, Unipath, Hampshire.

Potter, M.E. (1996) Risk assessment terms and definitions. *J. Food Protect.*, **Suppl.**, 6-9.

Pruitt, K.M. & Kamau, D.N. (1993) Mathematical models of bacterial growth, inhibition and death under combined stress conditions. *J. Ind. Microbiol.*, **12**, 221.

Ramsay, G. (1998) DNA chips: state-of-the-art. *Nat. Biotechnol.*, **16**, 40-44.

Raso, J., Pagán, R., Condón, S. & Sala, F. (1998) Influence of temperature and pressure on the lethality of ultrasound. *Appl. Environ. Microbiol.*, **64**, 465-71.

Raymond, G.M., *et al.* (1997) Molecular assessment of the potential transmissibilities of BSE and scrapie to humans. *Nature*, **388**, 285-8.

Reddy, B.S. & Riverson, A. (1993) An inhibitory effect of *Bifidobacterium longum* on colon, mammary and liver carcinogenesis induced by 2-amino-3-methyl-imidazo[4,5-f] quinolone, a food mutagen. *Cancer Res.*, **53**, 3914-18.

Rees, C.E.D., Dodd, C.E.R., Gibson, P.T., Booth, I.R. & Stewart, G.S.A.B. (1995) The significance of bacteria in stationary phase to food microbiology. *Int. J. Food Microbiol.*, **28**, 263-75.

Rehbein, H., Martinsdottir, E., Blomsterberg, F., Valdimarsson, G. & Oehlens-chlaeger, J. (1994) Shelf life of ice-stored redfish, *Sebastes marinus* and *S. mentella. Int. J. Food Sci. Technol.*, **29**, 303-13.

Reid, G. (1999) The scientific basis for probiotic strains of *Lactobacillus. Appl. Env. Microbiol.*, **65**, 3763-6.

Reid, G., Millsap, K. & Busscher, H.J. (1994) Implantation of *Lactobacillus casei* var. *rhamnosus* into the vagina. *Lancet*, **344**, 1229.

Reiter, B. & Harnulv, B.G. (1984) Lactoperoxidase antibacterial system; natural occurrence, biological function and practical applications. *J. Food Protect.*, **47**, 724-32.

Rhodehamel, E.J. (1992) FDA concerns with sous vide processing. *Food Technol.*, **46**, 73-6.

Roberts, D., Hooper, W. & Greenwood, M. (1995) *Practical Food Microbiology*, 2nd edn. Public Health Laboratory Service, London.

Roberts, T.A. & Gibson, A.M. (1986) Chemical methods for controlling *Clostridium botulinum* in processed meats. *Food Technology*, **40**, 163-71.

Robinson, A., Gibson, A.M. & Roberts, T.A. (1982) Factors controlling the growth of *Clostridium botulinum* types A and B in pasteurized, cured, meats. V. Prediction of toxin production: non-linear effects of storage temperature and salt concentration. *J. Food Technol.*, **17**, 727-44.

Rollins, D.M. & Colwell, R.R. (1986) Viable but nonculturable stage of *Campylobacter jejuni* and its role in survival in the natural aquatic environment. *Appl. Env. Microbiol.*, **52**, 531-8.

Rondon, M.R., Goodman, R.M. & Handelsman, J. (1999) The Earth's bounty: assessing and accessing soil microbial diversity. *TIBTECH*, **17**, 403-9.

Ross, T. (1993) Belehardek-type models. *J. Indust. Microbiol.*, **12**, 180.

Ross, T. & McMeekin, T.A. (1994) Review paper. Predictive microbiology. *Int. J. Food Microbiol.*, **23**, 241-64.

Rowan, N.J., Anderson, J.G. & Smith, J.E. (1998) Potential infective and toxic microbiological hazards associated with the consumption of fermented foods. In: *Microbiology of Fermented Foods* (ed. B.J.B. Wood), pp. 263-307. Blackie Academic & Professional, London.

Rowland, I.R. (1990) Metabolic interactions in the gut. In: *Probiotics* (ed. R. Fuller), pp. 29-52. Chapman & Hall, New York.

Rowland, I. (1999) Probiotics and benefits to human health - the evidence in favour. *Environ. Microbiol.*, **1**, 375-82.

Russel, A.D. (1991) Mechanisms of bacterial resistance to non-antibiotics: food additives and food and pharmaceutical preservatives. *J. Appl. Bacteriol.*, **71**, 191-201.

Safarik, I., Safarikova, M. & Forsythe, S.J. (1995) The application of magnetic separations in applied microbiology. *J. Appl. Bacteriol.*, **78**, 575-85.

Sanders, M.E. (1993) Summary of conclusions from a consensus panel of experts on health attributes of lactic cultures: significance of fluid milk products containing cultures. *J. Dairy Sci.*, **76**, 1819-28.

Sanders, M.E. (1998) Overview of functional foods: emphasis on probiotic bacteria. *Int. Dairy J.*, **8**, 341-9.

Savage, D.C. (1997) Microbial ecology of the gastointestinal tract. *Ann. Rev. Microbiol.*, **31**, 107-33.

SCF (1997) Commission Recommendation 97/618/EEC concerning the scientific aspects of the presentation of information necessary to support applications for the placing on the market of novel foods and novel food ingredients and the preparation of initial assessment reports under regulation (EC) No 258/97 of the European Parliament and of the Council. *Off. J. Euro. Commun.*, L253, Brussels.

Schellekens, M. (1996) New research in sous-vide cooking. *Trends Food Sci. Technol.*, **7**, 256-62.

Schellekens, M., Martens, T., Roberts, T.A., *et al.* (1994) Computer aided microbial safety design of food processes. *Int. J. Food Microbiol.*, **24**, 1-9.

Schena, M., Heller, R.A., Theriault, T.P., *et al.* (1998) Microarrays: biotechnology's discovery platform for functional genomics. *TIBTECH*, **16**, 301-306.

Scheu, P.M., Berghof, K. & Stahl, U. (1998) Detection of pathogenic and spoilage microorganisms in food with the polymerase chain reaction. *Food Microbiol.*, **15**, 13-31.

Schiffrin, E., Rouchat, F., Link-Amster, H., Aeschlimann, J. & Donnet-Hugues, A. (1995) Immunomodulation of blood cells following the ingestion of lactic acid bacteria. *J. Dairy Sci.*, **78**, 491-7.

Schleifer, K.-H., *et al.* (1995a) Phylogenetics for the genus *Lactobacillus* and related genera. *Syst. Appl. Microbiol.*, **18**, 461-7.

Schleifer, K.-H., Ehrmann, M., Brockmann, E., Ludwig, W. & Ammann, R. (1995b) Application of molecular methods for the classification and identification of the lactic acid bacteria. *Int. Dairy J.*, **5**, 1081-94.

SCOOP (1998) Reports on tasks for scientific co-operation. Microbiological criteria: Collation of scientific and methodological information with a view to the assessment of microbiological risk for certain foodstuffs. Report of experts participating in Task 2.1, European Commission, EUR 17638. Office for Official Publications of the European Communities, Luxembourg.

Seera, J.A., Domenech, E., Escriche, I. & Martorelli, S. (1999) Risk assessment and critical control points from the production perspective. *Int. J. Food Microbiol.*, **46**, 9-26.

Sethi, D., Wheeler, J.G., Cowden, J.M., *et al.* (1999) A study of infectious intestinal disease in England: plan and methods of data collection. *Commun. Dis. Publ. Health*, **2**, 101-107.

Shapton, D.A. & Shapton, N.E. (1991) *Principles and Practices for the Safe Processing of Foods*. Butterworth-Heinemann, Oxford.

Silley, P. & Forsythe, S. (1996) A review. Impedance microbiology - a rapid change for microbiologists. *J. Appl. Bacteriol.*, **80**, 233-43.

Skirrow, M.B. (1977) *Campylobacter enteritidis:* a 'new' disease. *Br. Med. J.*, **2**, 9-11.

Skirrow, M.B. (1991) Epidemiology of *Campylobacter* enteritis. *Int. J. Food Microbiol.*, **12**, 9-16.

Smith, K., Besser, J., Hedberg, C., *et al.* (1999) Quinolone-resistant *Campylobacter jejuni* infections in Minnesota, 1992-1998. *N. Engl. J. Med.*, **340**, 1525-32.

Snyder, O.P. Jr (1995) HACCP-TQM for retail and food service operations. In: *Advances in Meat Research - Volume 10. HACCP in Meat, Poultry & Fish Processing* (eds A.M. Pearson & T.R. Dutson). Blackie Academic & Professional, London.

Sockett, P.N. (1991) Food poisoning outbreaks associated with manufactured foods in England and Wales: 1980-89. *Comm. Disease Rep.*, 1, Rev No 10, R105-109.

Sofos, J.N. & Busta, F.F. (1981) Antimicrobial activity of sorbate. *J. Food Protect.*, 44, 614-22.

Stabel, J.R., Steadham, E. & Bolin, C.A. (1997) Heat inactivation of *Mycobacterium paratuberculosis* in raw milk: are current pasteurisation conditions effective? *Appl. Environ. Microbiol.*, 63, 4975-77.

Stanley, G. (1998) Cheeses. In: *Microbiology of Fermented Foods* (ed. B.J.B. Wood), pp. 263-307. Blackie Academic & Professional, London.

Stickler, D. (1999) Biofilms. *Curr. Opin. Microbiol.*, 2, 270-75.

Stringer, M.F. (1993) Safety and quality management through HACCP and ISO 9000. *J. Food Protect.*, 56, 904.

Stutz, H.K., Silverman, G.J., Angelini, P. & Levin, R.E. (1991) Bacteria and volatile compounds associated with ground beef spoilage. *J. Food Sci.*, 55, 1147-53.

Sugita, T. & Togawa, M. (1994) Efficacy of *Lactobacillus* preparation Biolactis powder in children with rotavirus enteritis. *Jpn. J. Pediatr.*, 47, 899-907.

Surkiewicz, B.F., Johnson, R.W., Moran, A.B. & Krumm, G.W. (1969) A bacteriological survey of chicken eviscerating plants. *Food Technol.*, 23, 1066-9.

Sutherland, J.P. & Bayliss, A.J. (1994) Predictive modeling of growth of *Yersinia enterocolitica*: the effects of temperature, pH and sodium chloride. *Int. J. Food Microbiol.*, 21, 197-215.

Sutherland, J.P., Bayliss, A.J. & Braxton, D.S. (1995) Predictive modeling of growth of *Escherichia coli* O157:H7: the effects of temperature, pH and sodium chloride. *Int. J. Food Microbiol.*, 25, 29-49.

Sutherland, J.P., Bayliss, A.J. & Roberts, T.A. (1994) Predictive modeling of growth of *Staphylococcus aureus*: the effects of temperature, pH and sodium chloride. *Int. J. Food Microbiol.*, 21, 217-36.

Sweeney, R.W., Whitlock, R.H. & Rosenberger, A.E. (1992) *Mycobacterium paratuberculosis* cultured from milk and supramammary lymph nodes of infected asymptomatic cows. *J. Clin. Microbiol.*, 30, 166-71.

Tannock, G.W. (1995) *Normal Microflora*. Chapman & Hall, London.

Tannock, G.W. (1997) Probiotic properties of lactic-acid bacteria: plenty of scope for fundamental R & D. *Trends Biotechnol.*, 15, 270-74.

Tannock, G.W. (1998) Studies of the intestinal microflora: a prerequisite for the development of probiotics. *Int. Dairy J.*, 8, 527-33.

Tannock, G.W. (ed.) (1999a) *Probiotics: A Critical Review*, pp. 1-4. Horizon Scientific Press, Norfolk.

Tannock, G.W. (1999b) Identification of lactobacilli and bifidobacteria. In: *Probiotics: A Critical Review* (ed. G.W. Tannock), pp. 1-4. Horizon Scientific Press, Norfolk.

Tannock, G.W. (1999c) A fresh look at the intestinal microflora. In: *Probiotics: A critical Review* (ed. G.W. Tannock), pp. 5-14. Horizon Scientific Press, Norfolk.

Taormina, P.J. & Beuchat, L.R. (1999) Infections associated with eating seed sprouts: an international concern. *Emerg. Inf. Dis.*, 5.

Tatsozawa, H., Murayama, T., Misawa, N., *et al.* (1998) Inactivation of bacterial respiratory chain enzymes by singlet oxygen. *FEBS Lett.*, 439, 329-33.

Thomas, P. & Newby, M. (1999) Estimating the size of the outbreak of new-variant CJD. *Br. Food J.*, 101, 44-57.

Titbull, R.W., Naylor, C.E. & Basak, A.K. (1999) The *Clostridium perfringens* α-toxin. *Anaerobe*, **5**, 51-64.

Todd, E.C.D. (1985) Economic loss from foodborne disease and non-illness related recalls because of mishandling by food processors. *J. Food Protect.*, **48**, 621-33.

Todd, E.C.D. (1989) Preliminary estimates of costs of foodborne disease in the U.S. *J. Food Protect.*, **52**, 595-601.

Todd, E.C.D. (1996a) Risk assessment of use of cracked eggs in Canada. *Int. J. Food Microbiol.*, **30**, 125-43.

Todd, E.C.D. (1996b) Worldwide surveillance of foodborne disease: the need to improve. *J. Food Protect.*, **59**, 82-92.

Tomlinson, N. (1998) Worldwide regulatory issues: legislation and labelling. In: *Genetic Modification in the Food Industry* (eds S. Roller & S. Harlander), pp. 61-8. Blackie, London.

Tompkins, D.S., Hudson, M.J., Smith, H.R., *et al.* (1999) A study of infectious intestinal disease in England: microbiological findings in cases and controls. *Comm. Dis. Publ. Hlth*, **2**, 108-13.

Trucksess, M.W., Mislevec, P.B., Young, K., Bruce, V.E. & Page, S.W. (1987) Cyclopiazonic acid production by cultures of *Aspergillus* and *Penicillium* species isolated from dried beans, corn, meal, macaroni and pecans. *J. Assoc. Anal. Chem.*, **70**, 123-6.

Turnbull, P.C.B., Kramer, J.M., Jorgensen, K., Gilbert, R.J. & Melling, J. (1979) Properties and production characteristics of *B. cereus* strains isolated from food, food poisoning outbreaks and environment. *J. Antibact. Antifung. Agents*, **13**, 547-54.

Vallejo-Cordoba, B. & Nakai, S. (1994) Keeping quality of pasteurised milk by multivariate analysis of dynamic headspace gas chromatographic data. *J. Agric. Food Chem.*, **42**, 989-93, 994-9.

Van de Venter, T. (1999) Prospects for the future: emerging problems – chemical/biological. Conference on International Food Trade Beyond 2000: science-based decisions, harmonization, equivalence and mutual recognition. Food and Agriculture Organization of the United Nations, Melbourne, Australia.

Van Gerwen, S.J.C. & Zwietering, M.H. (1998) Growth and inactivation models to be used in quantitative Risk Assessments. *J. Food Protect.*, **61**, 1541-9.

Van Schothorst, M. (1997) Practical approaches to risk assessment. *J. Food Protect.*, **60**, 1439-43.

Van Schothorst, M. (1998) Principles for the establishment of microbiological food safety objectives and related control measures. *Food Control*, **9**, 379-84.

Vandamme, P., Falsen, E., Rossau, R., *et al.* (1991) Revision of *Campylobacter*, *Helicobacter* and *Wolinella* taxonomy: emendation of generic descriptions and proposal of *Arcobacter* gen. nov. *Int. J. System Bacteriol.*, **41**, 88-103.

Vanderzant, C. & Splittstoesser, D.F. (eds) (1992) *Compendium of Methods for the Microbiological Examination of Foods*, 3rd edn. American Public Health Association.

Vaughan, E.E., Mollet, B. & de Vos, W.M. (1999) Functionality of probiotics and intestinal lactobacilli: light in the intestinal tract tunnel. *Curr. Opin. Biotech.*, **10**, 505-10.

de Vos, W.M. (1999) Safe and sustainable systems for food-grade fermentations by genetically modified lactic acid bacteria. *Int. Dairy J.*, **9**, 3-10.

Vose, D.J. (1998) The applications of quantitative risk assessment to microbial food safety. *J. Food Protect.*, **61**, 640-48.

Voysey, P.A. (1999) Aspects of Microbiological Risk Assessment. *New Food*, **2**, 3-13.

Walker, S.J. (1994) The principles and practice of shelf-life prediction for micro-organisms. In: *Shelf-life Evaluation of Foods* (eds C.M.D. Man & A.A. Jones), pp. 40-51. Blackie Academic & Professional, London.

Wallraff, G., Labadie, J., Brick, P., *et al.* (1997) DNA sequencing on a chip. *Chemtech*, February, 22-32.

Wan J., Wilcock, A. & Coventry, M.J. (1998) The effect of essential oils of basil on the growth of *Aeromonas hydrophila* and *Pseudomonas fluorescence*. *J. Appl. Microbiol.*, **84**, 152-8.

Wang, S.L., Pai, C.S. & Shieh, S.T. (1997) Production, purification and characterization of the hen egg-white lysozyme inhibitor from *Enterobacter cloaceae* M-1002. *J. Chin. Chem. Soc.*, **44**, 349-55.

Wassenaar, T.M. (1997) Toxin production by *Campylobacter* spp. *Clin. Microbiol. Rev.*, **10**, 466-76.

Webb, N.B. & Marsden, J.L. (1995) Relationship of the HACCP system to Total Quality Management. In: *HACCP in Meat, Poultry and Fish Processing* (eds A.M. Pearson & T.R. Dutson), Vol. 10, Chapter 8. Advances in Food Research Series. Blackie Academic & Professional, Chapman & Hall, London.

Weirup, M. (1992) As quoted by Altekruse, S.F., Tollefson, L.K. & Bögel, K. (1993) Control strategies for *Salmonella enteritidis* in five countries. *Food Control*, **4**, 10-16.

Wells, J.M., Robinson, K., Chamberlain, L.M., Schofiled, K.M. & LePage, R.W. (1996) Lactic acid bacteria as vaccine delivery vehicles. *Ant. van Leeu.*, **70**, 317-30.

Wesley, I. (1997) *Helicobacter* and *Arcobacter*: potential human foodborne pathogens? *Trends Food Sci. Tech.*, **8**, 293-9.

Wheeler, J.G., Sethi, D., Cowden, J.M., *et al.* (1999) Study of infectious intestinal disease in England: rates in the community, presenting to general practice, and reported to national surveillance. *Br. Med. J.*, **318**, 1046-50.

Whiting, R.C. (1995) Microbial modeling in foods. *Crit. Rev. Food Sci. Nutr.*, **35**, 467-94.

WHO (1991) *Strategies for Assessing the Safety of Foods Produced by Biotechnology*. Report of joint FAO/WHO consultation. World Health Organisation, Geneva.

Wijtzes, T., van't Riet, K., in't Veld, J.H.J. Huis & Zwietering, M.H. (1998) A decision support system for the prediction of microbial food safety and food quality. *Int. J. Food Microbiol.*, **42**, 79-90.

Will, R.G., *et al.* (1996) A new variant of Creutzfeldt-Jakob disease in the UK. *The Lancet*, **347**, 921-5.

Wimptheimar, L., Altman, N.S. & Hotchkiss, J.H. (1990) Growth of *Listeria monocytogenes* Scott A, serotype 4 and competitive spoilage organisms in raw chicken packaged under modified atmospheres and in air. *Int. J. Food Microbiol.*, **11**, 205.

de Wit, J.N. & van Hooydonk, A.C.M. (1996) Structure, functions and applications of lactoperoxidase in natural antimicrobial systems. *Neth. Milk Dairy J.*, **50**, 227-44.

Wood, B.J.B. (ed.) (1998) *Microbiology of Fermented Foods*. Blackie Academic and Professional, London.

Yeh, P.L., Bajpai, R.K. & Iannotti, E.L. (1991) An improved kinetic model for lactic acid fermentation. *J. Ferment. Bioeng.*, **71**, 75.

Zoppi, G. (1998) Probiotics, prebiotics, synbiotics and eubiotics. *Pediatr. Med. Chirurg.*, **20**, 13-17.

Zottola, E.A. & Sasahara, K.C. (1994) Microbial biofilms in the food processing industry - should they be a concern? *Int. J. Food Microbiol.*, **23**, 125-48.

Zwietering, M.H., Wijtzes, T., de Wit, J.C. & van't Reit, K. (1992) A decision support system for prediction of the microbial spoilage in foods. *J. Food Protect.*, **55**, 973.

Zwietering, M.H., de Wit, J.C. & Notermans, S. (1996) Application of predictive microbiology to estimate the number of *Bacillus cereus* in pasteurised milk at the point of consumption. *Int. J. Food Microbiol.*, **30**, 55-70.

INDEX

12-D concept, 22, 109, 116, 275
16S rRNA, 10, 12, 136, 192, 212, 217
23S rRNA, 89

Aspergillus spp., 12, 100, 132, 168,
    183, 184, 188
  A. flavus 11, 35, 100
  A. niger, 132
  A. ochraceus, 12, 188
  A. oryzae, 27, 132, 133, 188
  A. parasiticus, 12, 35
  A. sojae, 132, 188
acceptable quality level (AQL), 298
accreditation, 194, 254
acetic acid, 31, 33, 45, 96, 99, 106, 117,
    122
Acetobacter spp., 120
acid tolerance, 39, 93, 119, 342
acidophilus milk, 138
Acinetobacter spp., 101, 102, 103
additives, 2, 34, 63, 113, 362, 367
adenovirus, 59, 77, 176, 177
aerobic plate count, 194, 219, 288, 291,
    304, 305, 316, 317, 319-22,
    325, 326, 328, 329, 332-4
Aeromonas spp., 58, 60, 99, 101-3,
    143, 168, 169
  A. hydrophila, 20, 27, 28, 36, 37,
    168, 169, 207, 272, 279
aflatoxins, 11, 35, 144, 185, 187
AIDS, 183
air, 39, 159, 189, 226, 271, 356
allicin, 34, 36
American Type Culture Collection, 219
amnesic shellfish poisoning, 178, 179
amoeba, 11
Anasakis simplex, 183

antibiotics (including resistance), 13,
    66, 69-72, 147, 150, 190, 246,
    262, 342, 377
antigen, 13, 15, 16, 135, 149, 199, 205,
    206
appendicitis, 158
archaea organisms, 12
Arcobacter spp., 189, 192, 204
Ascaris spp., 11, 77
astrovirus, 54, 57, 59, 176
ATP bioluminescence, 199, 203, 213,
    215,—17
attaching and effacing, 58, 85, 91, 92,
    151, 155
attribute plans, 298
auditing, 267, 288, 290

Bacillus spp., 11, 12, 33, 37, 40, 42, 48,
    58, 95, 99, 101, 109, 131, 163,
    246, 275
  B. anthrasis, 163
  B. cereus, 163, 246
    death parameter, 20
    detection, 217, 221, 253, *254*
    growth parameters, 20, 27, 37, 48,
    50, 51, 97
    infections, 54, 56, 58, 67, 81, 82,
    142, 144, 164, 165, 274, *340*,
    341, 344
    sampling plans, 327, 319, 327, 330
    sporulation, 32, 99, 101, 109
  B. coagulans, 20
  B. licheniformis, 20, 37, 120, 163,
    165, 246
  B. subtilis, 20, 37, 42, 101, 109, 135,
    163, 165, 207, 246
  B. stearothermophilus, 20, 27

B. thurinigenesis, 163
Bacteroides spp., 75, 77
bacterial growth, 26, 49, 97, 115, 286, 288, 289
bacteriocins, 36, 117, 119, 122
bacteriophages and phage-typing, 12, 40, 70, 83, 86, 126, 128, 133, 147, 149, 150, 152, 159
Baird-Parker medium, 236, 247, 359
Baranyi equation, 46, 47
beef, 1, 10, *19*, 105, 106, 112, 158, 168, 183, 279, 349, 350
beer, 11, 96, 99, 110, 120, 188
benzoic acid, 31, 32, 33, 108, 117
betaine, 41, 42
Bifidobacterium spp., 75, 77, 78, 136, 137
Bif. bifidium, 75, 77, 137, 138
bile, 74, 91, 137, 151, 197, 224, 225, 228, 230, 231
biocides, 38, 210
biofilm, 38, 138, 139, *140*, *141*, 210, 271
biological hazards, 261, 262, 282, 284, 336
biotyping, 147
biuret reaction, 216
blanching, 23, 24
blood, 16, 79, 82, 88, 94, 129, 136, 146, 151-6, 158, 160, 165-7, 169, 171, 188, 235, 236, 238, 244, 274
botulinum poisoning, 22, 23, 30, 34, 40, 45, 56, 61, 82, 102, 109, 116, 117, 118, 142, 162, 163, 276, 327, 359
bovine spongiform encephalopathy (BSE), 1, 12, 63, 69, 190, 370
bread and bakery products, 11, 32, 100, 104, 120, 160, 279, *281*, 319
brilliant green medium, 219, 220, 222, 224, 225, 230
British Standards Institute (BSI), 201, 219, 295
Brochothrix thermosphacta, 48, 101, 102
Brucella spp., 54, 56, 144
  Br. abortus, 110, 167
  Br. canis, 167
  Br. melitensis, 167
  Br. suis, 167

Buffered peptone water, 197, 219, 220, 222

caliciviruses, 59, *60*, 172, 176
CAMP test, 235, 244, *245*, *246*
Campylobacter spp., 11, 54, 56, 58, *60*, 61, 142, 146, 147, 148, 192, 226, 331, 333
C. coli, 146, 278
C. jejuni,
  antibiotic resistance, 72
  costs of illness, 61, 62
  death parameter, 20, 22
  detection, 217, 220, 227, 229-34
  growth parameter, 12, 20, 27, 47
  infections, 60, 61, 68, 69, 143, 146, 148, 189, 192, 274, 344
  routes of transmission, 102, 272, 278
  toxins, 80, 87, 147
  viable nonculturable, 198
C. lari, 146
C. upsaliensis, 146
cancer, 136, 169, 187, 188
Candida spp., 128, 130, 131
canned foods, 22-4, 26, 28, 44, 73, 99, 104, 109, 122, 163, 317
capsule, 16, 139
carbon dioxide, 11, 76, 115, 116, 117, 128, 146, 206, 207, 226
case, 56, 64, 144, 158, 219, 299, 301, 302, 316, 318, 319, 361
catalase, 78, 87, 94, 122, 235
cattle, 12, 70, 146, 154, 167, 169, 182, 183, 187, 190, 191, 192, 351
central nervous system, 153, 154
cereals (including barley, maize and wheat), 11, 113-15, 120, 188, 189, 273, 319
cestode worms, 10
challenge testing, 271
cheese, 28-31, 34, 120, 124, 125, 127, 128, 130, 154, 156, 165, 171, 188, 238, 303, 317, 318, 323, 328, 332, 356, 357
chelators, 34
chemical hazards, 261, 262, 282, 284
chicken (see also poultry) , 66, 72, 114, 160, 165, 183, 279, *280*, 282, 305, 347, 348, 349
chilled food, 104, 108, 115, 279, 320

Index

cholera, 4, 54, 66, 79, 80, 82, 86–8, 90, 91, 147, 148, 168
chromogenic media, 203, 212, 231
ciguatera, 177–9
ciprofloxacin, 71, 147
citric acid, 31, 34, 124
Clostridium spp., 11, 12, 37, 40, 75, 77, 78, 99, 116, 131, 136, 161, 275, 276
Cl. botulinum,
  death parameter, 20, 23, 35, 117, 163
  growth parameter, 20, 27, 30, 34, 48, 61, 109, 116, 117, 276
  infections, 54, 142, 144, 274, 344
  microbiological risk assessment, 359
  routes of transmission, 102, 163, 272, 321
  sampling plan, 327
  sporulation, 40, 289
  toxins, 45, 81, 83, 84, 162,
Cl. difficile, 58, 136, 162
Cl. innocuum, 75
Cl. perfringens, 61, 317, 321, 328, 329, 332
  costs of illness, 62
  death parameter, 20, 162
  detection methods, 221, 242, 244, 245, 249
  growth parameter, 20, 27, 28, 37, 77, 275
  infections, 54, 56, 58, 67, 144, 161, 274, 344
  routes of transmission, 102, 272
  sampling plans, 318, 320, 327, 330, 333
  toxins, 82, 164
Cl. thermosaccharolyticum, 20
coagulase, 238
Coxiella burnetii, 110
Codex Alimentarius Commission, 3, 112, 257, 258, 260, 261, 293, 297, 298, 327, 336, 337, 361–4, 365, 377, 379
cold adaptation, 40
cold shock, 37, 40, 41
coliforms,
  detection methods, 194, 204, 220, 228, 231, 235
  growth parameters, 4, 78, 143, 145

routes of transmission, 78, 101
sampling plans, 317, 318, 321, 323–9, 332
colon, 75, 83, 87, 88, 93, 134, 136, 150, 152, 153
consumer risk, 312, 313
controlled atmosphere packaging, 115
corrective action, 260, 265, 267, 268
costs of foodborne diseases, 62
coxsackie virus, 4
Creutzveld Jacob Disease (vCJD), 12, 69, 190, 191, 370
critical control point, 3, 44, 49, 51, 64, 215, 216, 256, 259, 260, 263, 265, 266, 268, 273, 277, 278, 284, 285, 371–3
critical limit, 260, 265, 268, 277, 282, 283, 286–92, 371
Crohn's disease, 158, 192
cross-contamination, 72, 73, 146, 148, 150, 154, 277–9, 353
Cryptosporidium parvum, 54, 57, 59, 68, 69, 144, 182, 189, 201, 272
curing and cured products, 31, 107, 108, 117, 118, 131, 318
Cyclospora cayetanensis, 54, 57, 68, 69, 144, 181, 182, 189
cytolethal distending toxin, 80, 88, 147
cytolysin, 167
cytotoxin, 58, 79, 148

D value, 18–20, 22, 23, 109, 116, 149, 275, 348, 352, 353, 358
dairy and dairy products, 11, 31, 38, 95, 99, 116, 117, 122, 134, 150, 160, 167, 168, 187, 191, 203, 228, 231, 318, 323, 327, 328, 323, 357
decimal reduction time, 18, 19, 349
decision tree, 372
defensins, 94
Desulfotomaculum nigrificans, 20, 99
detergent, 141, 216
diarrhoeic shellfish poisoning, 178, 179, 181
diatoms, 181
dinoflagellates, 177, 180, 181
Diphyllobothrium latum, 10
Direct Epifluorescent Filter Technique (DEFT), 199, 201–3
disinfectants, 141, 278

DNA, 10, 12, 34, 38, 58, 72, 83, 89, 109,
  132, 136, 177, 187, 192, 198,
  203, 210, 212-14, 217, 238, 343
DNA chips, 212, *213*
DNAse, 238, 248
domoic acid, 178, 181, *182*
dose-response, 337, *340-43, 350*, 374
drying, 13, 24, 42, 96, 107, 108, 131,
  146, 156, 160, 189, 193, 197,
  318
dysentery, 80, 92, 168

echoviruses, 4, 173
economic, 55, 60, 139, 258, 340, 345,
  364, 377
eggs and products, *19*, 28, 29, 65, 68,
  103, 110, 150, 160, 226, 272,
  303, 317, 351-6, 365, 366
elderly, 5, 63, 153, 155, 156, 166, 181,
  343, 355
ELISA, 198, 203, *205*, 206, 233, 242,
  246, 249
emerging pathogens, 68, 152, 191, 204
emetic toxin, 81, 95, 249
endotoxins, 79
end-product testing, 3, 64, 256, 257,
  259, 268, 270, 315
enrichment media, 195, *196*, 235
enrofloxacin, 72, 147
Enter-Net, 66, 70
Enterobacter spp., 103
E. aerogenes, 77, 220
E. sakazakii, 144, 189, 191
Enterobacteriaceae, 33, 78, 131, 145,
  148, 168, 228, 230, 231, 331,
  334
Enterococcus spp., 76, 78, 101, 109,
  137, 170
  E. faecalis, 76, 77, 101, 109, 116,
    120, 123, 170, 207
  E. faecium, 170
enterotoxins, 58, 79, 372
enterovirulent, 60
enteroviruses, 4, *22*, 77
epidemiological studies, 12, 68, 149,
  188, 190, 192, 226, 335
Escherichia coli, 12, 13, 19, 66, 77,
  78
  attaching and effacing, 58, 85, 91,
    92, 151, 155
  death parameters, 22

detection, 201, 204, 206, 207, 220,
  228, 230-35
diffusely adherent *E. coli* (DEAC),
  58
enteroaggregative *E. coli* (EAggEC),
  58, *89*, 90, 91, 151, 152 189
enterohaemorrhagic *E. coli* (EHEC),
  90, *92*, 143, 144, 151-3, 206,
  235, *236, 240*
enteroinvasive *E. coli* (EIEC), 58, 90,
  92, *93*, 143, 152
enteropathogenic *E. coli* (EPEC), 58,
  90-92, 143, 151, 152, 155
enterotoxigenic *E. coli* (ETEC), 54,
  56, 58, 89-91, 143, 151, 152,
  233
growth parameters, 27, 37, 40, 41,
  48, 102, 145, 228, 289
infections, 60, 79, 142, 144, 147,
  189, 274
routes of transmission, 11
serotyping, 16
sampling plans, 320-24, 326, 328-34
shiga toxin producing *E. coli* (STEC),
  54, 58, 63, 66, 69, 70, 80, 82, 83,
  151,152, 331
toxins, 85, 89, 91, 151-5, 189
viable but nonculturable, 198
verocytotoxin producing *E. coli*
  (VTEC, see shiga toxin
  producing *E. coli*)
E. coli O111, 151, 152, 155
E. coli O157, 19, 61, 66,
  costs of illness, 62
  death parameters, 21, 275
  detection methods, 68-70, 201, 232,
    237-9
  growth parameters, 21, 39, 42, 48,
    152
  infections, 1, 5, 54, 56, 58, 71, 80, 90,
    92, 151, 153, 154, 189, 190,
    274, 351
  microbiological risk assessment,
    349, 350
  routes of transmission, 154, 155,
    190, 272
  sampling plans, 333
  serology, 16
E. coli O26, 151, 155
essential oils, 36
ethylene oxide, 112

Eubacterium, 76, 77, 78
European Union and EU Directives,
 64-6, 295, 304, 320, 330, 331,
 334, 367-70, 378
exotoxins, 79, 80, 161
exposure assessment, 336-40, *350*,
 372
extrinsic parameters, 24, 25, 96, 104,
 107, 198, 335

F value, 18
factory layout, 4, 74
faecal organisms, 5, 194
faecal-oral route, 172, 175, 176
Fenton reaction, 34
fermented foods, 9, 74, 117, 119, 120,
 135, 137
fish and fish products, 10, 11, 26,
 28-31, 64, 95, 99, 103, 106, 114,
 120, 150, 156, 158, 163, 165,
 166, 168, 169, 177, 178, 180,
 183, 187, 321, 330, 331, 334,
 356, 357
flagella, 15, 139, 147, 148, 156, 157,
 162
floppy baby syndrome, 163
flow cytometry, 217, 218
fluorescein isothiocyanate, 217
Food and Drug Administration (FDA),
 9, 61, 68, 112, 113, 155, 195,
 219, *225*, 228, *230*, 235, *236*,
 240, 243, *245*, 250, 263, 271,
 319, 365, 370, 377-9
food chain, 2, 150, 171, 184, 192, 256,
 258-60, 262, 297, 339, 346,
 372-4
food code (USA), 361, 365, 379
food handlers, 159, 171, 173, 175, 176,
 259
food lot, 297, 299, 307, 308, 312, 313,
 316
food processing, 61, 118, 158, 174,
 188, 293, 335, 339
Food Safety and Inspection Service
 (FSIS), 68, 351, 355, 377
food safety objectives, 8, 259, 346
food spoilage, 36, *97*, *98*, 105-7, 116,
 139
FoodMicro model, 49
FoodNet 66, 68, 69, 379
Fraser broth, 235, 244

freezing, 24, 36, 41, 42, 96, 107, 116,
 156, 159, 162, 184, 197, 261,
 262
fructo-oligosaccharides, 136
fruit and fruit juices, 4, 11, 26, 29, 31,
 32, 34, 39, 69, 100, 110, 112,
 115, 131, 154, 174, 188, 272,
 279, 284, 303
fumonisins, 189, 249
functional foods, 3, 9, 119, 134, 136
fungi, 9, 10, 11, 12, 33, 34, 100, 110,
 131, 132, 184, 187, 189, 193,
 215, 261
Fusarium, 110, 132, 184, 188, 189,
 252
F. culmorum, 189
Fusobacterium spp., 76, 77

generally regarded as safe (GRAS), 119,
 134
generic HACCP, 279, 284, 286, 377
genes, 37, 83, 85, 86, 91, 93, 132, 152,
 158
genetically modified foods, 1, 9
genomics, 212, *214*
Giardia lamblia, 54, 57, 59, 271, 272,
 344
glass, 15, 151, 262, 279
Gompertz equation, 45, *46*, 48, 52
good hygienic practice (GHP), 3, 7, 8,
 256, 293, 299, 320, 346, 369
good manufacturing practice (GMP), 3,
 7, 8, 256, 285, 293, 305, 346
Gram stain, 13, *14*, 244
ground beef, 105, 106, 349
growth models, 348
growth rate, 45-8, 145, 352, 358
guidelines, 72, 112, 163, 257, 258, 277,
 295, 315, 319, 320, 327, 331,
 334, 336, 345, 361-4
Guillain-Barre syndrome (GBS), 68,
 146, 148
growth curve, *17*, *46*, *47*, *50*, *51*, 348
growth temperature, 27, 29, 30, 37, 160

H antigen, 15
haemolysin, 82, 90, 91, 95, 151, 164,
 166, 167, 171
haemolytic, 92, 146, 151, 153, 154,
 156, 163, 170, 235, *246*, 327,
 352

haemolytic uraemic syndrome (HUS),
68, 71, 90, 144, 153-5, *350*
haemorrhagic colitis (HC), 90, 151,
153, 168
Hazard Analaysis Critical Control Point
(HACCP)
food poisoning, 68
food safety objectives, 258
implementation, 299, 327, 346, 369
management tool (including TQM),
3, 4, 7, 8, 68, 256, 293-5
microbiological criteria and end-
product testing, 64, 268, 296,
311, 315
microbiological risk assessment, 335
monitoring, 215, 216
plans (including generic), 276-92
chilled foods, 279-83
dried meats, 286-92
milk, 277-8
orange juice, 279, 284-5
pigs, 278-9
predictive microbiology, 44, 49, 51,
52, 257
principles, 257, 259-69, 320, 368
sanitation standard operating
procedures, 279-82
world wide web addresses, 377
hazard characterisation, *336, 338,* 341,
342, 345, 375
hazard identification, *336, 338,* 374
heat-labile toxin (i.e. LT), 82, 83,
89-91, 147, 151, 206
heat shock, 37, 39, 40, 119, 134
heat stable toxin (i.e. ST), 80, 83,
89-91, 206
helminths, 11, 189
hepatitis A, 11, 54, 57, 113, 143, 144,
173, 174, 271, 272, 274
hepatitis E, 174, 189
high-acid foods, 116, 117
high risk foods, 173, 270, 340
high pressure processing, 36-8, 42, 43,
63, 108
high temperature short time (HTST),
107
histamine, 177, 178, 180, 321
human volunteers, 177, 180
humectant, 25
hurdle concept, 36,
hydrogen peroxide, 33, 34

hydrophobic grid membrane (HGMF),
199, 203
hygienic manufacture, 293

ileum, 87, 150
immune system, 138, 156, 167, 184,
189
immunocompromised, 166-8, 183,
357
immunodeficient, 182
immunomagnetic separation, 199-201,
219, 232, 233
impedance, 48, 203, 206-10
index organism, 143
indicator, 74, 78, 143, 145, 228, 235,
238, *302,* 306, 315, 316, 319,
327, 333
infant botulism, 163
infectious dose, 63, 73, 79, 88, 142,
146, 148, 150, 155, 157-9, 163,
166, 168, 169, 171-3, 175, 176,
181, 187, 271, 274, 315, 317,
337, 341-3, 349, 355, 358, 359
insects, 96, 111, 114
intermediate moisture foods, 117
International Commission on
Microbiological Specifications
for Foods (ICMSF), 4, 63, 74,
142-4, 258, 271, 296, 302,
305-7, 318, 327, 334, 338, 377
International Organization for
Standardization (ISO), 7, 201,
219, 223, 227, 247, 256, 260,
294, 295
international trade, 111, 190, 257, 258,
260, 296, 361, 362
intestinal tract, 9, 39, 56, 60, 74-6, 78,
79, 81, 83, 87, 88, 91, 93, 94,
102, 129, 134-8, 142, 143, 147,
150, 152-6, 159, 163, 169, 170,
172, 183, 192, 231, 278, 372
intimin, 92
intrinsic parameters, 25, 39, 40, 73, 96,
98, 99, 101, 104, 107, 116, 335
inulin, 136
irradiation, 1, 12, 19, 37, 38, 96, 107-9,
111-15, 377

Johne's disease, 191

kanagawa reaction, 166

Kaufmann-White scheme, 149
kidney, 81, 152-4, 187
kimchi, 120
koji, 132, *133*
kojic acid, 133
Klebsiella, 77, 78, 145, 177

lactic acid, 9, 32, 99, 101, 106, 117,
    119, 122-8, 131-5, 137
Lactobacillus spp., 27, 122, 123, 129,
    130, 131, 136, 137
    Lb. acidophilus, 77, 127, 131, 134,
        136-8
    Lb. brevis, 77, 123
    Lb. casei, 77, 131, 137, 138
    Lb. delbrueckii, 127
    Lb. helveticus, 127
    Lb. lactis, 117, 122, 124, 126, 127,
        131, 134, 137, 138
    Lb. plantarium, 77, 120, 121, 123,
        131, 137
    Lb. rhamnosus, 137, 138
    Lb. sake, 131
    Lactococcus spp., 117, 122, 124, 126
lactoperoxidase, 33, 34, 122
lactose, 74, 78, 122, 129, 133, 136, 145,
    148, 155, 166, 175, 197, 219,
    225, 228, 231, 232, 235
lantibiotic, 117, 122
large intestine, 74, 135, 232
latex agglutination, 203, 206, 233, 242,
    249
lactic acid bacteria, 9, *33*, 99, 101, 105,
    106, 110, 117, 119-22, 124-6,
    132-5, 137
legislation, 4, 9, 30, 61, 63, 270, 331,
    364, 367, 368
Leuconostoc mesenteroides, 120, 123,
    124, 131
Lior scheme, 147
lipid A *14, 15*
lipopolysaccharide (LPS), 13-15, 79,
    94, 147-9, 167
Listeria monocytogenes, 61, 246, 325
    costs of illness, 62
    death parameters, 22, 35, 191
    detection methods, 68, 201, 206,
        207, 217, 221, 235, 241, 243-5
    growth parameters, 21, 27, 28, 30,
        37, 40-42, 45, 48, 52, 116, 276,
        279

infections, 12, 54, 56, 143, 144, 157,
        189, 192
    microbiological risk assessment,
        356, 357
    routes of transmission, 69, 102, 131,
        156, 190, 272, 279
    sampling plans, 320, 321, 323, 324,
        328-30, 332, 333
    toxins, 81, 82, 86, 94
listeriolysin, 81, 82, 86, 94
locus of enterocyte effacement, 85, 91
low acid foods, 116, 163
low risk foods, 270, 357
luciferase, 138, 215, 216

MacConkey medium (including
    SMAC), 152, 220, 228, 231, 232,
    235, *237, 238, 240*
meat and meat products, 26, 66, 96,
    99-103, 106, 117, 131, 146,
    148, 154, 156, 160, 161, 163,
    171, 183, 190, 274, 278, 279,
    318, 330, 331, 348, 357, 368,
    371
membranes, 13, *14*, 32, 34, 38-42, 74,
    79, 80, 82, 91-4, 117, 122, 124,
    151, 153, 155, 162, 199, 202,
    203, 352
meningitis, 157
mesophile, 26, 29, 36
methylumbelliferyl compounds, 203,
    230
microarrays, 132, 212
microbial growth, 26, 27, 49, 73, 97,
    115, 286, 288, 290
microbial risk, 257
microbiological criteria, 63, 64, 142,
    258, 270, 286, 288, 291, 297,
    298, 302, 307, 316, 320, 327,
    330, 331, 334, 377
microbiological guidelines, 315, 320,
    327, 331
microbiological hazards, 258, 270, 273,
    276, 318, 320, 339, 361, 362
microbiological limits, 105, 297
microbiological risk assessment (MRA),
    335-60
    B. cereus, 340, 341
    C. jejuni antibiotic resistance, 72
    Cl. botulinum, 359, 360
    Codex Alimentarius, 336, 362

E. coli 0157, 349-51
food safety objectives, 5, 8, 259, 346, 347
L. monocytogenes, 356-358
management tool, 256, 259, 335
predictive microbiology, 18, 44
Salmonella spp., 343, 347-9, 351-6, 358, 359
world wide web sites, 377
microbiological specifications, 286, 291, 316
Micrococcus spp., 101, 102, 109, 130, 131
MicroFit model, 46, *47*, 377
microvillus, *76*, *92*, 151, 155
milk and milk products, 11, 28, 29, 33, 34, 45, 95-7, 99, 101, 106, 107, 109, 110, 119-22, 124, 125, 128, 129, 131, 137, 138, 146, 150, 154, 156-60, 165, 170-72, 174, 187, 191-3, 203, 216, 219, 225, 228, 244, 265, 272, *277*, 278, 303, 317, 318, 328, 330, 331, 332, 340, 356, 357, 368
minimal processed foods, 36
modified atmosphere packaging, 36, 97, 99, 107, 108, 115, 117
moisture, 25, 117, 265, 339
monitoring, 65, 72, 104, 105, 213, 215, 216, 218, 265, 267, 268, 270, 275, 279, 296, 320, 371, 375
Moraxella spp., 101-103
Morganella morganii, 78, 177
moulds, 10, 24-6, 30, 33, 96, 100, 105, 112, 132, 188, 203, 249, 319
Mucor spp., 100, 102, 121, 127, *128*, 132
mucosa, 74, *76*, *87-9*, 91-4, 135, 152, 184
mussels and bivalves, 178, 179, 321, 326, 331
Mycobacterium spp., 201
  M. paratuberculosis, 189, 191, 192
  M. tuberculosis, 4, 101, 107, 110, 277
mycotoxins, 11, 132, 184, 187-9, 262, 272

natamycin, 31, 108
National Advisory (NACMCF), 261, 265, 267, 377

National Collection of Type Cultures, 219, 236
nematodes, 183
neosaxitoxin, 178, *180*
neurotoxic shellfish poisoning, 178, 179, 181
neurotoxin, 80
newborns, 156
nisin, 31, 42, 108, 117-19, 122
nitrate and nitrite, 18, 24, 31, 49, 117, 118, 131, 289
nitrosomyglobin, 117
Norwalk-like viruses, 54, 56, 57, 113, 172, 173, 176, 189, 252, 331, 334
nucleic acid probes, 217

O antigen, *13-15*
ochratoxins, 12, 188
odours, 97, 103, 115, 286, 315
ohmic heating, 63
okadaic acid, 181
operating characteristic, 310-12
orange juice, 39, 279
organic acids, 32-4, 38, 43, 117, 119, 122, 206
osmotic pressure, 41, 198
osmotic shock, 41
Oxford agar, 235
oxygen, 25, 33, 38, 49, 82, 99, 106, 108, 115, 124, 146, 161, 226, 241
oysters, 167, 178

P value, 18, 276
packaging, 25, 34, 36, 97, 99, 104, 107, 115, 117, 262, 278, 287, 294, 339, 359, 361, 368, 370
PALCAM medium, 235, 241
parabens, 32, 108
paralytic shellfish poisoning, 178-80
paratyphoid, 149
parvovirus, 176, 177
pasteurisation, 24, 39, 61, 99, 101, 107-10, 113, 116, 118, *128*, *129*, 156, 170, 192, 265, *277*, 317, 352, 353
pasteurised milk, 99, 101, 109, 128, 156, 171, 192, 277, 278, 340, 356
Pathogen Modeling Program, 49-*51*, 377

pathogenicity islands, 83, 85
patulin, 184, 188
pediocin, 42, 117
Pediococcus spp., 99, 122, 123, 130
Penicillium spp., 12, 100, 102, 128,
    131, 184, 188
  P. camembertii, 130, 188
  P. roqueforti, 100, 128, 130, 188
  P. viridicatum, 12, 188
Penner, 147
Peptostreptococcus spp., 76, 78
personal, 270, 294
personnel, 72, 74, 171, 173, 195, 218,
    259, 261, 270, 271, 276, 293,
    316
pest control, 279, 293
phosphatase, 164, 245, 248
phospholipase, 82, 94, 169, 242, 246
physical hazards, 261, 262, 282, 284,
    292
phytosanitary measures, 257, 364
picornaviruses, 173
pigs, 158, 161, 162, 183, 188, *278*
plasmids, 83, 124, 133, 152, 158, 159
Pleisiomonas spp., 169
polioviruses, 4, 173
polymerase chain reaction (PCR), 174,
    198, 201, 203, 210, *211*, 212,
    252
pork, 10, 48, 106, 112, 115, 158, 168,
    183, 184, 188
poultry (see also chicken), 3, 6, 30, 65,
    68, 70, 99, 101, 102, 112, 113,
    146-8, 150, 156, 160, 168, 183,
    272, 303, 305, 330, 358, 365,
    366
prebiotics, 134
predictive microbiology, 17, 18, 44, 49,
    257, 340
predictive modeling, 42, 44, 45, *46*, 48,
    49, 52, 271, 352
pregnancy, 157, 346
pregnant women, 12, 156, 157, 355
preservation, 24, 26, 34, 36, 39, 40, 42,
    61, 100, 107, 113, 118
preservatives, 2, 23, 31, 32, 33, 39, 40,
    63, 100, 107, 117, 159, 261,
    262, 276
prions, 12, 189, 190, 191
probability, 22, 43, 49, 113, 166, 270,
    298, 304, 308-10, 312, 313,

315, 335, 337, 342-4, 349, 351,
    355, 358, 359, 374
probability of infection, 342-4, 355
probiotics, 9, 119, 134-8
processed foods, 2, 36, 61, 147, 150,
    167
producer risk, 312, *313*
Propionibacterium spp., 78, 128, 130
propionic acid, 31-3, 39, 108, 117,
    122, 135
protein detection, 216, *217*
Proteus spp., 15, 78, 102, 103, 177, 221
protozoa, 10, 77, 196, 262
Pseudomonas spp., 45, 101, 102, 103,
    105, 106, 115, 117, 187, 203
Pseudoterranova decipiens, 183, 189
psychrophile, 26, 29, 36, 101, 102, 103,
    109, 359
public health and PHLS, 147, 163, 189,
    238, 239, 298, 299, 327, 334,
    338, 367-9, 378
PulseNet, 69, 379

quality, 3, 23, 44, 256, 259, 293-5, 298,
    335, 374, 375
quality assurance, 4, 293-5, 347, 375
quality control, 4, 40, 197, 219, 220,
    282, 286-92, 375
quality management, 7, 260, 294, 375
quality systems, 7, 8, 294

rapid methods, 195, 198, 203, 213, 216
Rappaport Vassiliadis medium, 203,
    204, 220, 222, 224
raspberries, 182
ready-to-eat food, 72, 163, 165, 196,
    219, 327, 332, 334, 357, 360
reduced oxygen packaging, 115, 116
regulations, 258, 299, 327, 331, 356,
    362, 369, 378
representative sample, 70, 193-5, 316,
    350
residues (veterinary and pesticide), 13,
    94, 115, 141, 215, 216, 262,
    271, 284, 363, 366
resuscitation, 197, 198, 219, 231, 232,
    235, 237
reverse phase latex agglutination
    (RPLA), 206, 207, 233, 242, 245,
    246
rhinoviruses, 173

Rhizopus spp. 100, 102, 120, 132
Rhodococcus equi, 236
rice, 28, 95, 114, 120, 121, 132, *133*,
   151, 165, 168, 171, 319
risk,
   food poisoning, 32, 119, 169, 172,
      173, 190, 276
   microbiological risk assessment
      (MRA), 18, 72, 256-8, 335-60,
      362, 377
   perception, 1, 3, 5, 9
   probiotics, 136, 138
   producer and consumer, 270, 307,
      312
risk analysis, 8, 18, 336, 363, 377
risk assessment, 8, 49, 52, 72, 142, 258,
   294, 296, 298, 299, *336*, 337,
   338, 345, 346, 348, 349, *352*,
   359, 365, 375, 377
risk characterisation, *336-8*, 344, 345,
   375
risk communication, 8, *336*, 346, 375,
   377
risk management, 336, 345, 358, 375
risk profiling, 336, 345
rotavirus, 54, 56, 57, 59, *60*, 77, 138,
   143, 175-7
roundworm, 184
routes of transmission, 5, 54, 56, 57,
   154, 170, 172, 174, 182, 190,
   192, 346, 361

Saccharomyces spp., 27, 100, 120
S. cerevisiae, 21, 121, 130, 132, 201
Salm-Net, 66, 69, 70
Salmonella spp., 61, 78, 224, 225, 347
   antibiotic resistance, 70, 71
   costs of infections, 62, 64, 66
   death parameters, 107, 156
   detection methods, 64, 68, 69, 72,
      196, 201, 203, 206, 207, 217,
      219, 220, 222, 223
   growth parameters, 27, 28, 37, 50,
      51
   infections, 12, 54, 56, 58, 60, 67,
      144, 149, 150, 274, 344
   microbiological risk assessment,
      347-9, 355, 358, 377
   routes of transmission, 4, 5, 11, 101,
      131, 149, 271, 272, 273, 279
   sampling plans, 306, 317-33

serology, 15, 16, 148
toxins, 13, 78, 80, 83, 85, 94
viable but nonculturable, 198
S. agona, 71, 150
S. anatum, 16, 71
S. dublin, 71, 149
S. enteritidis,
   death parameters, 21
   detection methods, 65, 220
   growth parameters, 21
   infections, 94, 189
   microbiological risk assessment,
      351-6
   routes of transmission, 150
   sampling plan, 65
   serology, *14*, 16,
S. hadar, 71, 150
S. heidelberg, 149
S. livingstone, 71
S. minneapolis, 16
S. newington, 16
S. newport, 71
S. paratyphi, 16, 144, 150, 151
S. senftenburg, 21
S. stanley, 71
S. tosamanga, 71
S. typhi, 4, 16, 54, 56, 94, 149, 150,
   151
S. typhimurium (including DT104),
   16, 25, 39, 42, 47, 48, 63, 65, 70,
   71, 85, 94, 149, 150, 189, 190
   antibiotic resistance, 63, 70, 71,
      190
   death parameters, 26,
   detection methods, 65, 150
   growth parameters, 42, 47, 48
   infections, 39, 94, 149, 150, 189
   sampling plans, 65
   serology, 16, 149, 150
   toxins, 85
salmonellosis, 61, 66, 149, 150, 151,
   343, 344
salt, 24, 26, 37, 81, 118, 119, 131,
   137, 160, 192, 197, 237, 238,
   265
sampling plans,
   EU, 320-34
   EU dairy products, 323, 328
   EU egg products, 322
   EU minced meat and meat
      preaprations, 322

EU shellffish, molluscs and bivalves, 326
FDA, 321
ICMSF cereals and bakery products, 319
ICMSF cook-chill and cook freeze preparations, 320
ICMSF dairy products, 318
ICMSF egg products, 317
ICMSF processed meats, 318
PHLS ready-to-eat foods, 163, 332-4
Three class plans, 298-300, 306, 307, 309, 310, 313-15, 317-19
Two class plans, 298-300, 306, 308, 310-14, 319
Variable plans, 303-5
sanitary and phytosanitary measures (SPS), 175, 257, 258, 285, 362, 364
sanitation standard operating procedures (SSOP), 279, 285
sanitizer, 281
saxitoxin, 178, *180*
scarlet fever, 170, 171
scombroid poisoning, 177-80
scrapie, 12, 190
seafood, 11, 120, 177, 273, 321, 332, 366
seeds, 189
selective broths, *196,* 220, 225
selenite cystine broth, 224
septicaemia, 15, 157, 162, 166-8
serotyping, 16, 147, 152, 156, 159
seven principles of HACCP, 260
sewage, 5, 156, 157, 159, 174
sheep, 12, 70, 146, 154, 171, 172, 190, 191, 236, 244
shelf stable, 30, 116
shelf life,
  pasteurisation, 23
  predicting, 43
  processing, 18, 61, 104, 107, 109, 113-16, 276, 317, 335
  preservation, 31, 63
  sampling plans, 276, 300, 334
  spoilage and poisoning organisms, 97, 98, 101, 105, 116, 194, 271, 315, 357
shellfish, 11, 145, 156, 166, 167-70, 173, 174, 177, 179-81, 252, 261, 272, 326, 331, 334

Shewanella putrefaciens, 99, 102, 103
shiga toxin, 63, 66, 80, *85,* 88, 89, 92, 151, 152, *237*
shiga toxin producing *E. coli* (STEC), 54, 63, 66, 69, 70, 80, 151, 152, 331
Shigella spp.,
  detection methods, 68, 220, 233, 235
  growth parameters, 27, 234, 239,
  infections, 54, 56, 58, 344
  routes of transmission, 4, 11, 78, 271, 272
  sampling plans, 327
  toxins, 80, 83, 88, 92, 143, 144, 155, 156
Sh. dysenteriae,
  detection methods, 206, 233
  growth parameters, 234
  infections, 155, 156, 274
  toxins, 80, 88,
Sh. flexneri, 274
Sh. sonnei, 71, 220, 240
shigelloides, 169, 170
short chain fatty acids, 39, 93, 101, 135
sigma factors, 37
slime, 18, 97, 99, 103, 105, 138, 139
small round structured viruses (SRSV), 59, *60,* 172
small round viruses, 176
smoked, 131, 356, 357
sodium chloride, 348
software, 44, 46, 49, 377
soil, 4, 5, 64, 156, 158, 161, 164, 189, 191, 371
sorbic acid, 31-3,100, 117
sore throat, 158, 170, 171
sour, 23, 99
Sous-vide, 109, 116
Spices, 35, 112, 114, 226, 273, 290
spoilage, 23, 36, 45, 49, 63, 96-103, 105-12, 115-17, 139, 315-17
standards, 28, 111, 175, 254, 257, 258, 270, 294, 295, 305, 315, 316, 327, 330, 331, 334, 345, 355, 361-5, 368, 369, 377
statistical analysis, 270
statistical probability, 270, 298
stomach, 39, 54, 63, 74, 75, 88, 93, 134, 150, 161, 342
Staphylococcus aureus, 61, 246
  cell wall structure, 13, 14

costs of infection, 62
death parameters, 19, 21, 22, 36, 43
detection methods, 201, 207, 219,
   220, 237, 238, 242, 247, 248
growth parameters, 25, 27, 37, 42,
   47, 48, 159, 160, 279, 317
infections, 54, 56, 58, 66, 67, 144,
   274, 344
routes of transmission, 11, 131, 271,
   272
sampling plans, 318–33
toxins, 81, 82, 102, 129, 142, 159,
   160, 165, 236, 276, 279, 317
starter culture, 122, 128, 131
sterilization, 23, 108, 109
Streptococcus spp., 13, 21, 54, 56, 101,
   107, 120, 122, 127, 129, 137,
   145, 170, 171, 271, 327
St. avium, 170
St. bovis, 170
St. durans, 170
St. parasanguinis, 171, 189
St. pyogenes, 81, 170
St. thermophilus, 101, 107, 127, 128,
   129, 134, 137
stress, 13, 17, 33, 36–40, 42, 87, 93,
   107, 109, 119, 132, 134, 167,
   188, 197, 342
stringency, 304, 307, 308, 312, 315
sublethally injured cells, 39, 196, 198,
   201
sulphites and sulphur dioxide, 31, 32
surveillance, 56, 66–70, 192, 338, 345,
   365, *369*
susceptibility, 2, 63, 98, 157, 160, 187,
   188, 271, 298, 341, 342
synbiotics, 134
syruping, 23, 24

Taenia saginata, 183
Taenia sainata, 10, 144
Taenia solium, 10, 183
tapeworm, 183, 272
technical barriers to trade and TBT
   Agreement, 257, 258
tetanus, 82
tetrathionate broth, 224
thermal processing (see also high
   pressure processing), 26, 63
thermoduric, 97
thermophile, 26, 29, 99, 122, 127, 146

three class plan, 299-*301*, 304, 305,
   306, 313
Total Quality Management (TQM), 3, 7,
   256, 259, 277, 294, 295, 335,
   374, 377
Toxoplasma gondii, 11, 54, 56, 57, 61,
   62, 183, 189
training, 4, 44, 49, 72, 259, 293
Transmissible spongiform
   encephalopathies, 190
Trichenella spiralis, 11, 54, 57, 144,
   184, 272
trichothecenes, 189, 249
two class plan, 299-*301*, 306, 308, 310
typhoid, 94, 149, 150, 151

ultrasound, 39, 118
ultraviolet (UV), 12, 36, 37, 38, 111,
   187, 189, 210, 249
under-reporting estimates, *53-5*
United States (USA), 2, 9, 34, 54-6, 60,
   61, 72, 112, 146, 152, 155, 158,
   166, 190, 191, 259, 263, 279,
   365, 370, 379
United States Department of
   Agriculture (USDA), 49, 68, 112,
   244, 279, 356
URL (see also World Wide Web), 277,
   351, 370, 376, 377
UVM medium, 235, 241, 243

vacuum packaging, 36, 99, 105, 107,
   108, 116
validation, 51, 107, 254, 255, 275
vegetables and vegetable products, 4,
   11, 28, 29, 31, 34, 95, 99, 131,
   154, 272, 332
verification, 64, 260, 267, 270, 286-92,
   368, 375
verocytotoxins, 152, 233
verocytotoxin producing *E. coli* (VTEC,
   see shiga toxin producing *E.
   coli*)
very young, 343, 351
Vi antigen, 16
viable but nonculturable (VNC), 87,
   146, 198
Vibrio spp., 11, 54, 56, 58, 68, 80, 87,
   165-7, 198, 273
   V. cholerae, 27, 56, 80, 85-7, 91, 144,
   189, 190, 198, 274, 343, 344

V. parahaemolyticus, 27, 35, 37, 144, 165, 274, 321, 327, 333
V. vulnificus, 27, 54, 56, 166, 167, 189, 321
violet red bile agars, 220, 231
virulence, 13, 16, 39, 83, 85, 87, 91, 147, 152, 158, 159, 169, 341
viruses (*see also* specific groups), 10-12, 56, 68, 113, 119, 172, 173, 176, 177, 193, 196, 199, 217, 252, 261, 334
water,
diarrhoea, 79, 83, 86, 95, 147, 151, 153, 155, 156, 164, 175
mucosa cells, 81, 88
source of infection, 4, 5, 10, 11, 78, 103, 145, 146, 154-6, 158, 159, 164, 166-70, 172-4, 176, 177, 181, 182, 190, 198, 199, 201, 217, 228, 271, 273, 303, 327, 365, 366
water activity (a_w), 17, 24-8, 30, 38, 39, 41, 42, 47, 49, *50, 51*, 98-101, 107, 108, 116-18, 131, 146, 160, 164, 198, 258, 261, 262, 271, 276, 291, 317, 319, 338, 339

World Wide Web (see also URL), 46, 49, 72, 271, 277, 279, 291, 302, 347, 351, 361, 369, 370, 376
wine, 11, 96, 99, 121, 188, 365
World Health Organization (WHO), 9, 72, 258, 261, 293, 361-4, 379
World Trade Organization (WTO), 257, 258, 361, 362, 364

xylose lysine desoxycholate medium (XLD), 220, 222, 226

Yakult, 119, 137, 138
yeast, 10, 11, 24, 30, 33, 35, 45, 63, 100, 105, 110, 119, 132, 150, 215, 303, 330
Yersinia spp., 58, 60, 83, 157, 158, 201
   Y. enterocolitica, 28, 37, 40, 48, 54, 56, 68, 69, 143, 144, 157, 158, 272, 274, 276, 279, 327
   Y. pestis, 157
yoghurt, 121, 125, 127, *129*, 138, 154, 332

Z value, 18, 20, *22*, 23, 275
zearalenone, 189
zoonosis, 174
Zygosaccharomyces baillii, 21, 100